I0036966

Numerical and Evolutionary Optimization 2018

Numerical and Evolutionary Optimization 2018

Special Issue Editors

Adriana Lara
Marcela Quiroz
Efrén Mezura-Montes
Oliver Schütze

MDPI • Basel • Beijing • Wuhan • Barcelona • Belgrade

MDPI

Special Issue Editors

Adriana Lara
Instituto Politécnico Nacional (IPN)
Mexico

Marcela Quiroz
University of Veracruz
Mexico

Efrén Mezura-Montes
University of Veracruz
Mexico

Oliver Schütze
CINVESTAV-IPN
Mexico

Editorial Office
MDPI
St. Alban-Anlage 66
4052 Basel, Switzerland

This is a reprint of articles from the Special Issue published online in the open access journal *Mathematical and Computational Applications* (ISSN 2297-8747) from 2018 to 2019 (available at: https://www.mdpi.com/journal/mca/special_issues/NEO18).

For citation purposes, cite each article independently as indicated on the article page online and as indicated below:

LastName, A.A.; LastName, B.B.; LastName, C.C. Article Title. *Journal Name* **Year**, *Article Number, Page Range*.

ISBN 978-3-03921-816-5 (Pbk)
ISBN 978-3-03921-817-2 (PDF)

© 2019 by the authors. Articles in this book are Open Access and distributed under the Creative Commons Attribution (CC BY) license, which allows users to download, copy and build upon published articles, as long as the author and publisher are properly credited, which ensures maximum dissemination and a wider impact of our publications.

The book as a whole is distributed by MDPI under the terms and conditions of the Creative Commons license CC BY-NC-ND.

Contents

Oliver Cuate and Oliver Schütze

Variation Rate to Maintain Diversity in Decision Space   within   Multi-Objective   Evolutionary
Algorithms

About the Special Issue Editors

Adriana Lara is a tenured Full-Time Professor at the Mathematics Department of the Physics and Mathematics School (ESFM) at IPN in México City. Her current research interests include multiobjective optimization, bio-inspired algorithms, memetic techniques, hybrid algorithms, and data analysis. Her research has received the IEEE Transactions on Evolutionary Computation Outstanding Paper Award twice, in 2010 as well as in 2012. She also received the 2010 Engineering Award granted by Mexico City's Science and Technology Institute (ICyTDF). She received her B.Sc. degree in Physics and Mathematics from IPN in 2001, an M.Sc. degree in Electrical Engineering, and a Ph.D. degree in Computer Sciences from Centro de Investigación y Estudios Avanzados (Cinvestav—IPN), in Mexico City, in 2003 and 2012, respectively.

Marcela Quiroz is Full-Time Researcher at the Artificial Intelligence Research Center at the Universidad Veracruzana, MEXICO. Her research interests focus on combinatorial optimization, metaheuristics, and experimental algorithms. She is the publicity chair of the Numerical and Evolutionary Optimization (NEO) workshop series. She received her M.Sc. degree in computer science from the Instituto Tecnológico de Ciudad Madero, Mexico, in 2009, and her Ph.D. degree in computer science from the Instituto Tecnológico de Tijuana, Mexico, in 2014.

Efrén Mezura-Montes is Full-Time Researcher of the Artificial Intelligence Research Center at the Universidad Veracruzana, MEXICO. His research interests are in the design, analysis, and application of bio-inspired algorithms to solve complex optimization problems. He has published over 145 papers in peer-reviewed journals and conferences. He also has one edited book and 11 book chapters published by international publishing companies. Google Scholar reports over 5800 citations of his work. He is a member of the Editorial Board of the journals: *Swarm and Evolutionary Computation, Complex & Intelligent Systems, International Journal of Dynamics and Control*, and *Journal of Optimization*. Dr. Mezura-Montes is a member of the IEEE CIS Evolutionary Computation Technical Committee and the IEEE Systems Man and Cybernetics Society Soft Computing Technical Committee.

Oliver Schütze is Full Professor at the Cinvestav—IPN in Mexico City, Mexico, and currently holds the Dr. Rodolfo Quintero Chair at the UAM Cuajimalpa in Mexico City, Mexico. His main research interests are in numerical and evolutionary optimization. He is co-author of more than 140 publications including one monograph, 4 school textbooks, and 10 edited books. Two of his papers have received the IEEE Transactions on Evolutionary Computation Outstanding Paper Award (in 2010 and 2012). He is the founder of the Numerical and Evolutionary Optimization (NEO) workshop series. He is Editor-in-Chief of the journal *Mathematical and Computational Applications*, and is a member of the Editorial Board of the journals *Engineering Optimization, Computational Optimization and Applications*, and *Mathematical Problems in Engineering*.

Preface to "Numerical and Evolutionary Optimization 2018"

This volume was inspired by the 6th International Workshop on Numerical and Evolutionary Optimization (NEO) hosted by the Instituto Politécnico Nacional (IPN), located in downtown Mexico City, Mexico. The workshop was held from September 26 to 28, 2018, and was attended by a total of around 70 researchers and students.

Solving scientific and engineering problems from the real world is nowadays a very complicated task; that is why the development of powerful search and optimization techniques is of great importance. Two well-established fields focus on this duty; they are (i) traditional numerical optimization techniques and (ii) bio-inspired metaheuristic methods. Both general approaches have unique strengths and weaknesses, allowing researchers to solve some challenging problems but still failing in others. The goal of NEO is to gather people from both fields to discuss, compare, and merge these complementary perspectives. Collaborative work allows researchers to maximize strengths and to minimize the weaknesses of both paradigms. NEO also intends to help researchers in these fields to understand and tackle real-world problems like pattern recognition, routing, energy, lines of production, prediction, modeling, among others.

Papers one to four of this book are about exciting applications of the differential evolution (DE) heuristic. In the first paper, [https://www.mdpi.com/2297-8747/23/3/34] R. Akararungruangkul et al. consider a particular case of the location routing problem (SLRP). Here, the objective function is fuel consumption, while the problem faces variations in vehicle speeds and admits time constraints regarding the kind of the carried material. The second paper [https://www.mdpi.com/2297-8747/23/3/40] by U. Ketsripongsa et al. concerns the economic crop planning problem, considering transportation logistics to maximize the profit from cultivated activities. In this case, income comes from the selling price and production rate, while costs are due to operating and transportation expenses. In the third paper [https://www.mdpi.com/2297-8747/23/4/79], P. Sresracoo et al. aim to solve the U-shaped assembly line balancing problem Type 1 (UALBP-1). They apply DE for balancing production lines while minimizing the number of workstations. The fourth paper by P. Sriboonchandr et al. [https://www.mdpi.com/2297-8747/24/3/80] compares metaheuristics performance on the flexible job-shop scheduling problem (FJSP).

The following two papers focus on genetic programming (GP) based techniques. Starting with paper five [https://www.mdpi.com/2297-8747/24/3/78], by P. Juarez-Smit et al., it approaches GP combining a numerical local search method and a bloat-control mechanism within a distributed model for evolutionary algorithms. After that, paper six [https://www.mdpi.com/2297-8747/23/2/19] by R. Lopez et al. is about considering computational models that can score the behavior of a driver based on a risky–safety scale. Potential applications of these models include car rental agencies, insurance companies, or transportation service providers.

Next, paper seven [https://www.mdpi.com/2297-8747/23/2/25] by W. Limmun et al. is Construction of a Model-Robust IV-Optimal Mixture Designs Using a Genetic Algorithm (GA). Their obtained results show that the GA-generated designs studied are robust across a set of potential mixture models. It is followed by paper eight [https://www.mdpi.com/2297-8747/23/4/60], where J. Guerrero et al. present a shape optimization workflow. This proposal considers fault-tolerant and software agnostics, and allows asynchronous simulations with a high degree of automation. The test focuses on a practical maritime industry case, aiming to optimize the shape of a bulbous bow to minimize the hydrodynamic resistance.

The last three papers address some aspects of multiobjective optimization. First, in paper nine [https://www.mdpi.com/2297-8747/23/2/30], S. Peitz and M. Dellnitz present an overview of recent developments in accelerating multiobjective optimal control. The results are about complex problems, where either PDE constraints are present or where there is a necessity of archive feedback behavior.

Second, in paper ten [https://www.mdpi.com/2297-8747/23/3/51], J. M. Bogoya et al. extend the two-parameter-based performance indicator delta p,q to asses multiobjective optimization algorithms. This extension applies to bounded subsets of the n-dimensional space, which makes it suitable for applications in the scope of NEO.

Finally, in paper eleven [https://www.mdpi.com/2297-8747/24/3/82], O. Cuate et al. propose a selection strategy for multiobjective evolutionary algorithms that aims to maintain diversity both in objective space—which is the commonly used space—as well as in decision variable space. This is done since a variation in decision space may represent valuable information such as backup solutions for the decision-maker in some instances.

Finally, we thank all participants at NEO 2018 and hope that this book can be a contemporary reference regarding the field of numerical evolutionary optimization and its exciting applications.

Adriana Lara, Marcela Quiroz, Efrén Mezura-Montes, Oliver Schütze
Special Issue Editors

Mathematical and Computational Applications

MDPI

Article

Modified Differential Evolution Algorithm Solving the Special Case of Location Routing Problem

Raknoi Akararungruangkul [1],* and Sasitorn Kaewman [2]

[1] Department of Industrial Engineering, Faculty of Engineering, Khon Kaen University,
    Khon Kaen 40000, Thailand
[2] Faculty of Informatics, Mahasarakham University, Maha Sarakham 44000, Thailand; sasitorn.k@msu.ac.th
*   Correspondence: raknoi55@gmail.com

Received: 18 June 2018; Accepted: 1 July 2018; Published: 3 July 2018

Abstract: This research article aims to solve the special case of the location routing problem (SLRP) when the objective function is the fuel consumption. The fuel consumption depends on the distance of travel and the condition of the road. The condition of the road causes the vehicle to use a different speed, which affects fuel usage. This turns the original LRP into a more difficult problem. Moreover, the volume of the goods that are produced in each node could be more or less than the capacity of the vehicle, and as the case study requires the transportation of latex, which is a sensitive good and needs to be carried within a reasonable time so that it does not form solid before being used in the latex process, the maximum time that the latex can be in the truck is limited. All of these attributes are added into the LRP and make it a special case of LRP: a so-called SLRP (a special case of location routing problem). The differential evolution algorithms (DE) are proposed to solve the SLRP. We modified two points in the original DE, which are that (1) the mutation formula is introduced and (2) the new rule of a local search is presented. We call this the modified differential evolution algorithm (MDE). From the computational result, we can see that MDE generates a 13.82% better solution than that of the original version of DE in solving the test instances.

Keywords: location routing problem; rubber; modify differential evolution algorithm; vehicle routing problem

---

1. Introduction

More than 14% of the Thai GDP comes from the transportation and logistics sector. More than 60% of the transportation cost incurred in Thailand is by one actor: the agricultural supply chain. This is because Thailand is known as an agricultural country. The major agricultural industry is the rubber industry. More than 80% of rubber is produced in the southern part of Thailand, with the remaining part of rubber production mainly in the northeastern part of Thailand. The rubber industry in the South has a long history, and much suitable infrastructure has been developed. Aside from the northeastern part of Thailand, the infrastructure and knowledge, including technology, knowledge, instruments and so forth, are not yet sufficient for the high growth of the rubber industry. The transportation costs in the rubber industry are not paid by the government or the person who takes care of this issue; farmers still deliver latex themselves to sell at the rubber collecting points. This activity generates a very high transportation cost for the whole country. Our research team is engaged in a project to design the latex collection system. This system comprises of finding the location of the collection points and the transportation route of the vehicles. This problem is actually similar to the vehicle routing problem which was first proposed by Dantzig and Ramser [1] and has been proven to be an NP-hard problem [2]. Apart from the vehicle routing problem (VRP), this problem must also decide the collection point of the latex; thus, the VRP turns out to be a location routing problem (LRP). Location routing [3].

The proposed problem is not the general LRP that we found in the literature but has a few characteristics that make it a special case of LRP. These characteristics are as follows:

(1) The volume of latex which is available in the rubber field can be more or less than the capacity of the vehicles;
(2) The potential location and the rubber fields have different attributes such as different road conditions which can affect the speed used and the fuel consumption rate of the truck.
(3) The maximum duration or distances used for each vehicle are limited due to the latex's fast transformation to solid;
(4) We focus the objective function on minimizing the total fuel consumption instead of the total distance as we often see in the literature.

These four special attributes make this problem a special case of the LRP problem (SLRP). The SLRP has never been found in the literature due to its 4 special attributes explained above. This problem can be found in many real-world applications. Nakorn Panom Province is one of the cities that has the fastest growing rate of development in the agricultural industry. The rubber cultural area has increased by more than 150% in the last 5 years in Nakorn Panom Province. We will take Nakorn Panom Province as the case study. The method which is presented here is customized to the rubber field industry, and it is also applicable to other industries where it has the same attributes as mentioned above.

In this study, the effective modified differential evolution algorithm will be presented to solve the proposed problem. The article is organized as follows; Section 2 is the literature review; Section 3 is the problem definition, and current practice procedures will be presented; Section 4 presents the proposed heuristics; and Section 5 gives the conclusions and outlook of the article.

2. Literature Review

Dantzig and Ramser [1] introduced the vehicle routing problem (VRP) and Lenstra and Rinnooy [2] proved that it is an NP-hard problem. Various types of VRP have been presented in many research articles. Braekers and Ramaekers [4] have given an overview of the scenario and the problem's physical characteristics, extending the basic uncapacitated VRP. Besides the various types of VRP problems, there are also various types of methods for solving the VRP. The metaheuristic method is one of the solution methods that is very popular for solving VRP as it is fast and effective. The solution approaches presented in many articles are simulated annealing [5].

The location routing problem (LRP) is one of many various types of VRP. This problem has two decision variables which need to be solved. These two decision variables are (1) the suitable location to be used and (2) the routing of the field or the clients that are assigned to that location [6,7]. The optimal value of the suitable location can be determined before or during the assignment of the clients to the locations.

There are many researchers that have presented a methodology to solve the LRP. Most of the algorithms proposed in the literature are two-phase heuristics, which are (1) cluster first, (2) route second or (1) route first, (2) cluster second [8]. From the literature, both choices perform well in many test problems; thus, selecting one will not affect the solution quality. In our research, the proposed heuristics employ the idea of using two-phase heuristics. We select cluster first, route second as the starting point of our algorithm.

There are many researchers who have focused on developing metaheuristics to solve the problem such as tabu search [9] and the simulated annealing (SA) technique [10,11].

In this research, the differential evolution algorithm (DE) is presented as it is a fast and effective algorithm.

The differential evolution (DE) algorithm is a branch of the evolutionary algorithm developed by Rainer Storn and Kenneth Price [12] for optimization problems over continuous optimization. Thereafter, DE has been successfully used in combinatorial optimization such as in assembly line balancing [13,14], the location-allocation problem [15], machine layout [16,17], and the manufacturing problem [18].

The DE has been applied to VRP problem [19–25]. The design of DE to solve VRP problem has both the traditional [20] way or the slight modification of some mechanism to get a more effective algorithm. Key successes of DE are the good designing of vectors to represent the problem, the encoding and decoding methods, the setting of good predefined parameters in the DE mechanism and the effectiveness of the mutation, recombination, and selection procedure [23–25]. Moreover, with DE, it is also easy to add some procedure to increase its performance such as add local search [19], adjust decoding method [21], and self-adaptive some parameters [23]. Though there are many articles proposing DE to solve VRP, we cannot find an effective DE to solve the LRP [22]. So it is our contribution to present the effective DE to solve the special case of LRP. When all key successes have been introduced to DE, DE will be very effective compared with other metaheuristics proposed in the literature. Moreover, DE has the advantage that it is fast and efficient; thus, in this article, we will present the modified version of it to solve the problem.

This article is organized as follows: in the next section, the problem definition and the current practice heuristics is presented. Section 3 presents the proposed heuristics. Section 4 proposes the computational results and Section 5 is the conclusion of the article.

3. The Problem Definition and the Current Practice Heuristics

3.1. The Problem Definition

The case study is composed of 110 rubber fields and 30 potential areas that can be set as the collecting points. All connections between the location and fields have different road conditions, with different speed limits and road condition factors. The example of the case study is shown in Figure 1.

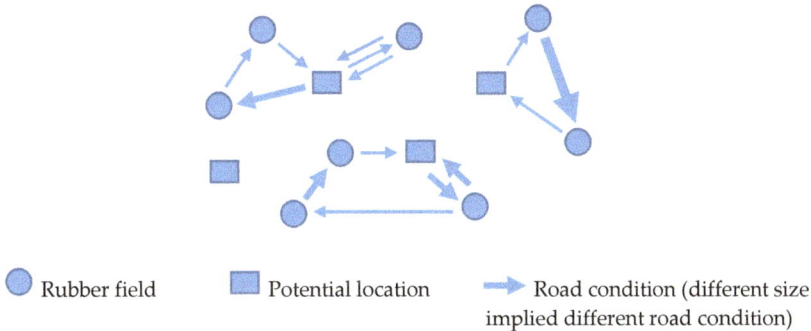

Rubber field   Potential location   Road condition (different size implied different road condition)

Figure 1. The problem definition.

From Figure 1, we can see that one field can be visited by direct shipping (which can be visited more than once by direct shipping) and routing or even combined direct shipping and routing. This makes the problem much harder than the normal LRP. The different size of the arrows used implies different road conditions and speed limits that are permissible to use, and the speed used on the road affects the fuel consumption of the vehicle. The objective function which is used here is the total fuel consumption, and the conditions that need to be satisfied are as follows:

(1)   The speed limit of the road need to be controlled;
(2)   All fields must be visited at least one time;
(3)   The maximum duration of traveling has to be under a limit;
(4)   The total working hours of the truck needs to be controlled.

The objective function and the constraints addressed above make the LRP harder to solve. The current practice method which is used in the Nakorn Panom Province will be explained in the next section.

The types of road are set to seven types, distinguished by using the average driving speed and the fuel consumption rate, as shown in Table 1. The consumption rate is calculated from https://en.wikipedia.org/wiki/Fuel_economy_in_automobiles.

Table 1. The road type and its consumption rate.

Road Type	Ave. Speed	Litre/km
S	30	0.118
T	40	0.107
V	50	0.112
W	60	0.090
X	70	0.098
Y	80	0.098
Z	90	0.102

Ave. Speed is the average speed of the vehicle that can be driven on the particular road types. In the case study, the historical data of the average speed used for that road is collected for two weeks. After that, the average speed is converted into road type and the fuel used using Table 1. The consumption rate is assigned to the road, then multiplied by the distance of the connection road. The average fuel used on that road will be obtained. For example, if the connection between fields 1 and 2 has a distance of 50 km and is a type C road. Thus, the fuel needed to travel on this road is $0.112 \times 50 = 5.6$ L. This mechanism will be applied to all connections. Finally, the metric of the fuel consumption of the road will be obtained.

3.2. The Current Practice Procedure

Previously, the practice procedure has not taken into account the fuel consumption rate and the speed limit. Thus, the current practice procedure will be decided based on the information given in the distance matric. The example used to explain the current procedure and the proposed heuristics composed of 5 potential locations (A, B, C, D, E) and 6 rubber fields. The traveling time of the locations to fields and fields to field is given in Tables 2 and 3 respectively. The road type of all locations and field connections is shown in Tables 6 and 7. The maximum traveling time per round of the 12-ton truck is 60 min and when the truck needs to travel more than one round, it cannot travel for more than 100 min. The procedure practice procedure can be explained as follows:

(1)   Select the current location to be used. Start by using the potential location that has the highest S_k when S_k is calculated by Formula (1):

$$S_k = \frac{C_k}{F_k} \tag{1}$$

where C_k is the capacity of location k and F_k is the operating cost per day of location k. For example, if a particular location has a capacity of 80 tons per day and an operating cost of 1500 baht per day, the S_k is 0.533. If we have five locations (A–E) with a capacity of 30, 20, 50, 40, and 50 tons and an operation cost of 1200, 1100, 1400, 1000, and 1200, then we get the S_k of each location as 0.025, 0.018, 0.035, 0.04, and 0.041, respectively. Location E will be selected to be the location first, then step 2 will be executed.

(2) Assign the field to the current location that is in use. This step simply assigns the fields according to the closest distance to the location that is in use. For example, if we have six fields, we have five locations; if we have six fields to travel, the traveling time between fields is calculated using the average speed of the road. The traveling times from locations to the fields are calculated and shown in Table 2, while the traveling time between the fields is shown in Table 3. Each field has a volume of latex of 5, 20, 40, 10, 5, and 13 tons, respectively. From step 1, location E is opened and it has a 50-ton capacity; thus, the closest fields to it that have a cumulative volume of latex of fewer than 50 tons are field 1 (5), 4 (10), 6 (13), and 5 (5). After all possible fields are assigned to location E and location E is full, then the next location will be opened (redo step 1). The second highest S_k is location D, which has a capacity of 40 tons. The remaining field will be assigned to location D. The closest field to D is field 2, which has a latex volume of 20, and we cannot assign field 3 to location D due to location D not having enough capacity. Step 1 will be performed again and we will open location C which has a capacity of 50. Field 3 will be assigned to this location. To conclude steps 1 and 2, the result of the assignment of the fields to the locations is shown in Table 4.

Table 2. The traveling times of the locations to the fields.

-	1	2	3	4	5	6
A	12	12	14	19	12	17
B	19	6	10	21	20	21
C	21	19	19	17	7	18
D	17	10	15	27	5	19
E	13	22	20	14	19	16

Table 3. The raveling times between fields.

-	1	2	3	4	5	6
1	0	11	15	14	6	9
2	11	0	13	9	6	16
3	15	13	0	27	11	15
4	14	9	27	0	16	25
5	6	6	11	16	0	14
6	9	16	15	25	14	0

Table 4. The result of steps 1 and 2 in the current practice procedure.

Location	Field	Full Cap.	Used Cap.
E	1, 4, 6, 5	50	33
D	2	40	20
C	3	50	40

(3) Route the fields in the opened locations using the nearest neighbor heuristics. Please note that a vehicle can travel more than one route, as long as it has enough time to travel in one day. If the truck has a capacity of 12 tons, the time limit per round is 60 min (including the loading-in and out time of (0.5 + 0.5 = 1) minutes per ton), and the maximum time per day that the truck can be in operation is 100 min. The loading time of latex to the collecting point is 1 min per ton.

Routing location E: A truck that has a capacity of 12 tons has to travel less than 60 min per round and the cumulative time used per day of that truck must not exceed 100 min. This truck will be sent to pick up the latex. The route is formed based on the nearest neighbor heuristics (the shortest route is preferred). Please note that the traveling time is different from the distance. Truck 1 will go out to pick up 5 tons from field 1 and 7 tons from field 4; then, it comes back to location E.

This route carries 12 tons and uses 41 min of traveling time and 12 min of loading time, thus the total time used in the first route is 53 min. Truck 1 still has a total time of less than 100 min, and so truck 1 will then go out to pick up 3 tons from field 4 and 5 tons from field 5. This route uses 57 min and carries 8 tons. The first truck has a total time of 98 min. This means that the second truck needs to start working. Truck 2 will be sent to field 6 two times due to field 6 having a latex volume of 13 tons, which cannot be carried back to E in only one route. The result of the current practice for the example is shown in Table 5.

From Table 5, we can see that the total time used is 438 min. This has a loading time of 93 min; thus, the traveling time total is 345 min.

The type of road for each connection is shown in Table 6 (location–field) and Table 7 (field–field).

Table 5. The result of the routing of the example.

Loc.	Truck	Route	T.U.	V.L.
	1	E-1-4-E	41	12
	1	E-4-5-E	57	8
E	2	E-6-E	44	12
	2	E-6-E	44	1
D	1	D-2-D	32	12
	1	D-2-D	28	8
	1	C-3-C	50	12
	1	C-3-C	50	12
C	2	C-3-C	50	12
	2	C-3-C	42	4
	Total		438	93

Remark: Loc. is location, T.U. is time used in the route, V.L. is volume carried in that route.

Table 6. The road type from location to the field.

-	1	2	3	4	5	6
A	S	T	V	W	S	V
B	V	T	S	V	W	V
C	W	X	V	V	W	V
D	T	S	X	T	S	X
E	X	X	T	Y	Y	V

Table 7. The road type from field to field.

-	1	2	3	4	5	6
1	-	S	W	V	T	W
2	S	-	S	X	T	S
3	W	S	-	V	X	W
4	V	X	V	-	X	S
5	T	T	X	X	-	X
6	W	S	W	S	X	-

The traveling distance of each connection is shown in Tables 8 and 9.

Table 8. The distance (km) from locations to fields.

-	1	2	3	4	5	6
A	6	8	11.7	19	6	14.2
B	15.8	4	5	17.5	20	17.5
C	21	22.2	15.8	14.2	8.17	15
D	11.3	5	17.5	18	2.5	22.2
E	15.2	25.7	13.3	18.7	25.3	13.3

Table 9. The distance (km) from field to field.

-	1	2	3	4	5	6
1	0	5.5	15	11.7	4	9
2	5.5	0	6.5	10.5	4	8
3	15	6.5	0	22.5	12.8	15
4	11.7	10.5	22.5	0	18.7	12.5
5	4	4	12.8	18.7	0	16.3
6	9	8	15	12.5	16.3	0

From the distances given in Tables 8 and 9, when we combine these distances with the road types of each connection given in Tables 6 and 7 and use the fuel consumption rate shown in Table 1, the fuel use in the traveled connections in Table 5 can be calculated. For example, for location E, route 1 of truck 1, the route is E-1-4-E, which has three connections. These are E-1, 1-4, and 4-E, which have distances of 15.2 km, 11.7 km, and 18.7 km, respectively. These connections have fuel consumptions of 0.098(X), 0.112(V), and 0.098(Y) liters per kilometer, respectively. Therefore, the fuel used in these three connections total 1.4896, 1.3104, and 1.8326 L, respectively. In total, in location E, route 1, truck 1 consumes 4.6324 L. The same mechanism will be performed with all trucks, routes, and locations. The result of fuel consumption using the current practice procedure is shown in Table 10.

Table 10. The fuel consumption of all routes.

Loc.	Truck	Route	Fuel Consumption (L)
E	1	E-1-4-E	4.6324
	1	E-4-5-E	6.1446
	2	E-6-E	2.9792
	2	E-6-E	2.9792
D	1	D-2-D	1.18
	1	D-2-D	1.18
C	1	C-3-C	3.5392
	1	C-3-C	3.5392
	2	C-3-C	3.5392
	2	C-3-C	3.5392
Total			33.2522

From Table 10, the total fuel consumption for all locations is 33.2522 L per day.

4. The Proposed Heuristic

The proposed heuristic is designed to solve the problem. Many metaheuristics are available in the literature. The differential evolution algorithm (DE) is selected to solve this problem because it is a fast and effective heuristic. Generally, DE is composed of four steps:

(1) Randomly generate the initial vector or solution;
(2) Use the mutation process;
(3) Use the recombination process;
(4) Use the selection process.

Steps (2) to (4) will be iteratively executed. Details of each step are explained in Section 4.1.

4.1. Randomly Generated Vector or Solution

The solution can be obtained by generating a set of vectors. Each vector is composed of D positions, where D is the number of fields. The number of the population is 5 (or NP = 5). Table 11 shows an example of the values in each position of five vectors which are randomly generated in the first iteration.

When the vectors are generated, obtaining the solution of the problem, the decoding method needs to be executed. There are five steps of decoding (transferring) the vector into the problem's solution, which are explained in Sections 4.1.1–4.1.4.

Table 11. The example of five random vectors.

Farmer		1	2	3	4	5	6
	Volume	5	20	40	10	5	13
Vector							
NP1		0.39	0.10	0.42	0.35	0.31	0.89
NP2		0.48	0.33	0.81	0.26	0.86	0.64
NP3		0.41	0.93	0.77	0.02	0.37	0.18
NP4		0.01	0.57	0.62	0.17	0.53	0.42
NP5		0.98	0.14	0.64	0.76	0.12	0.17

Figure 1: Assigning random code numbers to each vector in each position.

4.1.1. Finding the Order of the Fields

In this step, we need to sort the value of the position of each vector to get the order of the fields. The position's value is sorted according to the increasing order. For example, for vector 4, the original value of positions 1, 2, 3, 4, 5, and 6 are 0.01, 0.57, 0.62, 0.17, 0.53, and 0.42. The order of the value in each position according to increasing order is 0.01, 0.17, 0.42, 0.53, 0.57, and 0.62. This order of value generates an order of position for vector 4 of 1, 4, 6, 5, 2, and 3.

4.1.2. Randomly Select the Location

The method of selecting the opened location is explained in the following example. The probability of selecting the location can be calculated using Formula (2):

$$p_k = \frac{S_k}{\sum_{k=1}^{K} S_k} \qquad (2)$$

where p_k is the probability of location k and S_k is the heuristics information as explained in Section 3.2.

From the example, the probabilities p_k of locations A, B, C, D, and E are 0.16, 0.11, 0.22, 0.25, and 0.26, respectively. In the next step, the cumulative probability of each location is 0.16, 0.27, 0.49, 0.74, and 1. If the random number is 0.57, then location D is selected to be the first opened location. We can apply this mechanism to get the order of the location. The order of the locations in this step is D, E, A, B, and C; then, the next step, the decoding method, will be executed.

4.1.3. Assigning the Field to the Location According to Their Place in the Order

For example, the field order is 1, 4, 6, 5, 2, and 3 and the location order is D, A, E, B, and C. We start to assign the field to the location by adding the field in the very first order until this location is full, and then we move forward to the next possible location and continue doing this until all fields are assigned to exactly one location.

In this example, field 1 will be assigned to location D, and the capacity of D is updated whenever an assignment has been done. The location will close when it loads a volume of latex in the field reaching the full capacity level. The result of the assignment is shown in Table 12.

Table 12. The result of the decoding method of the differential evolution (DE) algorithm.

Location	Field	Cap	Load
D	1, 4, 6, 5	40	33
A	2	30	20
E	3	50	40

From Table 12, locations D, A, and E will be in use while the remaining locations will be closed.

4.1.4. Routing All Fields in the Locations

In this step, we construct the route of the fields that are assigned to the locations according to the capacity of the truck per round, the longest distance or traveling time per round and the total distance or time that the truck can travel per day. The result of Section 4.1.1 is the same as in Table 10 because the routing phase (Section 4.1.1) is the same as the routing phase of the current practice procedure and the order of the fields and locations are the same as in the current practice example.

4.2. Mutation Process

4.2.1. The Original Mutation Process

In this step, after we get the initial solution, we apply the mutation process formulas. In the proposed heuristic, the mutation formula is Formula (3):

$$V_{i,j,G} = X_{best,j,G} + F(X_{r_1,j,G} - X_{r_2,j,G}) \tag{3}$$

where $F = 1.5$, $X_{i,j,G}$ is a randomly selected target vector and $X_{best,j,G}$ is the vector that has the best solution compared among all target vectors.

4.2.2. The Modified Mutation Process

Originally, all vectors were random; we take them from the current iteration target vectors, and the best solution is the set of best vectors obtained from the best among all target vectors. The modified version of the mutation process is one in which we add one set of the best vectors. The number of the best vectors in this set is equal to the number of the population in the normal DE method. For example, if there are 10 target vectors (NP = 10), there will be 10 vectors in the best vectors set as well. The set of best vectors is obtained by collecting the best vectors that are found during the simulation. These vectors are not the same as the normal target vectors due to the collection of all good vectors, even if they come from the same vector number. The target vector of vector 1 at the current iteration is the best vector that obtains the best solution for vector 1. The second best for vector 1 is the vector from the last iteration if the current iteration found a new solution. This vector will forget the last best solution whenever it finds a better solution, even if the last best solution is better than that of the other vector in the set of target vectors. In the modified version of the mutation process, the second-best vector of a vector can be kept in the best vectors set if it is better than that of the other target vectors. The mutant vector used in the modified mutation process is generated from both the normal target vector and the set of best vectors, as shown in Figure 2.

From Figure 2, the mutant vector of vector 2 is generated from the best vector (assuming that vector 1 is the best vector) and the two random vectors are vector 3 (from the set of target vectors) and 5 (from the set of best vectors) and the result is given in Figure 2. The switching between using the set of best vectors or the set of target vectors can be executed by randomly choosing the using probability. Each vector is randomly chosen before the mutant process is executed. The probability function of the use of the set of best vectors is shown in Formula (4):

$$C = 1 - exp^{-[\,(It - \frac{MaxIt}{2})^2\,]} \tag{4}$$

If the random number is less than or equal to C, then the random vector used in the mutant process is drawn from the set of best vectors instead of the set of target vectors.

The updating of the set of best vectors can be done by getting rid of the worst vector in the set and putting the new best vector in if the new best vector is better than at least one vector in the set of best vectors. This is done whenever a new best vector (better than at least the worst vector in the set of best vectors) is found. The result of the mutant process is mutant vector $V_{i,j,G}$, where i is the number of the vector, j is the position number, and G is the current iteration.

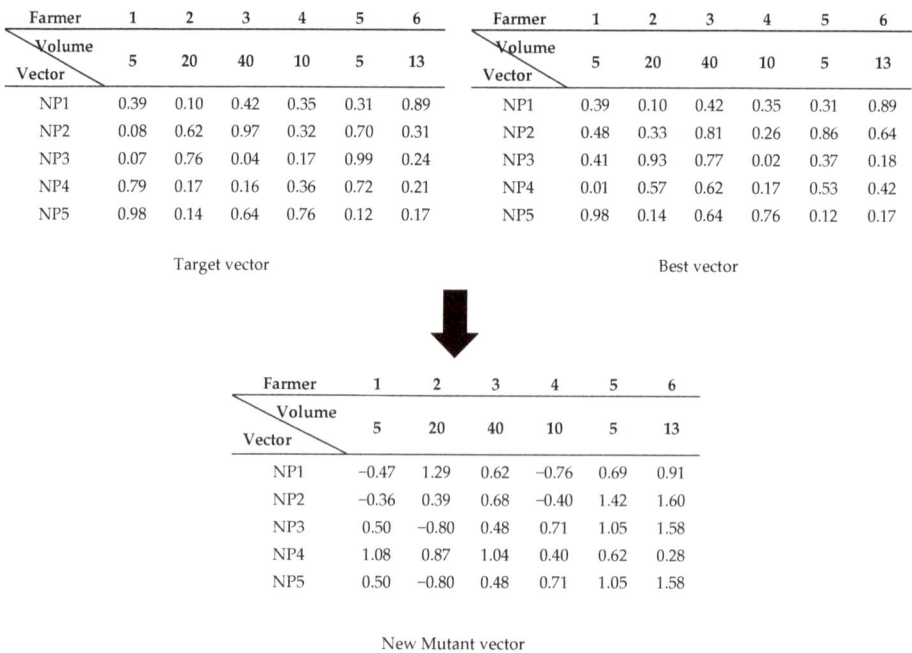

Farmer	1	2	3	4	5	6
Volume Vector	5	20	40	10	5	13
NP1	0.39	0.10	0.42	0.35	0.31	0.89
NP2	0.08	0.62	0.97	0.32	0.70	0.31
NP3	0.07	0.76	0.04	0.17	0.99	0.24
NP4	0.79	0.17	0.16	0.36	0.72	0.21
NP5	0.98	0.14	0.64	0.76	0.12	0.17

Target vector

Farmer	1	2	3	4	5	6
Volume Vector	5	20	40	10	5	13
NP1	0.39	0.10	0.42	0.35	0.31	0.89
NP2	0.48	0.33	0.81	0.26	0.86	0.64
NP3	0.41	0.93	0.77	0.02	0.37	0.18
NP4	0.01	0.57	0.62	0.17	0.53	0.42
NP5	0.98	0.14	0.64	0.76	0.12	0.17

Best vector

Farmer	1	2	3	4	5	6
Volume Vector	5	20	40	10	5	13
NP1	−0.47	1.29	0.62	−0.76	0.69	0.91
NP2	−0.36	0.39	0.68	−0.40	1.42	1.60
NP3	0.50	−0.80	0.48	0.71	1.05	1.58
NP4	1.08	0.87	1.04	0.40	0.62	0.28
NP5	0.50	−0.80	0.48	0.71	1.05	1.58

New Mutant vector

Figure 2. The modified mutant process.

4.3. Recombination Process

The recombination process can be executed by randomly assigning one number for each position in a vector; if this random number is less than or equal to the predefined parameter *CR*, then the value in position j of vector i is taken from the value of position j in vector i of the mutant vector. On the other hand, if the random number is higher than that of CR, the value in that position is taken from the target vector. The formula of the original version of DE is shown in Formula (5):

$$U_{i,j,G} = \begin{cases} V_{i,j,G} \ if \ rand_{i,j} \leq CR \\ X_{i,j,G} \ if \ rand_{i,j} > CR \end{cases} \quad (5)$$

When it is executed, the set of trial vectors ($U_{i,j,G}$) will be obtained.

In this article, a new recombination formula is presented in Formula (6).

$$U_{i,j,G} = \begin{cases} V_{i,j,G} \ if \ rand_{i,j} \leq CR_1 \\ X_{i,j,G} \ if \ CR_1 \ < rand_{i,j} \leq CR_2 \\ B_{i,j,G} \ \geq CR_2 \end{cases} \quad (6)$$

Predefined parameters CR_1 and CR_2 have to be set first and lie between 0 and 1. If the random number $rand_{i,j}$ is less than CR_1, then the value in position j of vector i is drawn from the mutant vector. If it is greater than CR_1 but less than CR_2, the value in position j, vector i will be taken from the target vector. Finally, if the random number is higher than CR_2, that position will take the value from the set of best vectors ($B_{i,j,G}$). After the trial vector is obtained, the local search will be applied for all vectors.

4.4. The Local Search

The local search that we use in the proposed heuristic is the SWAP algorithm. The swap algorithm will be executed in the order of the fields and the locations generated in the decoding method. For example, in Figure 3, the current interchange of the fields is order 1 and 2, which is now fields 5 and 4. The exchange is performed and obtains the order of fields as 4, 5, 1, 2, 6, and 3, respectively. The new order will use the decoding method using the location order E, A, D, C, and B, which is the exchange order between the locations B and C. The order of location has to exchange all positions, such as E exchanging with A, D, C, and B. After the exchange, the new order will be used to decode with the current order of the fields being 4, 5, 1, 2, 6, and 3.

After all possible exchanges of location are performed with the current order of fields, then the new order of the fields is generated by exchanging the next position in the field order.

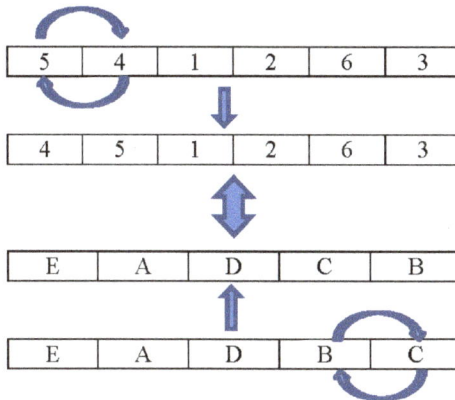

Figure 3. An example of the local search.

4.5. Selection Process

The selection process is according to Equation (7).

$$X_{i,j,G+1} = \begin{cases} U_{i,j,G} \ if \ f(U_{i,j,G}) \leq f(X_{i,j,G}) \\ X_{i,j,G} \qquad otherwise \end{cases} \tag{7}$$

When the selection process is finished, we will get the target vectors of the next generation (iteration).

The procedure of the proposed heuristics is shown in Figure 4.

Set NP, CR, CR1, CR2, F, D (size of vector)
Generate Initial Solution
Begin
For G = 1 to Gmax **when G = iterations and G**max **= Maximum iteration**
   For N = 1 to NP
      Random Generate Random Number r_1, r_2, r_3: r_1, r_2, $r_3 \in [1..NP]$, $r_1 \neq r_2 \neq r_3 \neq N$
      Produce Mutant Vector N (Mutation Process)
      For i = 1..D
         $V_{i,j,G} = X_{best,j,G} + F(X_{r_1,j,G} - X_{r_2,j,G})$
      Next Produce Trial Vector N (Recombination Process)

      $U_{i,j,G} = \begin{cases} V_{i,j,G} \ if \ rand_{i,j} \leq CR \\ X_{i,j,G} \ if \ rand_{i,j} > CR \end{cases}$ (optional)

      $U_{i,j,G} = \begin{cases} V_{i,j,G} \ if \ rand_{i,j} \leq CR_1 \\ X_{i,j,G} \ if \ CR_1 < rand_{i,j} \leq CR_2 \\ B_{i,j,G} \geq CR_2 \end{cases}$ (optional)

   Next Produce New Target Vector (Selection Process)
      Apply Decoding Method (4.1.1–4.1.4)
      Apply SWAP
      Selection Process
      Collect Best Vector $B_{i,j,G}$
End

Figure 4. The pseudo code of the modified DE.

5. Computational Results

The computational results are obtained by testing the DE (original version), the modified DE (MDE), and the current practice procedure with 10 test instances and one case study. The test instances are named N-1 to N-10 and the case study is named Case. The detail of the number of locations (#location) and the number of fields (#field) of each test instance is shown in Table 13.

Table 13. Parameter setting of test instances.

Test Instance	#Location	#Field
N-1	10	15
N-2	10	15
N-3	10	25
N-4	20	30
N-5	20	40
N-6	20	60
N-7	30	80
N-8	30	90
N-9	30	90
N-10	30	100
Case	30	110

CR is set to 0.6, F is set to 2 [13], CR_1 is set to 0.6, and CR_2 is set to 0.8. The simulation has been executed five times, and the best solutions among all five results are drawn to be representative of the algorithm. The simulation was performed in a Computer notebook Core™ i5-2467M CPU 1.6 GHz.

The stopping criterion which is used in the first test is the runtime limitation, which is set to 10 min. The computational result of the 11 test instances is shown in Table 14.

Table 14. The comparison of the current practice and the modified DE (MDE) results.

-	Fuel Used (L)		
Instance	Current Practice	DE	MDE
N-1	20.3	18.47	16.2
N-2	21.4	19.2	16.1
N-3	39.4	37.1	32.1
N-4	45.8	46.3	40.2
N-5	55.4	55.6	49.2
N-6	70.3	69.4	60.1
N-7	80.4	82.5	71.3
N-8	90.4	94.3	80.2
N-9	95.4	94.2	80.5
N-10	102.1	109.4	92.4
CASE	112.3	110.1	95.4

From Table 14, DE and MDE outperform the current practice procedure. To prove that they are outperforming it statistically, the result of the statistical test using the Wilcoxon signed-rank test with a 95% confident interval is shown in Table 15.

Table 15. The statistical result of the proposed heuristics.

-	Current Practice	DE	MDE
Current Practice	-	>	>
DE	<	-	<
MDE	<	<	-

From Table 15, DE and MDE generate a significantly better solution than that of the current practice. The solution obtained from MDE is statistically better than that of DE.

Now, the performance of DE and MDE will be tested to draw the contribution of the proposed heuristics. The proposed heuristics are split into 4 sub-heuristics, which are DE (original DE), MDE-1 (DE + new mutation Formula (6)), MDE-2 (DE + local search), and MDE-3 (DE + new mutation

Formula (6) + local search). The stopping criterion in this test is the runtime limitation, which is set to 10 min. The results of all proposed heuristics are shown in Table 16.

Table 16. The results of the proposed heuristics.

-	Fuel Used (L)			
Instance	DE	MDE-1	MDE-2	MDE-3
N-1	18.47	17.82	17.98	16.2
N-2	19.2	18.2	18.03	16.1
N-3	37.1	35.3	35.63	32.1
N-4	46.3	45.02	44.5	40.2
N-5	55.6	54.61	52.3	49.2
N-6	69.4	67.31	66.5	60.1
N-7	82.5	80.57	79.14	71.3
N-8	94.3	88.9	84.3	80.2
N-9	94.2	92.3	89.5	80.5
N-10	109.4	104.2	102.56	92.4
CASE	110.1	108.76	106.85	95.4

The significance test has been executed for all results using the Wilcoxon signed-rank test with a 95% confident interval as shown in Table 17.

Table 17. The statistic test for the proposed heuristics.

-	DE	MDE-1	MDE-2	MDE-3
DE	-	>	>	>
MDE-1	-	-	>	>
MDE-2	-	-	-	>
MDE-3	-	-	-	-

From Table 17, all MDE algorithms outperform the original version of DE, and MDE-3 outperforms MDE-1 and MDE-2. This means that both the local search and the new mutation formula should be combined to get a better solution. The local search alone can improve the solution better than that of DE using the new mutation formula. The new mutation formula works well in the DE due to it statistically improving the solution from the original DE. The percentage difference of each proposed heuristic to the best solution (MDE-3) is shown in Table 18.

Table 18. The percentage difference between the proposed heuristics to the best-proposed heuristics (MDE-3).

-	Fuel Used (L)			
Instance	DE	MDE-1	MDE-2	MDE-3
N-1	14.012	10.000	10.988	0.000
N-2	19.255	13.043	11.988	0.000
N-3	15.576	9.969	10.997	0.000
N-4	15.174	11.990	10.697	0.000
N-5	13.008	10.996	6.301	0.000
N-6	15.474	11.997	10.649	0.000
N-7	15.708	13.001	10.996	0.000
N-8	17.581	10.848	5.112	0.000
N-9	17.019	14.658	11.180	0.000
N-10	18.398	12.771	10.996	0.000
CASE	15.409	14.004	12.002	0.000
Ave.	16.056	12.116	10.173	0.000

Math. Comput. Appl. **2018**, 23, 34

From Table 18, we can see that the best heuristic (MDE-3) has a percentage difference of fuel usage less than that of MDE-2, MDE-1, and DE of 10.173%, 12.116%, and 16.056%, respectively.

6. Conclusions

This research presents solution approaches to solving a special case of the location routing problem (SLRP). The SLRP could have a volume of latex more or less than the capacity of the vehicle. The objective function of SLRP is to minimize the fuel usage, and the fuel usage depends on the road condition and the distance of the road.

We developed a differential evolution algorithm to solve the problem. The new mutation formula is presented in the article. The new rule of applying SWAP is presented and is used as the local search in the proposed heuristics.

The heuristics proposed in the article are split into four sub-heuristics, which are DE (original DE), MDE-1 (DE + new mutation Formula (6)), MDE-2 (DE + local search), MDE-3 (DE + new mutation Formula (6) + local search). The best heuristic is MDE-3, which has a 16.056% difference from the original version of DE and is 10.173 and 12.116 better than MDE-2 and MDE-1, respectively.

From the computational result, we can see that using the new mutation formula and the local search is beneficial to the original DE. The new mutation formula and the local search is designed based on the idea of increasing the intensification of DE. This makes MDE-1, MDE-2, and MDE-3 outperform the original DE.

Author Contributions: S.K. collected the important data and designed the algorithm; R.A. designed and performed the experiment and analyzed the result.

Acknowledgments: We would like to thank Department of Industrial Engineering, Faculty of Engineering, Khon Kaen University and Faculty of Informatics, Mahasarakham University for funding this research.

Conflicts of Interest: The authors declare no conflicts of interest.

References

1. Dantzig, G.; Ramser, J. The truck dispatching problem. *Manag. Sci.* **1959**, 6, 80–91. [CrossRef]
2. Lenstra, J.K.; Rinnooy Kan, A.H.G. Complexity of vehicle routing and scheduling problems. *Networks* **1981**, 11, 221–227. [CrossRef]
3. Watson-Gandy, C.; Dohrn, P. Depot location with van salesmen—A practical approach. *Omega* **1973**, 1, 321–329. [CrossRef]
4. Braekers, K.; Ramaekers, K.; Van Nieuwenhuyse, I. The vehicle routing problem: State of the art classification and review. *Comput. Ind. Eng.* **2015**, 99, 300–313. [CrossRef]
5. Hasanpour, H.A.; Mosadeghkhah, M.; Tavakoli Moghadam, R. Solving a Stochastic Multi-Depot Multi-Objective Vehicle Routing Problem by A Simulated Annealing. *J. Ind. Eng.* **2009**, 43, 25–36.
6. Min, H. Consolidation terminal location-allocation and consolidated routing problems. *J. Bus. Logist.* **1996**, 17, 235–263.
7. Nagy, G.; Salhi, S. Heuristic algorithms for single and multiple depot vehicle routing problems with pickups and deliveries. *Eur. J. Oper. Res.* **2005**, 162, 126–141. [CrossRef]
8. Barreto, S.; Ferriera, C.; Paixao, J.; Santos, B.S. Using clustering analysis in a capacitated location-routing problem. *Eur. J. Oper. Res.* **2007**, 179, 968–977. [CrossRef]
9. Tuzun, D.; Burke, L.I. A two-phase tabu search approach to the location routing problem. *Eur. J. Oper. Res.* **1999**, 116, 87–99. [CrossRef]
10. Lin, C.K.Y.; Chow, C.K.; Chen, A. A location-routing-loading problem for bill delivery services. *Comput. Ind. Eng.* **2002**, 43, 5–25. [CrossRef]
11. Lin, C.K.Y.; Chow, C.K.; Chen, A. Multi-objective metaheuristics for a location-routing problem with multiple use of vehicles on real data and simulated data. *Eur. J. Oper. Res.* **2006**, 175, 1833–1849. [CrossRef]
12. Storn, R.; Price, K. Differential Evolution—A Simple and Efficient Heuristic for Global Optimization over Continuous Spaces. *J. Glob. Optim.* **1997**, 11, 341–359. [CrossRef]

13. Pitakaso, R. Differential evolution algorithm for simple assembly line balancing type 1 (SALBP-1). *J. Ind. Prod. Eng.* **2015**, *32*, 104–114. [CrossRef]

14. Pitakaso, R.; Sethanan, K. Modified differential evolution algorithm for simple assembly line balancing with a limit on the number of machine types. *Eng. Optim.* **2016**, *48*, 253–271. [CrossRef]

15. Thongdee, T.; Pitakaso, R. Differential Evolution Algorithms Solving a Multi-Objective, Source and Stage Location-Allocation Problem. *Ind. Eng. Manag. Syst.* **2015**, *14*, 11–21. [CrossRef]

16. Storn, R.; Price, K. *Differential Evolution—A Simple and Efficient Adaptive Scheme for Global Optimization over Continuous Spaces*; Technical Report TR-95-012; ICSI: New Delhi, India, 1995.

17. Nearchou, A.C. Meta-heuristics from nature for the loop layout design problem. *Int. J. Prod. Econ.* **2006**, *101*, 312–328. [CrossRef]

18. Lampinen, J.; Zelinka, I. Mechanical engineering design optimization by differential evolution. In *New Ideas in Optimization*; Corne, D., Dorigo, M., Glover, F., Eds.; McGraw-Hill: London, UK, 1999; pp. 127–146.

19. Boon, E.T.; Ponnambalam, S.G.; Kanagaraj, G. Differential evolution algorithm with local search for capacitated vehicle routing problem. *Int. J. Bio-Inspired Comp.* **2015**, *7*, 321–342.

20. Huan, X.; Jiechang, W. Differential Evolution Algorithm for the Optimization of the Vehicle Routing Problem in Logistics. In Proceedings of the Eighth International Conference on Computational Intelligence and Security, Guangzhou, China, 17–18 November 2012; pp. 48–51.

21. Mingyong, L.; Erbao, C. An improved differential evolution algorithm for vehicle routing problem with simultaneous pickups and deliveries and time windows. *Eng. Appl. Artif. Intell.* **2010**, *23*, 188–195. [CrossRef]

22. Drexl, M.; Schneider, M. A survey of variants and extensions of the location-routing problem. *Eur. J. Oper. Res.* **2015**, *241*, 283–308. [CrossRef]

23. Zhi-Zhong, L.; Yong, W.; Shengxiang, Y.; Ke, T. An Adaptive Framework to Tune the Coordinate Systems in Nature-Inspired Optimization Algorithms. *IEEE Trans. Cybern.* **2018**, *99*, 1–14.

24. Yong, W.; Zhi-Zhong, L.; Jianbin, L.; Han-Xiong, L.; Jiahai, W. On the selection of solutions for mutation in differential evolution. *Front. Comp. Sci.* **2018**, *12*, 297–315.

25. Yong, W.; Zhi-Zhong, L.; Jianbin, L.; Han-Xiong, L.; Gary, G.Y. Utilizing cumulative population distribution information in differential evolution. *Appl. Soft Comput.* **2016**, *48*, 329–346.

© 2018 by the authors. Licensee MDPI, Basel, Switzerland. This article is an open access article distributed under the terms and conditions of the Creative Commons Attribution (CC BY) license (http://creativecommons.org/licenses/by/4.0/).

Mathematical and Computational Applications

MDPI

Article

An Improved Differential Evolution Algorithm for Crop Planning in the Northeastern Region of Thailand

Udompong Ketsripongsa [1], Rapeepan Pitakaso [1,*], Kanchana Sethanan [2] and Tassin Srivarapongse [3]

1　Metaheuristics for Logistics Optimization Laboratory, Department of Industrial Engineering, Ubon Ratchathani University, Ubon Ratchathani 34190, Thailand; udompong.jo@gmail.com
2　Department of Industrial Engineering, Faculty of Engineering, Khon Kaen University, Khon Kaen 40000, Thailand; skanch@kku.ac.th
3　Department of Economics, Faculty of Business Administration, Rajamangala University of Technology Thanyaburi, Patumthani 10900, Thailand; tassin66@hotmail.com
*　Correspondence: rapeepan.p@ubu.ac.th; Tel.: +66-85-921-0826

Received: 14 July 2018; Accepted: 9 August 2018; Published: 10 August 2018

Abstract: This research aimed to solve the economic crop planning problem, considering transportation logistics to maximize the profit from cultivated activities. Income is derived from the selling price and production rate of the plants; costs are due to operating and transportation expenses. Two solving methods are presented: (1) developing a mathematical model and solving it using Lingo v.11, and (2) using three improved Differential Evolution (DE) Algorithms—I-DE-SW, I-DE-CY, and I-DE-KV—which are DE with swap, cyclic moves (CY), and K-variables moves (KV) respectively. The algorithms were tested by 16 test instances, including this case study. The computational results showed that Lingo v.11 and all DE algorithms can find the optimal solution eight out of 16 times. Regarding the remaining test instances, Lingo v.11 was unable to find the optimal solution within 400 h. The results for the DE algorithms were compared with the best solution generated within that time. The DE solutions were 1.196–1.488% better than the best solution generated by Lingo v.11 and used 200 times less computational time. Comparing the three DE algorithms, MDE-KV was the DE that was the most flexible, with the biggest neighborhood structure, and outperformed the other DE algorithms.

Keywords: differential evolution algorithm; crop planning; economic crops; improvement differential evolution algorithm

---

1. Introduction

A variety of important, world-class economic crops are planted throughout the different regions in Thailand, including rice, cassava, sugarcane, rubber, and palm. Factors including land, soil, water, and weather influence the decisions of Thai farmers as to where such crops are planted [1]. Currently, these high-value crops play an important role in the economic growth of the country. Crop planting in the northeastern region of Thailand is valued at 63.84 million rai or approximately 23.27 rai per household, indicating that this region has high planting crop numbers. There is an imbalance, however, between the high rate of supply and the demand, which has caused a variety of problems. Rice farmers have been invading Bangkok to protest delayed payments for the rice subsidy program, sugarcane farmers and Para rubber farmers have closed a main road to protest low prices. The main causes of these problems appear to be unsuitable crop planting, high investment with low profits, unbalanced supply and demand, poor plantation knowledge, unsuitable marketing, and inconvenient transportation.

Thus, crop allotment, supply and demand, cost and profit, marketing, and transportation distance are crucial factors that need to be considered when planting crops.

Thailand is an agricultural country; most people make a living from selling their agricultural products. Therefore, their lives depend on the income generated from their product and the amount of product they can produce. Higher productivity can be obtained by growing the right plants in the right place, thus reducing the transportation cost by having the plants closer to the secondary producer. Therefore, selecting the ideal agricultural types to grow in a suitable area will generate higher productivity with lower transportation costs. A suitable area to grow a certain type of vegetation means having an available water supply, the correct earth type, and growing temperatures, for example. These can cause different production rates per area for each type of plant.

Thailand is composed of four main regions: the north, northeastern, middle, and south regions. The northeastern region is the main area where most of the agricultural products are grown. The three main plants produced to sell are rubber, rice, and sugarcane. These three types of plants have been cultivated widely in the whole northeastern part of Thailand. Some farmers grow rice because they are familiar with it, grow sugarcane for its high income, or grow rubber for the government subsidies. These planting choices can cause the problems mentioned above, such as the selling price of rice dropping due to oversupply in some areas, the sugarcane delivery is too far from the mill, or the latex quality is not good enough to produce a high-end product. These issues could be due to the farmer not growing a crop in the appropriate area. This study's research focuses on placing the correct plants in suitable areas to achieve high productivity and low transportation costs to provide the highest profit to the farmer.

The article has been organized as follows: Section 2 describes the intensive literature review. Section 3 outlines the mathematical model formulation of the problem, whereas Sections 4 and 5 are used to explain the proposed heuristics. Section 6 shows the computational results. The last section, Section 7, concludes the article.

2. Literature Review

Many different methods have been proposed to solve the crop planning model. Research has focused on the forecasting of agricultural products and some have attempted to match demand for the product to the supply to be planted. In this study, to address the crop planning problem, we focused on matching the earth types (different types of soil contributes to varying production rates in assorted types of plants), level of rain (water supply), experience of the farmer, and the transportation of the goods to the secondary producer in the agricultural chain. The required data were collected from the historical data from the area of interest. The proposed problem is similar to the special case of the generalized assignment problem, a location-allocation problem that has been widely investigated by researchers such as Thongdee and Pitakaso [2], who presented a multi-objective location-allocation problem to solve the agricultural transportation in Northeastern Thailand. Sethanan and Pitakaso [3] introduced an improved differential evolution algorithm to solve the generalized assignment problem.

Metaheuristics are methods widely used to solve difficult problems, such as crop planning transportation problems, or production challenges. The well-known metaheuristics include the Krill Herd (KH) algorithm [4–6], the Cuckoo algorithm [7,8], and the Monarch Butterfly Optimization (MBO) [9,10], which can improve efficiency by using the Lévy flight operator and fly straight operator [11], the Hybridizing Harmony Search [12], the Bat algorithm [13], the Elephant Herding Optimization (EHO) [14], or the Earthworm optimization algorithm (EWA) [15].

Metaheuristics are applicable to many types of problems and researchers are continuously improving its quality and capability to perform better searches. Many researchers paid attention to refining the quality of the original version. Wang published many articles about enhancing the capabilities of the original versions of many of the metaheuristics, especially Krill Herd (KH) [16]. There are many ways to improve the solution quality given by KH, such as adding new attributes to the algorithm [17], using the KH hybrid with other methods [18–21], exchanging information between top krill during the motion calculation process [22], using the best parameters [23], and adding local searches to improve search ability [24]. Aside from KH being applicable to many types of problems, it is also valid for function optimization [25]. An excellent review of the KH method was published by Want et al. [26].

The newly developed heuristics that have been used to solve real world problems include the parallel hurricane optimization algorithm [27], firefly-inspired krill herd (FKH) [28], Moth Search (MS) algorithm [29], Monarch Butterfly Optimization [30–32], Across Neighborhood Search (ANS) [33], Chaotic particle-swarm krill herd (CPKH) [34], Chaotic cuckoo search (CCS) [35], self-adaptive probabilistic neural network [36], and Differential Evolution algorithm (DE) [37].

Price and Storn [37] introduced the DE algorithm in 1995. Furthermore, using DE to solve logistics problems has attracted the attention of researchers and a program was developed in 2009 by Erbao and Mingyong [38] to solve vehicle routing problems (VRPs) and fuzzy demands by using DE. The program was used to design models in stochastic simulations and to create an algorithm using hybrid intelligence. The result of the study was an index of dispatcher settings returning the best value by using crossover parameters to develop different answers from the local optimum. Later, Dechampai et al. [39] presented a DE algorithm for a VRP, which had a vehicle capacity limit for the poultry industry, using two heuristics. Both heuristics were used for grouping customers before arranging transportation routes, which provided 7.59–31.28% lower costs than the present method and was better than the first heuristic by 0.84–13.15%. A year later, Sethanan and Pitakaso [40] presented a DE algorithm for scheduling raw milk transportation. The purpose was to find the lowest fuel cost, cleaning cost, and cost of disinfecting raw milk tanks in vehicles by presenting a modification of five DE methods with a two-step emerging and survival process called the re-born vector. The results were shorter routes and less truck transportation. Later the same year, Sethanan and Pitakaso [3] presented the development of DE in solving general operations by using three local search techniques to find better answers. Those three techniques were developed into seven other methods. They also measured the effectiveness of each method to find the best method to compare them with the BEE and Tabu algorithms in an experimental set of Gapa-Gape. DE was found to be better than the other two methods.

During the literature review, we studied the following problems related to DE for crop planning: the crop planning problem [41], plant production patterns [42], cropping patterns focused on the selection of crops and the allocation of cultivated areas using DE and gradient-based methods [43,44], the use of DE to solve crew rostering problems [45], the use of DE for multi-objective crop planning [46], the use of strategies of differential evolution algorithms (EPSDE) using parameters, mutation, and crossover [47], the development of multi-objective algorithms for optimal crop planning [48], and the use of DE for crop planning single and multi-objective optimization models [49]. Furthermore, we examined an application of DE presented by Pant et al. [50] to determine an optimal crop plan for the Pamba-Achankovil-Vaippar (PAV) link project area. Finally, the authors reviewed a study by Yi et al. [51], who used three improved hybrid metaheuristic algorithms for engineering design optimization. It can be seen from the abovementioned studies that DE is a highly effective algorithm for finding answers, which was an important part of the current authors' decision to use DE.

Improving the capability of DE involves many options. The hybrid DE has been widely used to improve the solution quality of the original DE search, like hybrid DE with Bat Algorithm [52], KH [53], or Particle Swarm Optimization [54]. DE uses some techniques to enhance the search capacity, such as using multi-population [55] or variable reduction [56] strategies.

Metaheuristics algorithms are effective for both single objectives [57–60] and multi-objectives [61–64].

Due to the success of metaheuristics in many numerical and combinatorial optimization problems such as those above, DE is a type of metaheuristics that is easy and powerful. It uses less computational time, fewer parameters, improves solution quality, and generates excellent results for many problems. Therefore, DE was selected to implement with this study's crop planning problem.

The contribution of this research is three-fold: (1) The transportation logistics are integrated into the model as well as selecting the suitable area to grow an appropriate plant, enabling higher productivity and reducing logistics costs; (2) The DE is modified by enhancing the diversification and intensification of the original DE, adding three local search techniques to the original DE; (3) It presents an excellent decoding method that can perform the assignment and transportation decision in the same code.

3. Mathematical Model for Crop Planning

During planting of the three economic crops, each farmer was assigned to plant only one crop and each farmer was allocated a transport vehicle, considering the lowest transport costs to maximize profit. The products of each farmer were sent to the factory or purchased on-site according to the type of crop (Figure 1).

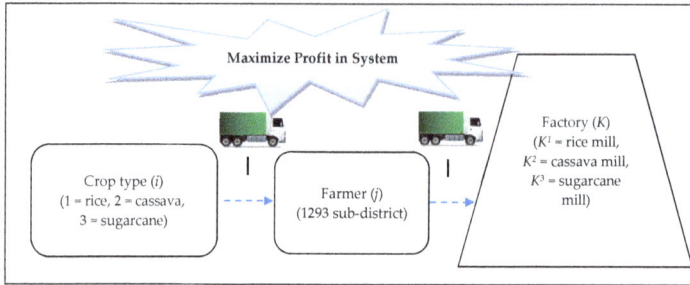

Figure 1. Model of transport of economic crops from the farmer to the factory.

The mathematical model for economic crop planning follows.

3.1. Indices

i stands for crop type (1 = rice, 2 = cassava, 3 = sugarcane)
j stands for planning area/famer ($j = 1, 2, \ldots, J$)
K stands for factory ($K = K^1, \ldots, K^i$, when K^1 = rice mill, K^2 = cassava mill, K^3 = sugarcane mill)

3.2. Parameters

P_{ij} stands for crop price i planted by farmer j (Baht/Kilogram)
C^1_{ij} stands for cost of planning i planted by farmer j (Baht/Rai)
B_{ij} stands for rate of crop yield i planted by farmer j (Ton/Rai)
A_j stands for planning area in each district (Rai)
D_{jk} stands for distance from planning area j to factory k (kilometer)
C^2_i stands for transportation costs of each crop i (Baht/Kilometer)
C_K stands for factory purchase capacity (Ton)
C^3_{ij} stands for cost of crop cultivation i planted by farmer j (Baht/Rai)
V_j stands for transportation capacity (Ton)
M stands for maximum production capacity

3.3. Decision Variables

$$X_{ijk} = \begin{cases} 1, \text{ if there is transportation } i \text{ from farmer } j \text{ to factory } k \\ 0, \text{ other cases} \end{cases}$$

$$Y_{ij} = \begin{cases} 1, \text{ if there is assignment of planting crop } i \text{ by farmer } j \\ 0, \text{ other cases} \end{cases}$$

H_{ijk} = Quantity factory k is given from crop i from farmer j
T^1_{ij} = Number of transport cycles, which must be an integer (Round)
T^2_{ij} = Number of crop transportation i by farmer j (Round)

3.4. Objective Function

We designed and developed a mathematical model to maximize the profits for the farmers. The related factors considered were crop price, cost of each crop's cultivation, yield rate of each crop, planning area in each district, transportation distance from planning area to factory, cost of each crop's transportation, amount of each crop's transportation, and cost of cultivation, which are expressed as follows:

$$\text{Maximize } Z = \sum_{i=1}^{I}\sum_{j=1}^{J}(P_{ij} - C^1_{ij})Y_{ij}B_{ij}A_j - \sum_{i=1}^{I}\sum_{j=1}^{J}\sum_{k=1}^{K}X_{ijk}D_{jk}C^2_{i}T^2_{ij}$$
$$- \sum_{i=1}^{I}\sum_{j=1}^{J}Y_{ij}C^3_{ij} \tag{1}$$

The objective function focused on economics to maximize profits for the farmers. Equation (1) consists of three main sequences: Sequence (1) is a function of raw materials cost, which depends on the purchase price of each crop, the cost of planting each crop, the size of planting area, and the yield rate; Sequence (2) is a function of transportation cost, which depends on the amount of raw materials, the transportation distance to factory, the cost of transportation for each crop, and the number of transportation rounds for each crop; Sequence (3) is a function of the cost of raising crops, which depends on the budget for raising each crop.

3.5. Constraints

$$\sum_{i=1}^{I}Y_{ij} = 1\, j \in J\, \forall j \in J \tag{2}$$

where the constraints function consists of Equation (2), which limits each farmer to planting only one crop.

$$\sum_{j=1}^{J}Y_{ij} \geq 1\, \forall i \in I \tag{3}$$

where Equation (3) is a limit requiring each crop to be assigned to at least one farmer.

$$\sum_{K \in K^i}H_{ijk} = Y_{ij}B_{ij}A_j\, \forall i \in I\, \forall j \in J \tag{4}$$

where Equation (4) is a limit requiring that the amount of crop yield to be delivered to the factory must be equal to the number of crops planted by each farmer.

$$H_{ijk} \leq MX_{ijk}\, \forall i \in I\, \forall j \in J\, \forall k \in K \tag{5}$$

where Equation (5) is a limit requiring that the amount of crop yield to be delivered to the factory must not exceed the number of crops planted by each farmer.

$$\sum_{k=1}^{K}X_{ijk} \leq 1\, \forall i \in I\, \forall j \in J \tag{6}$$

where Equation (6) is a limit requiring that a farmer can only use one transport route to the factory.

$$\sum_{j=1}^{J} H_{ijk} \leq C_K \; \forall i \in I \; \forall K \in K^i \tag{7}$$

where Equation (7) is a limit requiring that the number of crops must not exceed the purchase capacity of the factory.

$$T^2_{ij} \geq T^1_{ij} \; \forall i \in I \; \forall j \in J \tag{8}$$

where Equation (8) limits the number of transportation rounds, which must come from the yield rate multiplied by the crop area and divided by the capacity of the transport vehicle.

$$T^1_{ij} = \frac{Y_{ij} B_{ij} A_j}{V_j} \; \forall i \in I \; \forall j \in J \tag{9}$$

where Equation (9) is a limit requiring that the number of transportation rounds be an integer (round).

$$Y_{ij} M \geq \sum_{K \in K^i}^{K} X_{ijk} \; \forall i \in I \; \forall j \in J \tag{10}$$

where Equation (10) is a limit requiring that the maximum yield rate delivered when a farmer is assigned crop planting must not be less than the amount of yield rate sent from the farmer to the factory.

$$Y_{ij}(bin) \; \forall i \in I \; \forall j \in J \tag{11}$$

where Equation (11) is a limit requiring that a farmer who plants each crop be assigned a value of 0 or 1 only; 1 for plant and 0 for others.

$$X_{ijk}(bin) \; \forall i \in I \; \forall j \in J \; \forall k \in K \tag{12}$$

where Equation (12) represents the decision variables for when farmer j transports each crop to the factory k; value 0 or 1 only.

$$H_{ijk}(gin) \; \forall i \in I \; \forall j \in J \; \forall k \in K \tag{13}$$

where Equation (13) is a limit requiring that, for the amount of crop i from farmer j to the factory k, the value is an integer.

$$T^2_{ij}(gin) \; \forall i \in I \; \forall j \in J \tag{14}$$

where Equation (14) is a limit requiring that the number of transportation rounds be an integer.

4. Original Differential Evolution Algorithm

We used a DE algorithm to find the solution for the crop planning problem. There were four steps involved, which are outlined below.

4.1. Initial Population

The population number (NP) is determined by a process of random sample selection from the population under a certain limit. The population group is calculated for the answer and is called the fitness value, which entails creating the preliminary answer by using the initial population. The vector is designed based on the problem statement. The proposed crop planning problem involves deciding which famers will grow which type of plant. Considering this case study, three types of plants were used: (1) rice, (2) cassava, and (3) sugarcane. Table 1 shows the vector generated by the condition of having 10 farmers, 3 types of plants, and 5 populations for each iteration.

Table 1. Five vectors of the size 1 × 10.

Farmer Vector	1	2	3	4	5	6	7	8	9	10
1	0.84	0.39	0.92	0.56	0.06	0.72	0.85	0.19	0.27	0.09
2	0.71	0.45	0.40	0.63	0.78	0.07	0.49	0.81	0.71	0.30
3	0.55	0.20	0.63	0.34	0.39	0.14	0.50	0.43	0.69	0.17
4	0.80	0.75	0.49	0.76	0.74	0.55	0.11	0.34	0.65	0.51
5	0.22	0.20	0.12	0.42	0.74	0.90	0.49	0.44	0.73	0.40

Table 1 shows the example of 5 vectors (NP = 5), with each vector having a dimension of 1 × 10 due to there being 10 farms for which to make decisions. There is one rice mill (which has a capacity of 120 tons), two sugar mills (each sugar mill has a capacity of 80 tons), and one tapioca starch mill (which has a capacity of 90 tons). Ten fields have an expected production output (in tons) if they grow different types of plants, as shown in Table 2.

Table 2. Amount of product produced from each field if they grow different types of plants (ton).

-	1	2	3	4	5	6	7	8	9	10
Rice	15	20	18	8	15	13	19	21	19	15
Sugarcane	18	18	20	10	12	16	23	21	23	15
Cassava	25	16	29	9	14	20	30	19	23	16

The distance between each field to the factory is given in Table 3.

Table 3. Distance from fields to factories. RM is rice mill, SM1 and SM2 are the two sugar mills, and TS is the tapioca starch mill.

Factory Field	RM	SM1	SM2	TS
1	279	210	148	141
2	319	186	252	332
3	107	316	199	234
4	305	323	202	272
5	285	150	158	308
6	200	245	203	179
7	189	283	173	301
8	185	122	102	326
9	291	130	141	229
10	320	125	111	141

4.2. Decoding Method

Tables 1 and 2 as well as the information given above, were used to decode the continuous number to get the problem's solution by the following steps:

(1) Arrange the numbers in the vectors (value in position of each vector) in increasing order. For example, for Vector 1, the result of sorting is shown in Table 4.

Table 4. Sorting result of Vector 1.

Farmer Vector	5	10	8	9	2	4	6	1	7	3
1	0.06	0.09	0.19	0.27	0.39	0.56	0.72	0.84	0.85	0.92

Taken from Table 3, it can be seen that the order of the fields is 5, 10, 8, 9, 2, 4, 6, 1, 7, and 3, which is the order according to the sorting result of the value in the position of Vector 1.

(2) Assign the type of plant to a field and assign the field to the factory. The roulette wheel idea was used to assign the plants to the field. The probability of each plant to be selected can be determined by any idea, such as: (1) equal probability (0.333 each); (2) average productivity rate, such as rice, sugarcane, and cassava having average productivity rates of 19, 17, and 23 tons/km^2, with probabilities of 0.32, 0.29, and 0.39, respectively; (3) price per kilogram of the plants, for example, if the prices of rice, sugarcane, and cassava are 1000, 1200, and 900 baht per ton, the probability of each plant is 0.32, 0.39, and 0.29, respectively; (4) the ratio of factories that will take all the products. There is one rice mill, which has a capacity of 120 tons. There are two sugar mills, each of which has a capacity of 80 tons and, thus, a combined total capacity of 160 tons. There is one tapioca starch mill, which has a capacity of 90 tons. Therefore, each type of plant has a probability of 0.32, 0.43, and 0.25, respectively. During this research, we used the price per kilogram of the plants as the probability to select the plants to maximize the profits generated from the algorithm. Taken from the probability above (using price as the probability), a plant will be assigned to a field according to the value in the position of vector and that field will be assigned to the closest factory, as long as it has enough capacity. The assignment process is shown in Steps (a)–(d).

(a) Calculate the probability of assigning the plants to the fields. Using the proposed algorithm, we applied the price of the plants to calculate this. The probabilities of assigning rice, sugarcane, and cassava were 0.32, 0.39, and 0.29, respectively, for the current example.

(b) Calculate the cumulative probability of each type of plant. Taken from Step (2), the cumulative probabilities of rice, sugarcane, and cassava were 0.32, 0.71, and 1.0, respectively. Use the value in the vector position to decide which plants to grow. The Vector 1 position, which is Field 5, had value of 0.06. This value (0.06) falls in the rice area of the roulette wheel; therefore, Field 5 is assigned to grow rice. Then, Field 5 is assigned to the rice factory (using the information shown in Table 3). This rice factory has a 120-ton capacity and Field 5 can produce 15 tons. Thus, the rice mill has 105 tons remaining to receive rice from other fields.

(c) Redo Step (b) until all fields are assigned to grow exactly one type of plant. This step is needed to evaluate whether a factory has enough capacity to receive the product from all fields that fall into the area of that type of plant. When the factory does not have enough capacity, the field that has a higher value in that position needs to be changed to grow other types of plants. Fields 6, 1, 7, and 3 grow cassava because the values in the Vector 1 positions for these fields were 0.72, 0.84, 0.85, and 0.92, which fall in the area of cassava, for example. However, if all addressed fields grow cassava, this will generate 104 tons of cassava. The tapioca starch mill has only a 90-ton capacity, thus the last field needs to be changed to produce other types of plants. Field 3 needs to change to produce rice in this case. The total cassava that will be produced will decrease from 104 to 75 tons and the amount of rice that will be produced in the plan will increase from 60 to 78 tons (using the information given in Table 2). The assignment of the field to the factory, in case there is more than one factory to select, can be executed by selecting to deliver the product to the factory from that field. Field 4, which grows sugarcane, needs to deliver the sugarcane to the sugar mill and there are two sugar mills, for example. Field 4 has a distance to SM1 and SM2 of 323 and 202 km, respectively (using the information given in Table 3). Therefore, Field 4 will deliver sugarcane to SM2 at a distance of 202 km.

(d) Calculate all profit and cost terms according to the assignment obtained from Step (b).

The results of the assignment phase (Steps (a) and (b)) are shown in Table 5.

Table 5. The assignment results.

Factory	Field	Plant
RM	5, 10, 8, 9, 3	Rice
SM1	2	Sugarcane
SM2	4	Sugarcane
TS	6, 1, 7	Cassava

4.3. Mutation Process

The mutation process was executed using Equation (15), where $V_{i,j,G}$ is the mutant vector of i position j iteration $G + 1$; $X_{r1,j,G}$, $X_{r2,j,G}$, and $X_{r3,j,G}$ are random target vectors 1, 2, and 3, respectively; and F is the predefined scaling factor.

$$V_{i,j,G} = X_{r1,j,G} + F\left(X_{r2,j,G} - X_{r3,j,G}\right) \tag{15}$$

The value of the weighting factor (F) can range from 0 to 2 and was set to $F = 2.0$ [65]. The value in the vector coordinate was changed by using random numbers and then entered into the mutation process (mutant vector). This conventionally is called DE/rand/1/bin.

4.4. Crossover or Recombination Process

This is a mixed species process that produces new species of better or worse results for the selection of decision variables. The result is the trial vector ($U_{i,j,G}$).

Set $CR = 0.8$ [7,19], then enter the value exchange with Equation (16):

$$U_{i,j,G} = \begin{cases} V_{i,j,G} & if\ Rand_{i,j} \leq CR \\ X_{i,j,G} & if\ Rand_{i,j} > CR \end{cases} \tag{16}$$

4.5. Selection Process

This is the selection process for the best answer between the target vector and the trial vector using Equation (17), which was accomplished by comparing the function value or the cost value of the trial vector with the target vector. When the function value of the trial vector was better than the target vector, it was replaced by the trial vector in the next generation.

$$X_{i,j,G+1} = \begin{cases} U_{i,j,G} & if\ f\left(U_{i,j,G}\right) \leq f\left(X_{i,j,G}\right) \\ X_{i,j,G} & otherwise \end{cases} \tag{17}$$

Then, the answers are adjusted in each NP to determine if a better answer can be found. The answer (objective) was found from the calculation, compared against others, and the best answer was chosen from the entire population.

5. Improved Differential Evolution Algorithm

This section outlines developing and improving algorithms with DE and using Dev-C++ for testing three different-sized problems. DE is used with local search to develop the algorithm by adding a specific search step after the value exchange process in the recombination. The algorithms are developed to provide better results. There were three methods: (1) the swap algorithm, adapted from the method of Diaz and Fernandez [66]; (2) the cyclic move algorithm; and (3) the K-variable move algorithm. Figure 2 illustrates how these methods improved the steps of DE. The DE that is used in this article is conventionally called DE/rand/1/bin.

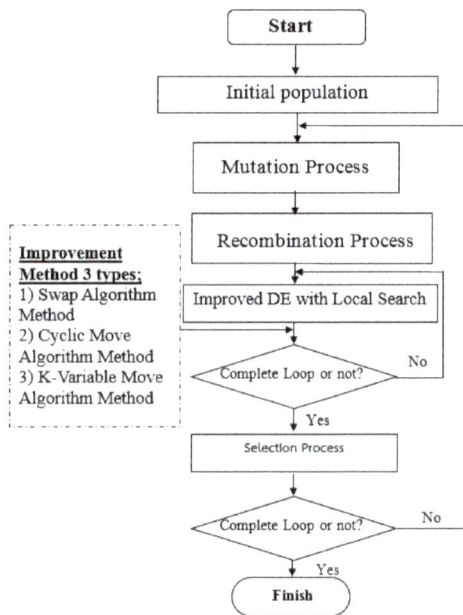

Figure 2. Steps using improved Differential Evolution (DE) algorithm with three local search methods.

5.1. Swap Algorithm

The swap algorithm is a method used to improve a heuristic-based solution by switching pair positions between groups of members. Assuming originally that Farmer 10 is assigned to plant rice, the algorithm will switch this farmer with Farmer 24 who plants cassava to make a greater profit. Then, it will switch Farmer 8, who planted cassava, with Farmer 30, who planted sugarcane, relocating all the positions using the same process. The amount of output that the farmer sells must not exceed the capacity of the factory. The switched pairs increase the profit, as seen in Figure 3.

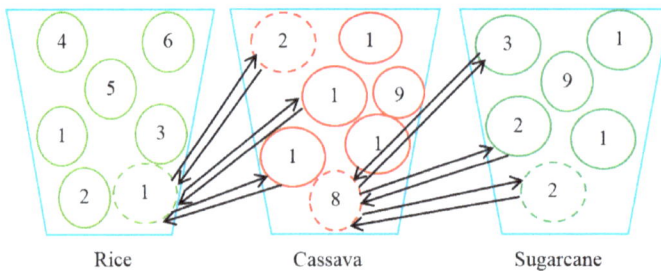

Figure 3. An example of developing an answer by the swap algorithm.

5.2. Cyclic Move Algorithm

This method selects one farmer from each group then switches each farmer in a circle. Farmer 6, who is originally assigned to plant rice, for example, is changed to cassava. Next, Farmer 8, who plants cassava, is changed to sugarcane. Additionally, Farmer 11, who plants sugarcane, is changed to rice, relocating all the positions using the same method by randomizing all rounds. Once again, the switched pairs increase the profit, as seen in Figure 4.

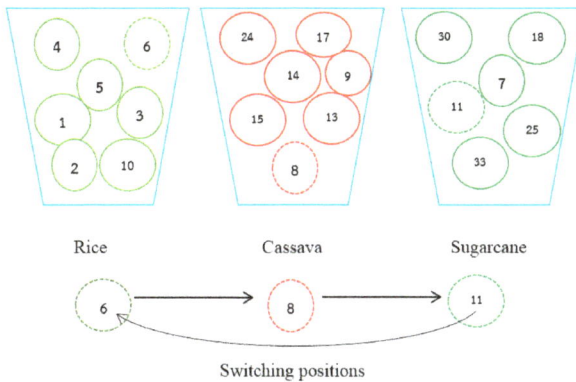

Figure 4. An example of developing an answer by the cyclic move algorithm.

5.3. K-Variable Move Algorithm

To use this algorithm, K = 5 (from randomly testing all rounds), then one crop is chosen from each group, and the farmer is switched to another crop. Farmer 17, for example, is originally assigned to plant cassava and then changed to sugarcane. Next, Farmer 33, who plants sugarcane, is changed to cassava. Farmer 8, who plants cassava, is changed to sugarcane. Farmer 24, who plants sugarcane, is changed to rice. Farmer 1, who plants rice, is changed to cassava. All the positions are relocated in the same process, but the amount of output that the farmers sell must not exceed the capacity of the factory. Once again, the switched pairs increase the overall profit, as seen in Figure 5.

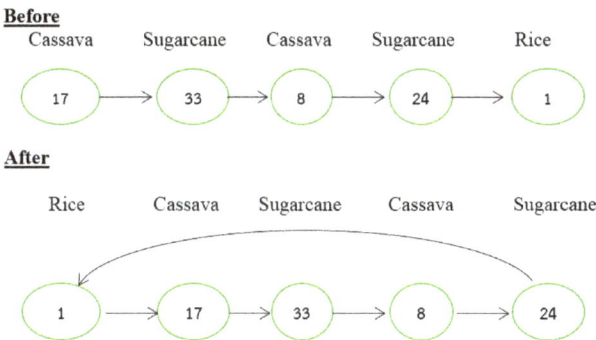

Figure 5. An example of developing an answer by the K-variable move algorithm.

6. Computational Experiment and Results

The computational results are divided into two parts. First, the result of the comparison of the proposed method (DE) with the result generated by Lingo v.11 is presented to check if the proposed heuristics are reliable and trustable. Second, a simulation is used to check if the improved DE (I-DE) is better than that of the original DE to determine if the contribution of adding three local search methods to the original DE has any benefit compared to the original algorithm.

6.1. Experimental Results of Differential Evolution Algorithm (DE) Compared with Lingo V.11

We used a general DE and a developed DE to apply to and solve problems by using Dev-C++ in design algorithms, which were calculated to compare the outcome with the processing unit (Intel® CoreTM i5-2410M 2.3 GHz and 4 GB memory). DE ran five replicates and the best solution among all five runs was the solution represented in Table 6. The problem instances were categorized into three groups: (1) small problem group, 5–20 farmers; (2) medium problem group, 40–70 farmers; and (3) large problem group, 80–500 farmers. Four test instances were randomly generated for the small-size instances, whereas three test instances were generated for medium- and large-size test instances. One real case study was used in the simulation test. The authors had 11 test instances in total. The stopping criterion for Lingo was set to when it found the optimal solution, with a maximum duration of 250 h. The stopping criterion of DE was the number of iterations, with a threshold of 500 iterations. Set NP = 50, F = 2.0, and CR = 0.8 [7,19]. The results of DE compared with Lingo v.11 are shown in Table 6.

Table 6. Experimental results of the Differential Evolution (DE) algorithm compared with Lingo v.11.

Problem Group	Farmer or Subdistrict	Rice Number of Factories	Cassava Number of Factories	Sugarcane Number of Factories	Methods			
					Lingo		Differential Evolution Algorithm (DE)	
.	.	.	.	.	Solution (Baht)	Time (s)	Solution	Time (s)
1 (Small Size)	5	3	3	4	153,040	00:00:02	153,040	00:00:01
	10	3	3	4	538,909	00:00:05	538,909	00:00:03
	15	3	3	4	704,463	00:00:27	704,463	00:00:05
	20	7	7	5	1.1175×10^6	00:00:46	1.1175×10^6	00:00:14
2 (Medium size)	40	16	30	5	2.4585×10^6	00:03:39	2.4585×10^6	00:00:23
	60	30	50	5	3.73617×10^6	00:06:50	3.73617×10^6	00:01:38
	70	35	55	3	4.31865×10^6	00:10:25	4.31865×10^6	00:03:37
3 (Large Size)	80	45	60	5	4.3341×10^6	250 h *	4.45924×10^6	00:02:13
	80	50	70	7	4.3378×10^6	250 h *	4.75384×10^6	00:03:26
	80	55	70	10	4.3467×10^6	250 h *	4.68498×10^6	00:03:69
	100	60	70	10	4.3593×10^6	250 h *	4.54516×10^6	00:10:16
	100	60	80	10	4.4355×10^6	250 h *	4.61473×10^6	00:13:10
	100	60	80	10	4.4355×10^6	250 h *	4.51473×10^6	00:13:10
	500	70	85	13	5.13148×10^6	250 h	5.18148×10^6	00:16:34
Case study	1293	95	98	19	1.39234×10^7	250 h *	1.41761×10^7	00:21:82

Remarks: * Best solution generated in 250 h.

Looking at the results in Table 6, it can be seen that the small- and medium-sized test instances of DE can find the same solution as Lingo v.11, using lower computational time. Regarding the large-size instances (including the case study), DE generated a better solution than that of Lingo v.11, which ran for 250 h, whereas DE ran for a maximum of 21.82 min to obtain the result. DE used less computational time in all test instances than Lingo v.11. The Wilcoxon signed-rank test was applied using a 95% confidence interval to compare the results of DE and Lingo v.11, which provided significant evidence to conclude that the results generated by Lingo v.11 and DE were different (p-value = 0.004 and α = 0.05). Therefore, we concluded that DE can find a better solution than Lingo v.11 when Lingo v.11 has a computational limit time set to 250 h (p-value = 0.043 and α = 0.05).

6.2. Experimental Results of DE Compared with Modified DE

The authors had three improved DE algorithms (I-DE), which were I-DE-SW (DE + swap algorithm), I-DE-CY (DE + cyclic move algorithm), and I-DE-KV (DE + K-variable move algorithm). The stopping criterion used in this session was a 2 h simulation time. Five simulation runs were performed and the best solution among all runs is shown in Table 7.

Table 7 shows the experiment used to compare the efficiency of the original DE with the I-DEs using the same duration (two hours). I-DE-KV provided the most effective answer and most comprehensive range of answers compared with the other methods, especially for the large-size problems and the case study for 1293 farmers. The best answer for the case study generated by I-DE-KV was 14,596,430 baht per crop cycle. I-DE-KV could also show the factory location and areas for economic crop planning.

Table 7. Overall profitability for the general DE and Improved DE (I-DE) methods with equal duration.

Problem Size	N (A-B-C)	Solution (Profit: Baht)			
		DE	I-DE-SW	I-DE-CY	I-DE-KV
Small size	5 (3-3-4)	153,040	208,933	153,040	208,933
	10 (3-3-4)	538,909	565,109	554,117	565,109
	15 (3-3-4)	704,463	704,463	704,219	704,219
	20 (7-7-5)	1.1175×10^6	1.10857×10^6	1.11875×10^6	1.11877×10^6
Medium size	40 (16-30-5)	2.4585×10^6	2.4585×10^6	2.4585×10^6	2.4585×10^6
	60 (30-50-5)	3.73617×10^6	3.73617×10^6	3.73617×10^6	3.73711×10^6
	70 (35-55-3)	4.31865×10^6	4.31865×10^6	4.31865×10^6	4.31865×10^6
Large size	80 (45-60-5)	4.55924×10^6	4.55924×10^6	4.55924×10^6	4.56032×10^6
	80 (50-70-7)	4.47249×10^6	4.65214×10^6	4.69214×10^6	4.71241×10^6
	80 (55-70-10)	4.78129×10^6	4.79241×10^6	4.79201×10^6	4.84542×10^6
	100 (60-70-10)	4.55681×10^6	4.60241×10^6	4.61042×10^6	4.72124×10^6
	100 (60-80-10)	4.64981×10^6	4.76292×10^6	4.75124×10^6	4.78419×10^6
	100 (60-80-10)	4.65924×10^6	4.71105×10^6	4.70924×10^6	4.72149×10^6
	500 (70-85-13)	5.23148×10^6	5.36866×10^6	5.3548×10^6	5.37866×10^6
Case study	1293 (95-98-19)	1.427614×10^7	1.457129×10^7	14.32278×10^6	14.59643×10^6

Note: **N** stands for Farmers, **A** stands for rice mill, **B** stands for cassava mill, and **C** stands for sugarcane mill.

The pair t-test has been executed using significant level $\alpha = 0.05$ and the result is shown in Table 8. The statistical test was performed only with the large instances. Considering the small and medium instances, it can be seen that DE and the proposed algorithm were not different (all results are the same).

Table 8. Pair *t*-test results (sign (*p*-value)).

Method	I-DE-SW	I-DE-CY	I-DE-KV
DE	$\leq$(0.02 *)	$\leq$(0.012 *)	$\leq$(0.009 *)
I-DE-SW	-	=(0.403)	$\leq$(0.038 *)
I-DE-CY	-	-	$\leq$(0.036 *)

Note: * means significantly different.

Viewing the statistical test shown in Table 8, I-DE-SW and I-DE-CY were significantly different to I-DE-KV. Thus, the K-variable move improved the solution quality of the DE algorithm. Additionally, I-DE-SW and I-DE-CY performance was not significantly different, which can be interpreted that when a local search in DE is added, it can improve the efficiency of DE (all proposed methods were significantly different from DE). This might be changed if a local search does not perform differently, especially when the neighborhood size is unaltered.

The last experiment tested I-DE-KV (computational time is set to two hours), which was the best proposed algorithm with the upper bound (maximization problem), and the best solution generated from Lingo v.11 within 200, 250, 300, 350, and 400 h. The result is shown in Table 9. The simulation considers only the large problem and the case study.

Table 9. Result of Lingo v.11 (Upper Bound and Best Objective) and I-DE algorithm using K-variable move (I-DE-KV).

Time / Instance	Method	200 h	250 h	300 h	350 h	400 h
80 (45-60-5)	Upper Bound	4.58894×10^6	4.58894×10^6	4.58894×10^6	4.57591×10^6	4.57591×10^6
	Best Objective	4.55145×10^6	4.55145×10^6	4.55981×10^6	4.55981×10^6	4.55981×10^6
	I-DE-KV			4.56032×10^6		
80 (50-70-7)	Upper Bound	4.80512×10^6	4.80512×10^6	4.78214×10^6	4.78214×10^6	4.76542×10^6
	Best Objective	4.56821×10^6	4.56995×10^6	4.58912×10^6	4.60156×10^6	4.60156×10^6
	I-DE-KV			4.71241×10^6		
80 (55-70-10)	Upper Bound	4.90124×10^6	4.90259×10^6	4.899128×10^6	4.899128×10^6	4.899128×10^6
	Best Objective	4.71249×10^6	4.71249×10^6	4.71249×10^6	4.71249×10^6	4.71249×10^6
	I-DE-KV			4.84542×10^6		
100 (60-70-10)	Upper Bound	4.98241×10^6	4.85215×10^6	4.79242×10^6	4.79242×10^6	4.79242×10^6
	Best Objective	4.61983×10^6	4.62488×10^6	4.62488×10^6	4.62488×10^6	4.63245×10^6
	I-DE-KV			4.72124×10^6		
100 (60-80-10)	Upper Bound	4.88812×10^6	4.88812×10^6	4.85289×10^6	4.85289×10^6	4.81457×10^6
	Best Objective	4.62388×10^6	4.62388×10^6	4.62388×10^6	4.64514×10^6	4.64514×10^6
	I-DE-KV			4.73419×10^6		
100 (60-80-10)	Upper Bound	4.8190×10^6	4.8190×10^6	4.8024×10^6	4.8024×10^6	4.79842×10^6
	Best Objective	4.6786×10^6	4.69782×10^6	4.70113×10^6	4.70113×10^6	4.70113×10^6
	I-DE-KV			4.71105×10^6		
500 (70-85-13)	Upper Bound	5.46991×10^6	5.46991×10^6	5.46892×10^6	5.46892×10^6	5.44412×10^6
	Best Objective	5.34917×10^6	5.35178×10^6	5.35178×10^6	5.35980×10^6	5.35991×10^6
	I-DE-KV			5.37866×10^6		
1293 (95-98-19)	Upper Bound	14.7919×10^6	14.7919×10^6	14.7919×10^6	14.7814×10^6	14.7814×10^6
	Best Objective	14.56891×10^6	14.56891×10^6	14.57129×10^6	14.57643×10^6	14.57610×10^6
	I-DE-KV			14.59643×10^6		

Concerning Table 9, "Upper Bound" is the upper bound found by Lingo v.11 within the predefined computational times (200, 250, 300, 350, and 400 h) and "Best Objective" is the best objective found by Lingo v.11 within the predefined computational times. The I-DE-KV result is the outcome of the proposed heuristics using two hours as the computational time. Taken from Table 9, it can be simplified into %$diff$, which can be calculated from Equation (18):

$$\%diff. = \frac{Result\ from\ Candidate\ Method - Result\ from\ I-DE-KV}{Result\ from\ I-DE-KV} \times 100\% \tag{18}$$

Table 10 shows the %$diff$ of the Upper Bound and the Best Objective found by Lingo v.11 using different computational times.

Table 10 shows that the Upper Bound has all positive %$diff$ values, and Best Objective has negative %$diff$ values. This means that Upper Bound generates more profit to the system than I-DE-KV, but Best Objective has lower profits than I-DE-KV. Upper Bound can be a feasible or an infeasible solution. Therefore, it generates a better solution than the DE algorithm. The Best Objective is a feasible solution and it is the best solution found during the simulation. Taken from this, we concluded that the DE algorithm can find a better solution than Lingo v.11, and Lingo v.11 uses a computational time of 200–400 h and DE runs for only two hours. The average %$diff$ of I-DE-KV and Upper Bound generated by Lingo v.11, when using 200, 250, 300, 350, and 400 h of computational time were 2.232, 1.891, 1.523, 1.479, and 1.265%, respectively. The %$diff$ of Best Objective and I-DE-KV ranged from -1.488 to -1.196%. Further, I-DE-KV had a gap of only 0.342–5.532% from the Upper Bound, from which we concluded that it is an effective algorithm. The pair t-test was used to see if I-DE-KV performed differently than Upper Bound and Best Objective found by Lingo v.11. The result is shown in Table 11.

Table 10. %*diff* of the candidate method and I-DE-KV.

Time	Method	200 h	250 h	300 h	350 h	400 h
Instance						
80	Upper Bound	0.628	0.628	0.628	0.342	0.342
(45-60-5)	Best Objective	−0.195	−0.195	−0.011	−0.011	−0.011
80	Upper Bound	1.967	1.967	1.480	1.480	1.125
(50-70-7)	Best Objective	−3.060	−3.023	−2.616	−2.352	−2.352
80	Upper Bound	1.152	1.180	1.108	1.108	1.108
(55-70-10)	Best Objective	−2.743	−2.743	−2.743	−2.743	−2.743
100	Upper Bound	5.532	2.773	1.508	1.508	1.508
(60-70-10)	Best Objective	−2.148	−2.041	−2.041	−2.041	−1.881
100	Upper Bound	3.251	3.251	2.507	2.507	1.698
(60-80-10)	Best Objective	−2.330	−2.330	−2.330	−1.881	−1.881
100	Upper Bound	2.291	2.291	1.939	1.939	1.855
(60-80-10)	Best Objective	−0.689	−0.281	−0.211	−0.211	−0.211
500	Upper Bound	1.697	1.697	1.678	1.678	1.217
(70-85-13)	Best Objective	−0.548	−0.500	−0.500	−0.351	−0.349
1293	Upper Bound	1.339	1.339	1.339	1.267	1.267
(95-98-19)	Best Objective	−0.189	−0.189	−0.172	−0.137	−0.139
Average	Upper Bound	2.232	1.891	1.523	1.479	1.265
	Best Objective	−1.488	−1.413	−1.328	−1.216	−1.196

Table 11. Statistical test of I-DE-KV and Upper Bound and Best Objective found by Lingo v.11.

	Method	Best Objective	I-DE-KV
Lingo	Upper Bound	≥(0.001 *)	≥(0.003 *)
	Best Objective	-	≤(0.003 *)

Note: * means significantly different.

Looking at Table 11, it can be seen that Upper Bound has a significantly higher benefit than Best Objective and I-DE-KV. Upper Bound normally contains an infeasible solution. The Best Objective has significantly lower benefit than I-DE-KV with a 95% confidence interval.

7. Conclusions

This study aimed to solve the economic crop planning allotment problem for farmers in eight provinces in the lower northeast region of Thailand by improving mathematical models and algorithms and by considering the economic value of maximizing profits for the farmers. The study focused on three economic crops—rice, cassava, and sugarcane—and used a DE algorithm to solve the problems to maximize profits for the farmers. The outcomes and running times of the DE algorithms were compared with Lingo v.11. When comparing the performance of all algorithms using DE in small-size problem simulations, DE had the best performance and enhanced the chance of finding a better outcome.

Three improved DE algorithms were proposed in this article: I-DE-SW, I-DE-CY, and I-DE-KV. Viewing the computational results, I-DE-KV generated the best method compared to all other methods proposed, including the original DE algorithm. All the I-DE algorithms found better solutions than the original DE. We concluded that a DE algorithm that includes local search can improve the efficiency of the original version. The swap algorithm limited the number of entities involved to two, the cyclic move algorithm had three entities involved, and the K-variable move algorithm can have two, three, four, or more depending on the random value of K. This attribute makes the K-variable move algorithm freer and more flexible, thus enabling it to generate the best solution among all proposed heuristics while using the same computational time.

Comparing I-DE-KV with Upper Bound, the proposed method resulted in a 1.265–2.232% difference from the Upper Bound. I-DE-KV found a 1.196–1.488% higher profit than that of the best solution found by Lingo v.11 using 400 h computational time, whereas DE only required two hours.

Therefore, the proposed heuristic is an effective algorithm for finding a good solution to the crop planning problem. It uses more than 200 times less computational time than that of Lingo v.11 while generating a better solution. Future work should focus on using different kinds of attractiveness to assign the plants to the fields.

The advantage of the proposed heuristic is that it is fast and can find a better solution in comparison to the results generated from Lingo v.11 when the problem size is large. The computational times when the problem size is small are not much different from that of Lingo v.11. Due to DE working on the random environment using a guided search, sometimes the guided search leads to a bad search area, which can result in local optimization and generate a worse solution than that of Lingo v.11—the optimization tool which always finds the lowest cost. The decision-maker must clarify the problem size first before making the decision to use optimization tools or metaheuristics to obtain the optimal solution.

Author Contributions: U.K. and R.P. collected the data and designed the algorithm; T.S. performed the experiment while K.S. analyzed the result.

Acknowledgments: We would like to thank Metaheuristics for Logistics Optimization Laboratory, Department of Industrial Engineering, Faculty of Engineering, Khon Kaen University and Department of Economics, Faculty of Business Administration, Rajamangala University of Technology Thanyaburi for funding this research.

Conflicts of Interest: The authors declare no conflicts of interest.

References

1. Agricultural Information Office. *Agricultural Information*; Agricultural Economics, Ministry of Agriculture and Cooperatives: Bangkok, Thailand, 2016.
2. Thongdee, T.; Pitakaso, R. Differential Evolution Algorithms Solving a Multi-Objective, Source and Stage Location-Allocation Problem. *Ind. Eng. Manag. Syst.* **2015**, *14*, 11–21. [CrossRef]
3. Sethanan, K.; Pitakaso, R. Improved differential evolution algorithms for solving generalized assignment problem. *Expert Syst. Appl.* **2016**, *45*, 450–459. [CrossRef]
4. Wang, G.; Tan, Y. Improving Metaheuristic Algorithms with Information Feedback Models. *IEEE Trans. Cybern.* **2017**, *99*, 1–14. [CrossRef] [PubMed]
5. Wang, G.; Guo, L.; Gandomi, A.H.; Hao, G.; Wang, H. Chaotic Krill Herd algorithm. *Inf. Sci.* **2014**, *274*, 17–34. [CrossRef]
6. Wang, G.; Gandomi, A.H.; Alavi, A.H. An effective krill herd algorithm with migration operator in biogeography-based optimization. *Appl. Math. Model.* **2014**, *38*, 2454–2462. [CrossRef]
7. Cui, Z.; Sun, B.; Wang, G.; Xue, Y.; Chen, J. A novel oriented cuckoo search algorithm to improve DV-Hop performance for cyber-physical systems. *J. Parallel Distrib. Comput.* **2016**, *103*. [CrossRef]
8. Wang, G.; Alavi, A.H.; Zhao, X.J.; Hai, C.C. Hybridizing harmony search algorithm with cuckoo search for global numerical optimization. *Soft Comput.* **2016**, *20*, 273–285. [CrossRef]
9. Feng, Y.; Wang, G. Binary Moth Search Algorithm for Discounted {0–1} Knapsack Problem. *IEEE Access* **2018**. [CrossRef]
10. Wang, G.; Deb, S.; Cui, Z. Monarch Butterfly Optimization. *Neural Comput. Appl.* **2015**. [CrossRef]
11. Wang, G.; Guo, L.; Gandomi, A.H.; Cao, L.; Alavi, A.H.; Duan, H.; Li, J. Lévy-Flight Krill Herd Algorithm. *Math. Probl. Eng.* **2013**. [CrossRef]
12. Wang, G.; Guo, L.; Duan, H.; Wang, H.; Liu, L.; Shao, M. Hybridizing Harmony Search with Biogeography Based Optimization for Global Numerical Optimization. *J. Comput. Theor. Nanosci.* **2013**, *10*, 2312–2322. [CrossRef]
13. Wei, Z.J.; Wang, G. Image Matching Using a Bat Algorithm with Mutation. *Appl. Mech. Mater.* **2012**, *203*, 88–93. [CrossRef]
14. Wang, G.; Coelho, L.; Gao, X.Z.; Deb, S. A new metaheuristic optimisation algorithm motivated by elephant herding behavior. *Int. J. Bio-Inspired Comput.* **2016**, *8*. [CrossRef]
15. Wang, G.; Deb, S.; Coelho, L.D.S. Earthworm optimization algorithm: A bio-inspired metaheuristic algorithm for global optimization problems. *Int. J. Bio-Inspired Comput.* **2018**, *12*, 1–12. [CrossRef]

16. Wang, G.; Guo, L.; Wang, H.; Duan, H.; Liu, L.; Li, J. Incorporating mutation scheme into krill herd algorithm for global numerical optimization. *Neural Comput. Appl.* **2014**, *24*, 853–871. [CrossRef]

17. Wang, G.; Gandomi, A.H.; Alavi, A.H. Stud krill herd algorithm. *Neurocomputing* **2014**, *128*, 363–370. [CrossRef]

18. Wang, G.; Gandomi, A.H.; Yang, X.; Alavi, A.H. A new hybrid method based on krill herd and cuckoo search for global optimisation tasks. *Int. J. Bio-Inspired Comput.* **2016**, *8*, 286–299. [CrossRef]

19. Wang, H.; Yi, J.H. An improved optimization method based on krill herd and artificial bee colony with information exchange. *Memet. Comput.* **2017**, *10*, 177–198. [CrossRef]

20. Wang, G.; Gandomi, A.H.; Alavi, A.H.; Deb, S. A hybrid method based on krill herd and quantum-behaved particle swarm optimization. *Neural Comput. Appl.* **2016**, *27*, 989–1006. [CrossRef]

21. Wang, G.; Guo, L.; Gandomi, A.H.; Alavi, A.H.; Duan, H. Simulated Annealing-Based Krill Herd Algorithm for Global Optimization. *Abstr. Appl. Anal.* **2013**. [CrossRef]

22. Guo, L.; Wang, G.; Gandomi, A.H.; Alavi, A.H.; Duan, H. A new improved krill herd algorithm for global numerical optimization. *Neurocomputing* **2014**, *138*, 392–402. [CrossRef]

23. Wang, G.; Gandomi, A.H.; Alavi, A.H. Study of Lagrangian and Evolutionary Parameters in Krill Herd Algorithm. In *Adaptation and Hybridization in Computational Intelligence. Adaptation, Learning, and Optimization*; Fister, I., Fister, I., Jr., Eds.; Springer: Cham, Switzerland, 2015.

24. Wang, G.; Gandomi, A.H.; Alavi, A.H.; Deb, S. A Multi-Stage Krill Herd Algorithm for Global Numerical Optimization. *Int. J. Artif. Intell. Tools* **2016**, *25*. [CrossRef]

25. Wang, G.; Deb, S.; Gandomi, A.H.; Alavi, A.H.A. Opposition-based krill herd algorithm with Cauchy mutation and position clamping. *Neurocomputing* **2016**, *177*, 147–157. [CrossRef]

26. Wang, G.; Gandomi, A.H.; Alavi, A.H.; Gong, D. A comprehensive review of krill herd algorithm: Variants, hybrids and applications. *Artif. Intell. Rev.* **2017**. [CrossRef]

27. Rizk, M.; Rizk, A.; Ragab, A.; El-Sehiemy, R.A.; Wang, G. A novel parallel hurricane optimization algorithm for secure emission/economic load dispatch solution. *Appl. Soft Comput.* **2018**, *63*, 206–222. [CrossRef]

28. Wang, G.; Gandomi, A.H.; Alavi, A.H.; Dong, Y. A Hybrid Meta-Heuristic Method Based on Firefly Algorithm and Krill Herd. In *Handbook of Research on Advanced Computational Techniques for Simulation-Based Engineering*; IGI: Hershey, PA, USA, 2016; pp. 521–540.

29. Wang, G. Moth search algorithm: A bio-inspired metaheuristic algorithm for global optimization problems. *Memet. Comput.* **2018**, *10*, 151–164. [CrossRef]

30. Feng, Y.; Wang, G.; Deb, S.; Lu, M.; Zhao, X. Solving 0–1 knapsack problem by a novel binary monarch butterfly optimization. *Neural Comput. Appl.* **2017**, *28*, 1619–1634. [CrossRef]

31. Feng, Y. Solving 0–1 knapsack problems by chaotic monarch butterfly optimization algorithm with Gaussian mutation. *Memet. Comput.* **2016**. [CrossRef]

32. Wang, G.; Deb, S.; Zhao, X.; Cui, Z. A new monarch butterfly optimization with an improved crossover operator. *Oper. Res.* **2016**. [CrossRef]

33. Wu, G. Across neighborhood search for numerical optimization. *Inf. Sci.* **2016**, *329*, 597–618. [CrossRef]

34. Wang, G.; Gandomi, A.H.; Alavi, A.H. A chaotic particle-swarm krill herd algorithm for global numerical optimization. *Kybernetes* **2013**, *42*, 962–978. [CrossRef]

35. Wang, G.; Deb, S.; Gandomi, A.H.; Zhang, Z.; Alavi, A.H. Haotic cuckoo search. *Soft Comput. Fusion Found. Methodol. Appl.* **2016**, *20*, 3349–3362.

36. Yi, J.; Wang, J.; Wang, G. Improved probabilistic neural networks with self-adaptive strategies for transformer fault diagnosis problem. *Adv. Mech. Eng.* **2016**, *8*. [CrossRef]

37. Storn, R.; Price, K.V. Differential evolution: A simple and efficient adaptive scheme for global optimization over continuous spaces. *J. Glob. Optim.* **1995**, *11*, 341–359. [CrossRef]

38. Erbao, C.; Lai, M. A Hybrid differential evolution algorithm to vehicle routing problem with fuzzy demands. *J. Comput. Appl. Math.* **2009**, *231*, 302–310. [CrossRef]

39. Dechampai, D.; Tanwannichkul, L.; Sethanan, K.; Pitakaso, R. A differential evolution algorithm for the capacitated VRP with Flexibility for mixing pickup and delivery services and the maximum duration route in poultry industry. *J. Intell. Manuf.* **2017**, *28*, 1357–1376. [CrossRef]

40. Sethanan, K.; Pitakaso, R. Differential evolution algorithms for scheduling raw milk transportation. *Comput. Electron. Agric.* **2016**, *121*, 245–259. [CrossRef]

41. Das, S.; Suganthan, P.N. Differential Evolution: A Survey of the State-of-the-Art. *IEEE Trans. Evol. Comput.* **2011**, *15*, 4–31. [CrossRef]
42. Glen, J.J. Mathematical models in farm planning: A survey. *Oper. Res.* **1987**, *35*, 641–666. [CrossRef]
43. Itoh, T.; Ishii, H.; Nanseki, T. A model of crop planning under uncertainty in agricultural management. *Int. J. Prod. Econ.* **2003**, *81*, 555–558. [CrossRef]
44. Sarker, R.A.; Talukdar, S.; Haque, A. Determination of optimum crop mix for crop cultivation in Bangladesh. *Appl. Math. Model.* **1997**, *21*, 621–632. [CrossRef]
45. Santosa, B.; Sunarto, A.; Rahman, A. Using Differential Evolution Method to Solve Crew Rostering Problem. *Appl. Math.* **2010**, *1*, 316–325. [CrossRef]
46. Adeyemo, J.; Otieno, F. Differential evolution algorithm for solving multi-objective crop planning model. *Agric. Water Manag.* **2010**, *97*, 848–856. [CrossRef]
47. Mallipeddi, R.; Suganthan, P.N.; Pan, Q.K.; Tasgetiren, M.F. Differential evolution algorithm with ensemble of parameters and mutation strategies. *Appl. Soft Comput.* **2010**, *11*, 1679–1696. [CrossRef]
48. Adekanmbi, O.A.; Olugbara, O.O.; Adeyemo, J. A Comparative Study of State-of-the-Art Evolutionary Multi-objective Algorithms for Optimal Crop-mix planning. *Int. J. Agric. Sci. Technol.* **2014**, *2*, 8–16. [CrossRef]
49. Adeyemo, J.; Bux, F.; Otieno, F. Differential evolution algorithm for crop planning: Single and multi-objective optimization model. *Int. J. Phys. Sci.* **2010**, *5*, 1592–1599.
50. Pant, M.; Radha, T.; Rani, D.; Abraham, A.; Srivastava, D.K. Estimation Using Differential Evolution for Optimal Crop Plan. In Proceedings of the HAIS 2008 International Conference on Hybrid Artificial Intelligence Systems, Oviedo, Spain, 20–22 June 2008; pp. 289–297.
51. Yi, H.; Duan, Q.; Liao, T.W. Three improved hybrid metaheuristic algorithms for engineering design optimization. *Appl. Soft Comput.* **2013**, *13*, 2433–2444. [CrossRef]
52. Wang, G.; Cheng, H.; Chu, E.; Mirjalili, S. Three-dimensional path planning for UCAV using an improved bat algorithm. *Aerosp. Sci. Technol.* **2016**, *49*, 231–238. [CrossRef]
53. Wang, G.; Gandomi, A.H.; Alavi, A.H.; Hao, G. Hybrid krill herd algorithm with differential evolution for global numerical optimization. *Neural Comput. Appl.* **2014**, *25*, 297–308. [CrossRef]
54. Wang, G.; Gandomi, A.H.; Yang, Z.; Alavi, A.H. A novel improved accelerated particle swarm optimization algorithm for global numerical optimization. *Eng. Comput.* **2014**, *31*, 1198–1220. [CrossRef]
55. Wu, G.; Shen, X.; Li, H.; Chen, H.; Lin, A.; Suganthan, P.N. Ensemble of differential evolution variants. *Inf. Sci.* **2018**, *423*, 172–186. [CrossRef]
56. Wu, G.; Pedrycz, W.; Suganthan, P.N.; Lie, H. Using variable reduction strategy to accelerate evolutionary optimization. *Appl. Soft Comput.* **2017**, *61*, 283–293. [CrossRef]
57. Wang, R.; Purshouse, R.C.; Fleming, P.J. Preference-Inspired Coevolutionary Algorithms for Many-Objective Optimization. *IEEE Trans. Evol. Comput.* **2013**, *17*, 474–494. [CrossRef]
58. Wang, R.; Zhang, O.; Zhang, T. Decomposition-Based Algorithms Using Pareto Adaptive Scalarizing Methods. *IEEE Trans. Evol. Comput.* **2016**, *20*, 821–837. [CrossRef]
59. Wu, G.; Pedrycz, W.; Li, H.; Ma, M.; Liu, J. Coordinated Planning of Heterogeneous Earth Observation Resources. *IEEE Trans. Syst. Man Cybern. Syst.* **2016**, *46*, 109–125. [CrossRef]
60. Wang, G.; Guo, L.; Duan, H.; Wang, H. A New Improved Firefly Algorithm for Global Numerical Optimization. *J. Comput. Theor. Nanosci.* **2014**, *11*, 477–485. [CrossRef]
61. Wang, G.; Cai, X.; Cui, Z.; Min, G.; Chen, J. High Performance Computing for Cyber Physical Social Systems by Using Evolutionary Multi-Objective Optimization Algorithm. *IEEE Trans. Emerg. Top. Comput.* **2017**. [CrossRef]
62. Ke, L.; Gong, D.W.; Meng, F.L.; Chen, H.H.; Wang, G. Gesture segmentation based on a two-phase estimation of distribution algorithm. *Inf. Sci.* **2017**. [CrossRef]
63. Rizk, R.M.; El-Sehiemy, R.A.; Deb, S.; Wang, G. A novel fruit fly framework for multi-objective shape design of tubular linear synchronous motor. *J. Supercomput.* **2017**, *73*, 1235–1256. [CrossRef]
64. Wang, R.; Zhou, Z.; Ishibuchi, H.; Liao, T.; Zhang, T. Localized Weighted Sum Method for Many-Objective Optimization. *IEEE Trans. Evol. Comput.* **2018**, *22*, 3–18. [CrossRef]
65. Qin, A.K.; Huang, V.L.; Suganthan, P.N. Differential evolution algorithm with strategy adaptation for global numerical optimization. *IEEE Trans. Evolut. Comput.* **2009**, *13*, 398–417. [CrossRef]

Math. Comput. Appl. **2018**, 23, 40

66. Diaz, J.A.; Fernandez, E. A Tabu search heuristic for the generalized assignment problem. *Eur. J. Oper. Res.* **2001**, *132*, 22–38. [CrossRef]

© 2018 by the authors. Licensee MDPI, Basel, Switzerland. This article is an open access article distributed under the terms and conditions of the Creative Commons Attribution (CC BY) license (http://creativecommons.org/licenses/by/4.0/).

Mathematical and Computational Applications

MDPI

Article

U-Shaped Assembly Line Balancing by Using Differential Evolution Algorithm

Poontana Sresracoo [1,*], Nuchsara Kriengkorakot [1], Preecha Kriengkorakot [1] and Krit Chantarasamai [2]

[1] Industrial Engineering, Department, Faculty of Engineering, Ubon Ratchathani University, Ubon Ratchathani 34190, Thailand; ennuchkr@ubu.ac.th (N.K.); preecha.k@ubu.ac.th (P.K.)
[2] Manufacturing Engineering, Department, Faculty of Engineering, Mahasarakham University, Mahasarakham 44150, Thailand; krit@msu.ac.th
* Correspondence: poontana.teay@gmail.com; Tel.: +66-86-228-5472

Received: 9 November 2018; Accepted: 10 December 2018; Published: 12 December 2018

Abstract: The objective of this research is to develop metaheuristic methods by using the differential evolution (DE) algorithm for solving the U-shaped assembly line balancing problem Type 1 (UALBP-1). The proposed DE algorithm is applied for balancing the lines (manufacturing a single product within a fixed given cycle time), where the aim is to minimize the number of workstations. After establishing the method, the results from previous research studies were compared with the results from this study. For the UALBP, two groups of benchmark problems were used for the experiments: (1) For the medium-sized UALBP (21–45 tasks), it was found that the DE algorithm DE/best/2 to Exponential Crossover 1 produced better solutions when compared to the other metaheuristic methods: it could generate 25 optimal solutions from a total of 25 instances, and the average time used for the calculation was 0.10 seconds/instance; (2) for the large-scale UALBP (75–297 tasks), it was found that the basic DE algorithm and improved differential evolution algorithm generated better solutions, and DE/best/2 to Exponential Crossover 1 generated the optimal solutions and achieved the minimum solution search time when compared to the other metaheuristic methods: it could generate 36 optimal solutions from a total of 62 instances, and the average time used for the calculation was 4.88 seconds/instance. From the comparison of the DE algorithms, it was found that the improved differential evolution algorithm generated optimal solutions with a better solution search time than the search time of the basic differential evolution algorithm. The basic and improved DE algorithm are the effective methods for balancing UALBP-1 when compared to the other metaheuristic methods.

Keywords: U-shaped assembly line balancing; basic differential evolution algorithm; improved differential evolution algorithm; optimal solutions

---

1. Introduction

Nowadays, the degree of competition in many industries is very high. Therefore, organizations that respond quickly to changes in their customers' needs, require less effort to control the storage of their inventory, and spend less time in production will certainly achieve business advantages over their competitors. Moreover, organizations need to show continuous improvement and development of their products' values in order to respond to the needs of their customers by reducing costs and improving product quality during the production process. This has resulted in changes to the production system, such as the change from the push system to the pull system, which has reduced the volume of each batch size produced. The changes also include replacing the traditional production layout of straight lines with U-shaped production lines or U-lines. When compared to the traditional production layout of straight lines, it was found that using U-lines was more advantageous in terms of balanced production lines, improved worker visibility, better communications, fewer workstations, higher flexibility, shorter operation travel, and easier material handling, among other benefits. Relevant research studies on simple assembly line balancing using a single model and mixed model were first published in 1955, and the subject has been researched intensively since then. Meanwhile, research studies on U-shaped assembly line balancing have not received much interest from researchers, with few studies conducted. Therefore, there are many interesting areas under the topic of U-shaped assembly line balancing that require further research studies [1].

The assembly line balancing problem (ALBP) is considered a problem of the NP-hard class of combinatorial optimization problems [2], which are complicated to solve. If using mathematics with exact methods for finding optimal solutions, a lot of time will be spent on calculations and expenses, especially of large-scale problems with more variables and limitations. Hence, many heuristic methods for obtaining good solutions have been developed [3]. In the last 10 years, the development of the heuristic methods known as metaheuristic methods has been of great interest to many researchers because the methods were used to obtain results for the general assembly line balancing problem (GALBP) and simple assembly line balancing problem (SALBP) [4]. The assembly line balancing problem (ALBP) has come to be considered a classic problem that interests many researchers and research studies on this problem have been carried out since 1955. Many researchers have developed mathematics methods (exact methods) and heuristic methods, including metaheuristic methods. [5] A new hybrid GSA-GA algorithm is presented for the constraint nonlinear optimization problems with mixed variables. In it, firstly the solution of the algorithm is tuned up with the gravitational search algorithm and then each solution is upgraded with the genetic operators such as selection, crossover, and mutation [6]. The main objective of this paper is to present a hybrid technique named as a PSO-GA for solving the constrained optimization problems. In this algorithm, particle swarm optimization (PSO) operates in the direction of improving the vector while the genetic algorithm (GA) has been used for modifying the decision vectors using genetic operators. However, only a small number of metaheuristic methods have been developed for solving the problems discussed in this study. Therefore, it is of interest to develop metaheuristic methods for solving the ALBP and thus increase the chance of finding an effective solution of this problem [7]. The objective of this paper is to solve the reliability redundancy allocation problems of series parallel system under the various nonlinear resource constraints using the penalty guided based biogeography based optimization. In the same year, [8] The main goal of the present paper is to present a penalty based cuckoo search (CS) algorithm to get the optimal solution of reliability e redundancy allocation problems (RRAP) with nonlinear resource constraints.

Hence, this research is a study on U-shaped assembly line balancing for the manufacture of a single product with a given fixed cycle time (c), where the aim is to minimize the number of workstations (m). The U-line assembly line balancing problem type 1 (UALBP-1) was studied, and a new metaheuristic method was developed for finding the solution by using the differential evolution algorithm (DE) [9]. The aim of this new method is to generate good solutions or the optimal solutions to this classic problem.

The main contribution of this work includes: (1) background; (2) the assembly line balancing problem; and (3) objective of the work.

2. Literature Review

2.1. U-Shape Assembly Line Balancing by Using Other Metaheuristic Methods

The previous studies on the UALBP were reviewed and are summarized in this section. The first UALBP study in the literature was conducted in 1994 by [10]. In their study entitled "The U-line Line Balancing Problem", the mathematical formulation of the problem established by using dynamic programming for the single-model U-line to minimize the number of stations, and the RPWT (ranked positional weight technique) for solving a large-sized UALBP (111 tasks), which contained problems derived from previous research studies. These problems were more complicated than traditional problems because the tasks could be placed from forward to backward, from backward to forward, or from both directions simultaneously according to a sequential flowchart. A year later, [11] developed three mathematical exact algorithms to solve the UALBP. The dynamic programming (DP) formulation was used in the first algorithm, and the other two were breadth-first and depth-first branch and bound (B&B) algorithms. The results of the calculation revealed that B&B was more effective than the DP-based algorithm, and breadth-first spent less time on calculation than depth-first, but depth-first could find the optimal solutions faster. Later, [12] conducted research entitled "The Mixed-Model U-line Balancing Problem" and developed a heuristic method for solving the mixed-model UALBP with a precedence graph of each batch size with 25 tasks. In the same year, [13] presented a formulation for Integer Programming (IP) for finding the optimal balance to solve the UALBP. This formulation solved the large-scale problems better than the traditional ones. Later, [14] proposed a DP formulation for solving problems of numerous U-lines with a maximum of 22 tasks. The objective was to assign the tasks to workstations in various ways, aiming to minimize the number of workstations and time wasting [15,16]. This integrated model was solved with a black hole optimization based algorithm. The quality of the heuristic solution was checked with special data sets. A year later, proposed a black hole optimization (BHO) based algorithm dealing with a multi objective supply chain model is presented. The sensitivity of the enhanced algorithm is tested with benchmark functions. Numerical results with different datasets demonstrate the efficiency of the proposed model and validate the usage of Industry 4.0 inventions in first mile and last mile (FMLM) delivery.

The principles proposed could be effective methods for ALB and Rebalancing the UALBP. Later, [17] developed a heuristic method for solving the UALBP with nine U-lines, where there were 18 tasks in each U-line with the same cycle time. The time between work and the time between U-lines were used in the consideration. A year later, [18] produced a thesis called "Incorporating Ergonomics Criteria into Assembly Line Balancing" and developed three heuristic methods: (1) multiple ranking heuristic, (2) combinatorial genetic algorithm (GA), and (3) problem-space GA. These methods were developed by employing the criteria for ergonomic designs (such as a reduction in cycle time and the loss of grip strength due to fatigue at workstations) for the consideration of solutions of the I-shaped and U-shaped line ALB in order to obtain the lowest values. Consequently, many industrial factories benefited from the results of the study in terms of both production and ergonomics. Later, [19] carried out a study entitled "ULINO: Optimally balancing U-shaped JIT assembly lines" by implementing a B&B method that was developed from the simple assembly line balancing optimization method (SALOME). The SALOME method, which had been previously used for solving straight-line problems, was used for solving U-line problems with 297 tasks. This new method was called ULINO (U-line optimizer). The method was based on depth-first branch and bound and dominance rules. The purpose was to minimize the number of workstations, cycle time, or both. In 2003, a method for obtaining the optimal solution of UALB with parallel stations was developed by using multiple lower bounding and a new heuristic method for finding the upper bounding. The results from the two developed methods were improved over the traditional methods.

A year later, [20] applied the genetic algorithm (GA), which had been used for solving the SALBP, for solving the UALBP-1. Then, the optimal solutions from previous studies were compared. The results from the study showed that the GA could generate the optimal solutions or nearest solutions in the very first rounds of the experiment. Later, [21] published "A Shortest Route Formulation of Simple U-Type Assembly Line Balancing Problem", which applied the theory of the shortest route formulation. The theory was proposed by [22] for solving the SALBP. However, the principles of the theory were further developed to solve the UALBP. In addition, some examples of the calculations of this method were published in many research articles and, In the same year, [23] achieved the goal of developing an equation formulation for the UALBP that was based on the integer linear programming formulation developed for the UALBP, as well as an equation formulation for the SALBP. The principles were flexible, helping in decision making in the UALBP for cases with many conditions.

In the same year, [24] carried out research called "The Stochastic U-line Balancing Problem: A Heuristic Procedure". The authors explained the reason why the U-lines were popular: it was because the JIT (just-in-time) production system played a major role in production. Therefore, U-shaped assembly lines were replacing the traditional straight assembly lines. The issue that was emphasized as needing consideration in the UALBP was the unreliability of the task time due to workers and other factors with unpredictable timing.

The developed heuristic approach was divided into two parts: finding basic solutions and improving basic feasible solutions. The results of the calculation revealed the effectiveness of the method used. In another study, conducted by [25], simulated annealing (SA) was developed for solving the UALBP. This method is widely used at present. The aim of this method is to minimize the number of workstations (Type-I). The effectiveness of the principles was evaluated by a solution search for large-scale problems and by comparing the results with ULINO (U-Line optimizer), which was B&B based on the heuristic procedure. In addition, [26] the authors proposed recommendations for further research studies as follows: (1) SA methods should be used for more difficult problems, such as mixed-/multi-model lines, stochastic task time, and U-line with other characteristics; (2) the principles of other metaheuristic methods should be utilized for solving the UALBP; (3) exact methods for solving U-line problems at a large scale should be developed; (4) the principles of other metaheuristic methods should be used for solving problems of Type-II (min. c, given m). Since the study, many researchers have explored methods for solving the UALBP by developing GA metaheuristic methods. In [27], a GA was presented for solving the UALBP by using the Just-In-Time (JIT) approach for solving the UALBP. The results were better than those of the traditional approaches that were employed for solving the UALBP. Therefore, the GA was used for solving UALBP. The results from the verification showed that greater effectiveness was achieved. The criteria used for consideration were the number of workstations of assembly lines and the variation of the workloads. The results from the experiment showed that the proposed method was effective because the workstations were well grouped and the workload formats were improved.

Moreover, many researchers used metaheuristic methods for solving the UALBP. They proposed the ant colony system (ACS) algorithm for finding solutions and proposed four heuristic methods: the method of Kilbridge and Wester (K&W), ranked positional weight (RPW), maximum task time (Max. T. T.), and total maximum number of following tasks (Max. N.F.). The four methods were then compared to the metaheuristic method of ACS for finding solutions, and it was found that the ACS method generated better solutions than the four heuristic methods. Later, other researchers studying methods for solving the UALBP developed a heuristic method and metaheuristic method by using the max-min ant system (MMAS) of max. task time and min. task time, together with a local search method for solving the SALBP and UALBP. For the SALBP, a large-scale benchmark problem set with 45–111 tasks and the large-scale benchmark problem set of Lapierre's Tabu Search (TS) with 297 tasks were tested. For solving the UALBP, experiments were conducted on the benchmark problem set using Max. RPW and the medium-sized UALBP with 21–45 tasks, and the benchmark problem set with some given information consisted of a large-scale problem with 75–297 tasks. From the results of the study, it was concluded that the developed Max-Min Ant system was the most efficient method when compared to the heuristic method and Max. RPW method. From the experiments, Min. task time generated the worst solution. When compared to other methods, and when there was a change that increased cycle time, the results showed that there was no effect on the capacity of the Max-Min Ant system for finding the solution. Therefore, it was concluded that the developed method was highly efficient. After that, other methods for solving the UALBP were researched by other researchers.

2.2. U-Shape Assembly Line Balancing by Using Differential Evolution Algorithm

Assembly line balancing (ALB) by using the differential evolution algorithm for solving the ALBP has been studied by many researchers. In [28], the differential evolution algorithm was studied and proposed for solving the SALBP from the benchmark problem (retrieved from http://www.

assembly-line-balancing.de) with 7–111 tasks. The purpose was to find the minimum number of workstations with the condition that the time of each station must not exceed the production cycle time and the preceding conditions must be met. The results of the study revealed that the proposed method generated a good solution with a minimum search time. This generated an initial solution from the sampling of real numbers, and the solution was further optimized. Later, [29] carried out research by using the differential evolution (DE) algorithm for solving the SALPB-1. The proposed heuristic is composed of four main steps: (1) initialization, (2) mutation, (3) crossover, and (4) selection processes. In the study, the mutation and crossover processes were combined, and a new method was found. The computational results based on many tests using a set of standard instances showed that the proposed DE algorithm was very competitive for solving the SALPB-1.

2.3. Differential Evolution Algorithm for Solving Other Problems

Differential evolution (DE) algorithms have been used for solving other research problems which aim to generate solutions as follows. The mutation DE algorithm was used in [24] and involved the positions of the optimal vectors in the population of each batch and employed the crossover or recombination of positions which were exchanged between vectors through the comparison of the crossover rate (CR) with random positions. This method was found to be effective and, in 2011, [30] developed the DE algorithm for solving assignment problems by improving two main crucial parameters in the DE process: the weighting factor (F) and crossover rate (CR). The improved differential evolution (IDE) was used for allowing an adjustable F, and the CR changed in terms of taxonomy steps. The sample problems were compared with the solutions of opposition-based differential evolution (ODE) and adaptive differential evolution (JADE). The results revealed that the developed IDE generated better solutions than the other two methods in terms of cost reduction and increased performance in the system. A year later, [31] developed the Pareto Utility Discrete Differential Evolution (PUDDE) algorithm for handling operator allocation problems (OAP) in order to allocate jobs appropriately for the balance control of assembly lines when multiobjective functions and conditions were formulated, and a decision based on only a single objective cannot be made. The procedure, which included the discrete event simulation DES model, was used in the general simulation, and PUDDE was employed for solving OAPs by improving the operator condition in two ways: decreasing the number of operators or increasing the number of operators. The results from the experiment concluded that PUDDE could find the solutions effectively. However, this method was suitable for decreasing the number of operators. When compared to the traditional DE, the PUDDE algorithm had a much better performance when finding objectives of multi-assembly lines in the same problem.

From the literature and related research studies above, it can be concluded that the DE algorithm was the method able to find the optimal solutions for large-scale complicated problems within the possible solution area by using a short search time when compared to the other metaheuristic methods. It was reported by [32] that DE algorithms with an evolution algorithm procedure were a new technique for increasing the efficiency of and capacity for handling problems with nondifferentiable, nonlinear, and multimodal objective functions, particularly large-scale complicated problems. It was found that the speed of the solution search of DE was better than that of other methods. Therefore, the DE method can be used for solving the ALBP and increasing the efficiency of the solution search for the ALBP. Hence, DE was chosen for solving the UALBP-1 in the current study.

3. U-line Assembly Line Balancing Problem Pattern and Mathematical Model

Only the UALBP-1 was focused on in this study, and the details of the problem are as follows:

3.1. U-line Assembly Line Balancing Problem (UALBP)

The UALBP-1 can be divided into three problem versions which are similar to the divisions of the Simple Assembly Line Balancing Problem [10,19]:

1.   UALBP-1: Given the cycle time (*c*), minimize the number of stations (*m*);
2.   UALBP-2: Given the number of stations (*m*), minimize the cycle time (*c*).;
3.   UALBP-E: Maximize the line efficiency (*E*) for c and m being variable.

U-line Assembly Line Balancing Problem (UALBP) Pattern: The UALBP of each task can be located at only one station and is performed before its predecessor and after its successor tasks. The total time of each station cannot exceed more than the cycle time according to the conditions of the workstation assignment in the UALBP-1. The UALBP-1 can be explained using the benchmark problem called the Jackson Problem, shown in Figure 1, which is illustrated as a precedence diagram of the ALBP (11 tasks): the number inside each box represents the name of each task, and the number above each box represents the time of that task. When applying the example of Figure 1 to the UALBP, given *c* = 10, an optimal solution with *m* = 5 can be found, which is shown in Figure 2. The begging of process line balancing, the task 1 and task 11 are in station 1. It can generate a feasible line balance with a cycle time of *c* = 10 and with *m* = 5 stations given by the station loads S_1 = {1, 11}, S_2 = {2, 4, 5}, S_3 = {6, 7, 9}, S_4 = {3, 10}, and S_5 = {8}. While no idle time occurs in stations 1, 2, 3, and 4, station 5 shows an idle time of 4.

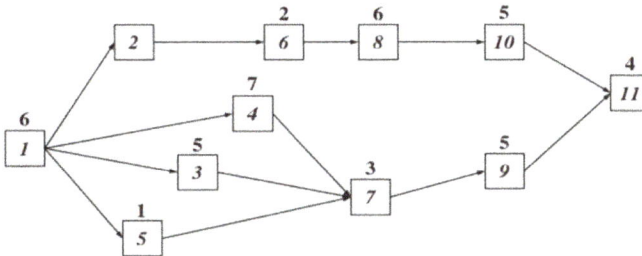

Figure 1. Precedence diagram with assembly network [33].

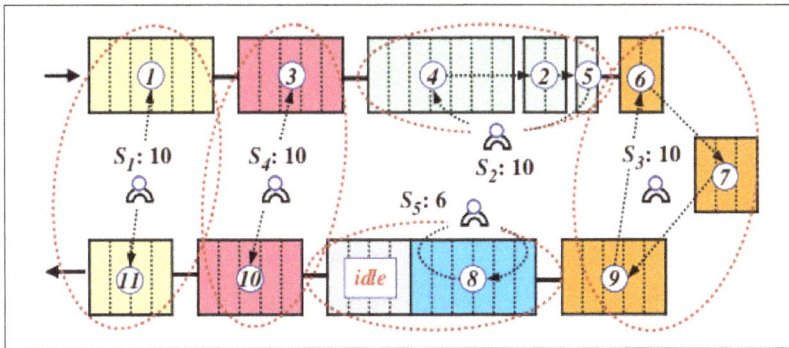

Figure 2. Completed line assignments for the U-shaped assembly line.

3.2. Mathematical Model of the U-Shaped Assembly Line

This section presents the mathematical model for the UALBP-1, adapted from Bowman [34]. The indices, parameters, and decision variables are as defined below.

3.2.1. Indices

n denotes the index of a task, where *n* = 1 .... *N*
k denotes the index of workstation *k*, where *k* = 1 ... *M*
N denotes the total number of tasks
M denotes the total number of workstations

3.2.2. Parameter

P_n denotes the processing time of task n

CT denotes the cycle time of a workstation

P_{nj} denotes the relationship of task n to task j

$$F_{nj} = \begin{cases} 1 & \text{if task } n \text{ is predecessor of task } j \\ 0 & \text{otherwise} \end{cases}$$

3.2.3. Decision Variables

$$X_{nk} = \begin{cases} 1 & \text{if task } n \text{ is assigned to station } k \\ 0 & \text{otherwise} \end{cases}$$

$$Y_k = \begin{cases} 1 & \text{if station } k \text{ is opened} \\ 0 & \text{otherwise} \end{cases}$$

Objective function:

$$Min \ Z = \sum_{k-1}^{M} Y_k \tag{1}$$

subject to

$$\sum_{k=1}^{M} X_{nk} = 1 \qquad \forall n = 1 \dots N, \tag{2}$$

$$\sum_{k=1}^{M} (K \times X_{jk}) - (K \times X_{nk}) \geq 0 \\ \forall n = 1 \dots N, \ k = 1 \dots M, \ F_{nj} = 1 \tag{3}$$

$$\sum_{n=1}^{N} X_{nk} \times P_n \leq CT \times Y_k \\ \forall k = 1 \dots M \tag{4}$$

$$Y_k \leq Y_{k-1} \ \forall k = 2 \dots M \tag{5}$$

Equation (1) represents an objective function of the model to minimize the number of stations. Equation (2) guarantees that task n must be assigned to exactly one workstation. Equation (3) ensures that the precedence constraints are not violated on the U-line. Equation (4) ensures that the total processing time used by all tasks assigned to a particular workstation must not exceed a prespecified cycle time (CT). Finally, Equation (5) ensures that the station will be opened successively according to the station number.

4. General Differential Evolution Algorithm

4.1. Differential Evolution Algorithm

DE is a population-based random search method where an initial population of size N of D-dimensional vector is randomly generated, and a new population is generated through the cycles of calculations. A solution in DE algorithm is represented by D-dimensional vector, and each value in the D-dimensional space is represented as a real number. The key idea behind DE which makes the algorithm from other evolutionary algorithms (EAs) is its mechanisms for generating new solutions, called trial vectors, by mutation and crossover operation. In DE, each vector is served as a target vector which is then combined with other vectors in the population to form a new vector, called a mutant vector. Next, the mutant vector is crossover with its corresponding occurs only if the trial vector outperforms its corresponding target vector. The evolution process of DE population continues through repeated cycles of three main operator; mutation, crossover, and selection until some stopping criteria are met [30].

The general procedure of DE consists of several steps: (1) construct a set of initial target vectors, (2) perform a mutation process, (3) perform a recombination process, and (4) perform a selection process. The design of the procedure application is shown in Figure 3.

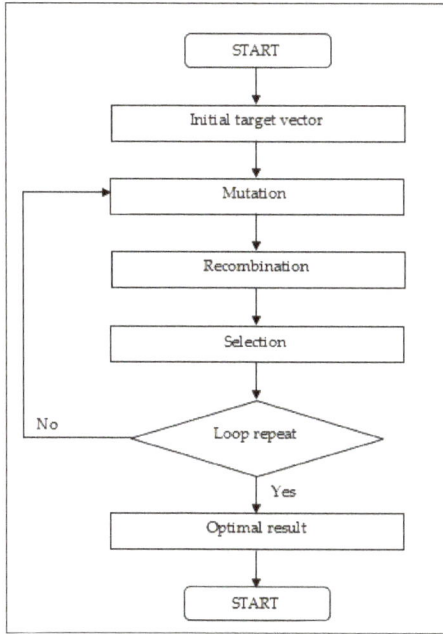

Figure 3. General Differential Evolution Algorithm Procedure.

From the general differential evolution algorithm procedure in Figure 3, an initial vector is generated and the designed vector is further optimized by mutation, recombination, and selection processes. The procedure can be illustrated using the Bowman problem as an example, as shown in Figure 4.

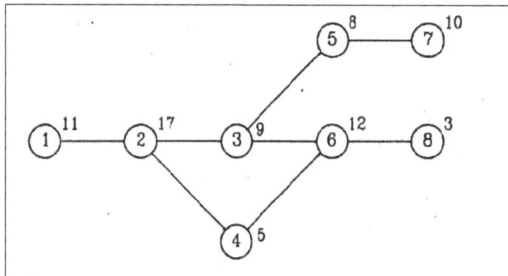

Figure 4. Precedence Diagram of the Bowman Problem [34].

4.2. Procedure of UALB-1 by Using Differential Evolution Algorithm

The procedure of UALB-1 using the general differential evolution algorithm from the Bowman problem set in Figure 4, which presents the precedence diagram of the ALBP (8 tasks), can be interpreted as follows. The number in each circle represents the name of each task and the number above each circle represents the time of that task. Before balancing assembly lines, the values used in the calculation must be set. The variables are as follows: R = Round, NP = Numbers of tasks, F = Scaling factor, and CR = Crossover rate. In this problem's calculation, these suitable variables were set from the experiment as $R = 1$, $NP = 5$, $F = 0.8$, $CR = 0.8$ [35], and CT (Cycle Time) = 20. The General Differential Evolution Algorithm "DE/rand/1" and binomial crossover were used in the calculation as follows.

4.2.1. Calculation Using the General Differential Evolution Algorithm DE/rand/1 and Binomial Crossover

(1) Initial population. In this step, a randomized real number between 0 and 1 for each task is obtained. This is the formulation of the target vector or initial solutions for the decision value in the workload allocation to the workstations, and the initial vector will be used further in Mutation and crossover, as shown in Table 1. In the table, the initial population calculation is presented by a randomized real number between 0 and 1 for each task. A target vector will be generated for each task in order to generate the initial solution and decision value in the workload allocation to the workstations. The workload allocation to the workstations must in line with the conditions of UALB by arranging the random numbers in ascending order for workstation allocation. From the table, $NP = 5$ vectors. The benchmark problem is illustrated in Figure 4, and it is described in Table 1.

Table 1. Result of initial population $NP = 5$ vectors.

	Station	1	2	3	4	5
Vector 1	Work	1, 8	6, 4	2	3, 5	7
	Time	11, 3	12, 5	17	9, 8	10
	Target Vector	0.30, 0.57	0.44, 0.61	0.72	0.53, 0.68	0.92
	Station	1	2	3	4	5
Vector 2	Work	7, 5	1, 8	6	3	4, 2
	Time	10, 8	11, 3	12	12	5, 11
	Target Vector	0.57, 0.32	0.74, 0.92	0.21	0.44	0.69, 0.82
	Station	1	2	3	4	5
Vector 3	Work	1, 8	7, 5	6, 4	3	2
	Time	11, 3	10, 8	12, 5	9	17
	Target Vector	0.51, 0.96	0.88, 0.67	0.84, 0.62	0.41	0.92
	Station	1	2	3	4	5
Vector 4	Work	1, 8	7, 5	2	3, 4	6
	Time	11, 3	10, 8	17	9, 5	12
	Target Vector	0.13, 0.26	0.58, 0.81	0.64	0.42, 0.21	0.69
	Station	1	2	3	4	
Vector 5	Work	8, 6, 4	7, 5	3, 1	2	
	Time	3, 12, 5	10, 8	9, 11	17	
	Target Vector	0.84, 0.41, 0.59	0.91, 0.76	0.32, 0.98	0.48	

(2) Mutation. In this step, a position of the vector is mutated to obtain new solutions that differ from the initial population number by targeting the mutation. The calculation for the mutant vector ($V_{i,j,G+1}$) is shown in Equation (6), and an example of a mutation is illustrated in Table 2.

$$V_{i,j,G} = X_{r1,j,G} + F(X_{r2,j,G} - X_{r3,j,G}) \tag{6}$$

where:

$V_{i,j,G}$ = Mutant Vector

$X_{r1,j,G}$, $X_{r2,j,G}$, $X_{r3,j,G}$ = Random vector from G round

F = Scaling factor (random real number between 0 and 2)

Table 2. Results of Mutation in Vector 1 by using DE/rand/1.

	Position	1	2	3	4	5	6	7	8
Vector 1	Work	1	8	6	4	2	3	5	7
	Time	11	3	12	5	17	9	8	10
	Target Vector	0.3	0.57	0.44	0.61	0.72	0.53	0.68	0.92
	Position	1	2	3	4	5	6	7	8
Vector 2	Work	7	5	1	8	6	3	4	2
	Time	10	8	11	3	12	12	5	11
	Target Vector	0.57	0.32	0.74	0.92	0.21	0.44	0.69	0.82
	Position	1	2	3	4	5	6	7	8
Vector 3	Work	1	8	7	5	6	4	3	2
	Time	11	3	10	8	12	5	9	17
	Target Vector	0.51	0.96	0.88	0.67	0.84	0.62	0.41	0.92
Mutant Vector 1		0.35	0.06	0.33	0.81	0.22	0.39	0.90	0.84

The results of the mutation in Vector 1 by using "DE/rand/1" are depicted in Table 2.

Table 2 presents the results of Mutation in Vector 1 by using the DE/rand/1 method, where $X_{r1,j,G}$, $X_{r2,j,G}$, and $X_{r3,j,G}$ are randomized to form Vectors 1, 2, and 3, respectively, and $F = 0.8$ is set so that at position 6, $X_{r1,j,G} = 0.53$, $X_{r2,j,G} = 0.44$, and $X_{r3,j,G} = 0.62$ substituted into the equation as $V_{i,j,G} = X_{r1,j,G} + F(X_{r2,j,G} - X_{r3,j,G})$ will be $V_{i,j,G} = 0.53 + 0.8(0.44 - 0.62) = 0.39$. Therefore, Mutant Vector 1's position 6 is 0.39. After that, the calculation for every position is done until all eight positions are calculated.

(3) Crossover or Recombination. In this step, vector positions are exchanged. New vectors, both better and worse, are generated. The Trial Vector $(U_{i,j,G+1})$ is formulated, and the Trial Vectors are compared and exchanged as in Equation (7). The examples are presented in Tables 3 and 4.

$$U_{i,j,G} = \begin{cases} V_{i,j,G} & \text{if } Rand_{i,j} \leq CR \text{ or } j = Irand \\ X_{i,j,G} & \text{if } Rand_{i,j} > CR \text{ or } j \neq Irand \end{cases} \tag{7}$$

where:

$V_{i,j,G}$ = Mutant Vector

$X_{i,j,G}$ = Target Vector

CR = Crossover Constant (real number in the range 0–1)

rand j [0,1) = random real number between 0 and 1 in every position, j = 1, 2, 3, ..., G (G = number of position).

Table 3. Results of Binomial Crossover in Vector 1 with the DE/rand/1 Method.

Vector	Position	1	2	3	4	5	6	7	8
	Work	1	8	6	4	2	3	5	7
1	Time	11	3	12	5	17	9	8	10
	Target Vector	0.3	0.57	0.44	0.61	0.72	0.53	0.68	0.92
Vector	Position	1	2	3	4	5	6	7	8
	Mutant Vector	0.35	0.06	0.33	0.81	0.22	0.39	0.90	0.84
1	rand(j)	0.40	0.40	0.06	0.96	0.47	0.40	0.94	0.33
	Trial Vector	0.35	0.06	0.33	0.61	0.22	0.39	0.68	0.84

Table 4. The results of UALB by using the Trial Vector from Table 3.

	Station	1	2	3	4	5
	Work	8, 6	1, 3	2	4, 5	7
Vector 1	Time	3, 12	11, 9	17	5, 8	10
	Trial Vector	0.06, 0.33	0.35, 0.39	0.22	0.61, 0.68	0.84

Table 3 shows the results of binomial crossover in Vector 1 if $CR = 0.8$; therefore, the value of the target vector will be used in the Trial Vector at positions 4 and 7. For other positions, the values of the mutant vector are used.

From Table 4, when employing the trial vector obtained for UALB according to the preceding conditions, the total time of each workstation must not exceed the production cycle time, which can be satisfied by considering a Trial Vector with low values before assigning the tasks to workstations. Therefore, a solution of 5 workstations is found.

(4) Selection. In this step, the next generation is selected ($G + 1$): only better solutions are selected by comparing the results of the Target Vector with the Trial Vector for cases in which the number of workstations of the Trial Vector is lower than/equal to that of the Target Vector. Therefore, the Trial Vector is selected as the next generation, as in Equation (8):

$$X_{i,j,G+1} = \begin{cases} U_{i,j,G} & if \ f(U_{i,j,G}) \leq f(X_{i,j,G}) \\ X_{i,j,G} & otherwise \end{cases} \tag{8}$$

where:

$U_{i,j,G}$ = Trial Vector

$X_{i,j,G+1}$ = Target Vector in the next generation, $i = 1,2,...n$

In sum, from the selection step for selecting the next generation, the Trial Vector of the 5 assigned workstations is found with the values higher than those of the Target Vector for 4 workstations. Therefore, the Target Vector is selected as the optimal solution, and the process is repeated for the next generation.

4.2.2. Procedure of UALB-1 by Using the Improved Differential Evolution Algorithm

(1) Improved DE. In mutation process, a mutant vector ($V_{i,j,G}$) will be calculated from one or more selected target vector ($X_{i,j,G}$). Traditionally, the mutation process of DE is performed using Equations (6) [30]. Improved DE was developed by applying the four Mutation Equations DE/best/1,

DE/rand-to-best/1, DE/best/2, and DE/rand/2, as seen from Equations (9), (10), (11) and (12), respectively, as follows:

$$V_{i,j,G} = X_{best,j,G} + F(X_{r1,j,G} - X_{r2,j,G}) \tag{9}$$

$$V_{i,j,G} = X_{i,j,G} + F\left(X_{best,j,G} - X_{i,j,G}\right) + F(X_{r1,j,G} - X_{r2,j,G}) \tag{10}$$

$$V_{i,j,G} = X_{best,j,G} + F(X_{r1,j,G} - X_{r2,j,G}) + F(X_{r3,j,G} - X_{r4,j,G}) \tag{11}$$

$$V_{i,j,G} = X_{r1,G} + F(X_{r2,G} - X_{r2,G}) + F(X_{r4,G} - X_{r5,G}) \tag{12}$$

Let $r1, r2, r3, r4,$ and $r5$ denote the vectors which are randomly selected from a set of target vectors j. represent the best vector found so far in the algorithm. F is a predefined integer parameter (scaling factor). In the proposed heuristics, F is set to 2; i is vector number which starts from 1 to NP, and j is position of a vector which run from 1 to D.

(2) Improved DE. The result of mutation process is a set of mutant vector $V_{i,j,G}$ (i run from 1 to NP). Then a mutant vector will apply recombination equations (13) and (14) to yield trial vector ($U_{i,j,G}$) as a product of recombination processes. In traditional DE for UALBP–1, a binomial recombination Equations (7) is applied in the basic differential evolution algorithm [30]. Improved DE was developed by applying two crossover or recombination equations: Exponential Crossover 1 position and Exponential Crossover 2 position, as seen in Equations (13) and (14), as follows:

$$U_{i,j,G} = \begin{cases} V_{i,j,G} \text{ when } randb_i \leq j \\ X_{i,j,G} \text{ if } randb_i > j \end{cases} \tag{13}$$

$$U_{i,j,G} = \begin{cases} V_{i,j,G} \text{ when } j \leq randb_{i,1} \text{ and } j \geq randb_{i,2} \\ X_{i,j,G} \text{ when } randb_{i,1} < j < randb_{i,2} \end{cases} \tag{14}$$

Let $randb_i$, to be random number between 0 and 1, and CR is recombination probability which is the predefined parameters in the proposed heuristics. $randb_i$, $randb_{i,1}$ and $randb_{i,2}$ are random integer numbers which are used to represent position of a vector and these random numbers ranges from 1 to D.

On the basis of the explanations in steps 1–4, the Improved Diff. is shown in Algorithm 1.

Algorithm 1. Pseudo-code of the DE for (UALBP-1)

Setup initial DE parameter
Do while from first iteration to final iteration
   Do while from first DE to final DE
     Setup initial parameters: cycle time, remaining time, station number
     Do while from first task to final task
       Find start/following task with task time is less than or equal to
       Remaining time, and proper precedence to data list
       Input scaling factor, crossover rate and NP to data list
       Select task randomly to list
       Update remaining time/station number
       Produce the four Mutation Equations

$$V_{i,j,G} = X_{best,j,G} + F(X_{r1,j,G} - X_{r2,j,G})$$

$$V_{i,j,G} = X_{i,j,G} + F\left(X_{best,j,G} - X_{i,j,G}\right) + F\left(X_{r1,j,G} - X_{r2,j,G}\right)$$

$$V_{i,j,G} = X_{best,j,G} + F\left(X_{r1,j,G} - X_{r2,j,G}\right) + F\left(X_{r3,j,G} - X_{r4,j,G}\right)$$

$$V_{i,j,G} = X_{r1,G} + F(X_{r2,G} - X_{r2,G}) + F(X_{r4,G} - X_{r5,G})$$

Developed by applying the two Crossover or Recombination Equations

$$U_{i,j,G} = \begin{cases} V_{i,j,G} \text{ when } randb_i \le j \\ X_{i,j,G} \text{ if } randb_i > j \end{cases}$$

$$U_{i,j,G} = \begin{cases} V_{i,j,G} \text{ when } j \le randb_{i,1} \text{ and } j \ge randb_{i,2} \\ X_{i,j,G} \text{ when } randb_{i,1} < j < randb_{i,2} \end{cases}$$

Produce new target vector (selection\process)

$$X_{i,j,G+1} = \begin{cases} U_{i,j,G} \text{ if } f(U_{i,j,G}) \le f(X_{i,j,G}) \\ X_{i,j,G} \qquad\qquad otherwise \end{cases}$$

     End do
    End do
   Select best solution from all DE in the iteration
  End do
Show/select best solution from all DE in all iteration

---

5. Analysis of the Results from the Experiment on DE for Solving UALBP

    The results obtained from the experiment on DE for solving the UALBP by using the basic DE algorithm and improved DE algorithm were analyzed. Six methods from the 15 methods for generating optimal solutions with the minimum search time were selected through the experiment. Then, the selected methods were compared. The results from the experiment were compared with other metaheuristic methods.

    DE for solving the UALBP: the basic DE algorithm and improved DE algorithm consisted of six methods as follows:

1. DE/rand/1 to Binomial Crossover (Basic)
2. DE/rand/1 to Exponential Crossover 1 Position (improved)
3. DE/rand/1 to Exponential Crossover 2 Position (improved)
4. DE/ rand-to-best/1 to Binomial Crossover (improved)
5. DE/Best/2 to Exponential Crossover 1 Position (improved)

6. DE/Best/2 to Exponential Crossover 2 Position (improved)

The ALBP can be solved by applying the Java program (operated on a computer with Core i3, 2.3 GHz, 2 GB RAM, and the operating system Windows 7). The metaheuristic methods developed, including the Basic UALB and Improved UALB for solving the U-shaped Assembly Line Balancing Problem Type 1 (UALBP-1), used the data retrieved from http://www.assembly-line-balancing.de

The variables used in the calculation for ALB were defined as follows: R = Round, NP = Number of tasks, F = Scaling factor, and CR = Crossover rate. In this problem's calculation, these variables were set at R = 30, NP = 30, F = 0.8, and CR = 0.8 [35]. Then, the experiment for finding the optimal solutions was conducted as follows.

1) The benchmark problems of Sawyer (30 tasks, 8 instances) and Arcus 1 (83 tasks, 16 instances) were used. The results of the experiment on the UALBP-1 by using the basic DE algorithm and improved DE algorithm are depicted in Table 5.

Table 5. The results of the experiment on the UALBP-1 by using the basic DE algorithm and improved DE algorithm with the Sawyer and Arcus 1 problems.

Problem	*c*	*m**	DE1*		DE2**		DE3***		DE4****		DE5*****		DE6******	
			m	Cal Time (s)	*m*	Cal Time (s)	*m*	Cal Time (s)	*m*	Cal Time (s)	*m*	Cal Time (s)	*m*	Cal Time (s)
Sawyer 30	25	14	14	1.87	14	0.67	14	0.11	14	1.87	14	0.07	14	0.08
	27	13	13	1.11	13	0.44	13	0.13	13	0.98	13	0.12	13	0.09
	30	11	11	1.23	11	0.83	11	0.42	11	0.77	11	0.08	11	0.10
	33	10	10	1.11	10	0.67	10	0.67	10	0.44	10	0.18	10	0.11
	36	10	10	1.34	10	0.55	10	0.55	10	0.34	10	0.05	10	0.15
	41	8	8	1.96	8	0.59	8	0.59	8	0.50	8	0.04	8	0.19
	54	6	6	1.61	6	0.78	6	0.38	6	0.61	6	0.09	6	0.38
	75	5	5	1.94	5	0.53	5	0.07	5	0.57	5	0.06	5	0.35
Arcus 183	3786	21	21	0.91	21	0.31	21	0.54	21	0.31	21	0.30	21	0.14
	3985	20	20	0.89	20	0.89	20	0.17	20	0.89	20	0.06	20	0.11
	3786	21	21	0.91	21	0.31	21	0.54	21	0.31	21	0.30	21	0.14
	4454	18	18	0.85	18	0.55	18	0.31	18	0.55	18	0.10	18	0.98
	4732	17	17	0.46	17	0.98	17	0.89	17	0.46	17	0.28	17	0.16
	5048	16	16	0.31	16	0.77	16	0.64	16	0.31	16	0.08	16	0.25
	5408	15	15	0.89	15	0.44	15	0.06	15	0.89	15	0.12	15	0.06
	5824	14	14	0.64	14	0.34	14	1.98	14	0.64	14	0.04	14	1.98
	5853	13	13	0.06	13	0.50	13	0.24	13	0.06	13	0.06	13	0.24
	6309	13	13	1.98	13	0.61	13	0.55	13	1.98	13	0.09	13	0.55
	6842	12	12	0.64	12	0.57	12	0.54	12	0.24	12	0.19	12	0.54
	6883	12	12	0.79	12	0.99	12	0.47	12	0.59	12	0.38	12	0.16
	7571	11	11	0.74	11	0.44	11	0.98	11	0.54	11	0.35	11	0.25
	8412	10	10	1.97	10	0.34	10	1.98	10	0.47	10	0.22	10	0.06
	8898	9	9	1.89	9	0.50	9	0.24	9	0.51	9	0.45	9	1.66
	10816	7	7	1.98	7	0.61	7	0.55	7	0.98	7	0.71	7	0.78
Total Optimal Solutions Found from 24 Problem Instances			24	1.16	24	0.61	24	0.53	24	0.67	24	0.17	24	0.40

Notes: *m* is the number of station (*m** is the optimal). *c* is the cycle time. DE1* = DE/rand/1 to Binomial Crossover. DE2** = DE/rand/1 to Exponential Crossover 1 position. DE3*** = DE/rand/1 to Exponential Crossover 2 position. DE4**** = DE/rand-to-best/1 to Exponential Crossover 1 position. DE5***** = DE/best/2 to Exponential Crossover 1 position. DE6****** = DE/best/2 to Exponential Crossover 2 position.

From Table 5, the results from the experiment on the UALBP-1 by using the basic method and improved DE algorithm with the Sawyer and Arcus 1 problems show that all six DE algorithms generated the optimal solution of 24 instances from a total of 24 instances. The DE or DE/Best/2

Exponential Crossover 1 position was the best method to calculate the optimal solutions with the minimum average times (0.17 seconds).

The comparison of the basic and improved DE algorithm using the optimal solution search time with the Sawyer problem (30 tasks, 8 instances) is shown in Figure 5.

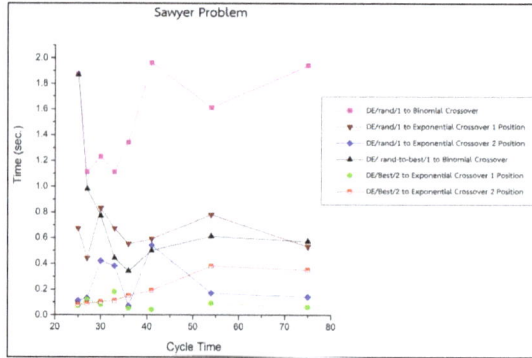

Figure 5. Comparison of Basic and Improved DE Algorithm Using Optimal Solution Search Time with Sawyer Problem.

Figure 5 shows that the improved DE algorithm DE/Best/2 to Exponential Crossover 1 with the Sawyer problem (30 tasks, 8 instances) was the best method for generating the optimal solutions with the minimum solution search time when compared to the other DE methods, and the average optimal solution search time was 0.09 seconds. The comparison between the basic and improved DE algorithms in terms of optimal solution search time with the Arcus 1 problem (83 tasks, 16 instances) is presented in Figure 6.

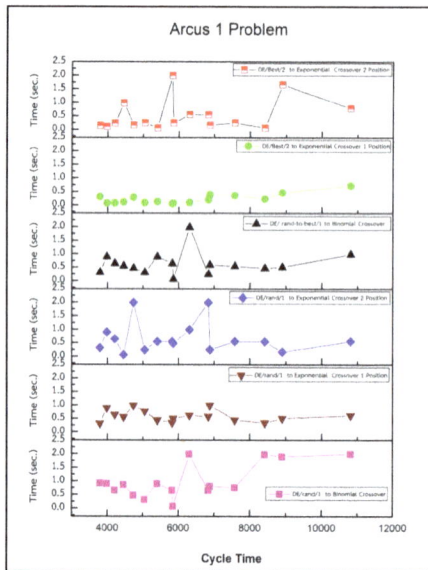

Figure 6. Comparison of Basic and improved DE Algorithm Using Optimal Solution Search Time with Arcus 1 Problem.

From Figure 6, it is seen that the Improved DE algorithm DE/Best/2 to Exponential Crossover 1 was the best method for generating optimal solutions with the minimum solution search time when compared to the other DE methods. The average optimal solution search time was 0.22 seconds.

6. The Results from the Comparison of DE Algorithm and Other Metaheuristic Methods

From the experimental results, it can be seen that the DE algorithm DE/Best/2 to Exponential Crossover 1 was the method for the optimal solution search when finding the number of workstations and minimizing the solution search time. Therefore, the DE algorithm DE/Best/2 to Exponential Crossover 1 was compared with other metaheuristic methods in order to determine the efficiency of the optimal solution search. MMAS (no local search) was used for the comparison [36] with the benchmark problems of the medium-sized UALBP-1 (21–45 tasks), which is presented in Table 3, and of the large-scale UALBP-1 (75–297 tasks), which is presented in Table 6.

Table 6. The Results of Using the Proposed DE Compared with Using MMAS (No Local Search) in UALBP-1 Benchmark Problems with the Medium-Sized UALBP-1 (21–45 tasks).

Problems	Size	Cycle Time	IP* Solution			MMAS**					DE1***		
			m^*	m	%	cal. Time (s)	E	m	%	cal. Time (s)	E		
Mitchell	21	14	8	8	0	1.50	93.75	8	0	0.05	93.75		
		15	8	8	0	1.63	87.75	8	0	0.03	87.75		
		21	5	5	0	5.43	100.00	5	0	0.01	100.00		
Heskiaoff	28	114	9	10	11.11	30.00	79.33	9	11.11	0.04	89.82		
		128	8	9	12.50	30.00	81.28	8	12.50	0.08	88.89		
		138	8	8	0	1.71	92.75	8	0	0.09	92.75		
		205	5	5	0	1.68	99.99	5	0	0.07	99.99		
		216	5	5	0	2.36	94.81	5	0	0.13	94.81		
		256	4	4	0	5.52	100.00	4	0	0.16	100.00		
		324	4	4	0	2.35	79.01	4	0	0.09	79.01		
		342	3	3	0	2.45	99.81	3	0	0.12	99.81		
Sawyer	30	25	14	14	0	1.22	92.57	14	0	0.07	92.57		
		27	13	13	0	0.96	92.31	13	0	0.12	92.31		
		30	11	11	0	7.48	98.18	11	0	0.08	98.18		
		33	10	10	0	20.15	98.18	10	0	0.18	98.18		
		36	10	10	0	2.08	90.00	10	0	0.05	90.00		
		41	8	8	0	14.67	98.78	8	0	0.04	98.78		
		54	6	6	0	5.56	100.00	6	0	0.09	100.00		
		75	5	5	0	2.97	86.84	5	0	0.06	86.84		
Kilbridge and Wester	45	57	10	10	0	1.33	96.84	10	0	0.11	96.84		
		79	7	8	14.28	30.00	90.67	7	0	0.12	99.82		
		92	6	7	16.67	30.00	89.28	6	0	0.14	100.00		
		110	6	6	0	1.48	83.64	6	0	0.10	83.64		
		138	4	4	0	4.08	100.00	4	0	0.14	100.00		
		184	3	3	0	6.73	100.00	3	0	0.15	100.00		
Total Optimal Solution (or Lower Bound) Found from 25 Problem Instances			21	2.18/ins	9.19/inst.		95.51/inst.	25	0.94/i1	0.10/inst.	97.84/inst.		

Notes: $E = \sum \frac{t}{mc} \times 100$; E = Efficiency of Balance. m^* is the optimal solution (data set) MMAS** (no local search). DE1*** = DE/best/2 to Exponential 1 position. m is the number of stations. % is the average relative deviation from the best-known solution.

From Table 6, the experimental results of the ULAB benchmark problems with the medium-sized UALBP-1 (21–45 tasks) by using the DE algorithm DE/Best/2 to Exponential Crossover 1 with

MMAS (no local search) (operated on a computer with Pentium 4, 3.0 GHz, 512 MB RAM, and the operating system Window XP) are presented. Four sets of the medium-sized UALBP (21–45 tasks) were tested: Mitchell's, Heskiaoff's, Sawyer's, and Kilbridge and Wester's problems, which were retrieved from http://www.assembly-line-balancing.de. Twenty-five instances resulted from calculating the minimum number of workstations, as depicted in the table. It was found that the DE algorithm DE/Best-2 to Exponential Crossover 1 can generate 25 optimal solutions from 25 instances. The method can find optimal solutions better than MMAS (no local search), which can generate 21 optimal solutions from 25 instances. The average solution search time of DE/Best-2 to Exponential Crossover 1 was 0.10 seconds, which is less than the solution search time of MMAS (no local search), which was 9.19 seconds.

Table 7 presents the experimental results of the ULAB benchmark problem with the large-scale UALBP-1 (75–297 tasks) by using the DE algorithm DE/Best-2 to Exponential Crossover 1 with MMAS (no local search) (operated on a computer with Pentium 4, 3.0 GHz, 512 MB RAM, and the operating system Window XP). Three sets of the large-scale UALBP (75–297 tasks) were used: Wee-mag's, Arcus 1's, and Scholl's problems were used, and the problems were retrieved from http://www.assembly-line-balancing.de. Sixty-two instances resulted from calculating the minimum number of workstations, as presented in the Table. It was found that the DE algorithm DE/Best-2 to Exponential Crossover 1 can generate 36 optimal solutions from 62 instances. The method can find optimal solutions better than MMAS (no local search), which can generate 35 optimal solutions from 62 instances. The average solution search time of DE/Best-2 to Exponential Crossover 1 was 4.88 seconds, which is less than the solution search time of MMAS (no local search), which was 5.70 seconds.

Table 7. The Results of Using the Proposed DE Compared to Using MMAS (No Local Search) [36] in UALBP-1 Benchmark Problems with the Large-Scale UALBP-1 (75–297 tasks).

Problems	Size	Cycle Time	IP* Solution			MMAS**				DE1***		
			m^*	m	%	cal. Time (s)	E	m	%	cal. Time(s)	E
		28	63	63	0	1.19	93.75	63	0	0.12	93.75
		29	63	63	0	1.19	87.75	63	0	0.14	87.75
		30	62	62	0	1.23	100.00	62	0	0.07	100.00
		31	62	62	0	1.14	89.82	62	0	0.05	89.82
		32	61	61	0	1.16	88.89	61	0	0.10	88.89
		33	61	61	0	1.22	92.75	61	0	0.28	92.75
		34	61	61	0	1.22	99.99	61	0	0.08	99.99
		35	60	60	0	1.22	94.81	60	0	0.12	94.81
		36	60	60	0	1.25	100.00	60	0	0.04	100.00
		37	60	60	0	1.23	79.01	60	0	0.06	79.01
Wee-mag	75	38	60	60	0	1.19	99.81	60	0	0.22	99.81
		39	60	60	0	1.14	92.57	60	0	0.08	92.57
		40	60	60	0	1.25	92.31	60	0	0.19	92.31
		41	59	59	0	1.19	98.18	59	0	0.10	98.18
		42	55	55	0	1.16	98.18	55	0	0.11	98.18
		43	50	50	0	1.30	90.00	50	0	0.15	90.00
		45	38	38	0	4.31	98.78	38	0	0.19	98.78
		46	34	34	0	3.59	100.00	34	0	0.38	100.00
		52	31	31	0	2.59	86.84	31	0	0.35	86.84
		54	31	31	0	1.17	96.84	31	0	0.22	96.84

Table 7. *Cont.*

Problems	Size	Cycle Time	IP* Solution	MMAS**				DE1***			
			m*	m	%	cal. Time (s)	E	m	%	cal. Time(s)	E
Wee-mag	75	56	30	30	0	1.55	99.82	30	0	0.45	99.82
Arcus 1	83	3786	21	21	0	1.03	100.00	21	0	0.30	100.00
		3985	20	20	0	1.00	83.64	20	0	0.06	83.64
		4206	19	19	0	1.08	89.82	19	0	0.07	89.82
		4454	18	18	0	1.03	88.89	18	0	0.10	88.89
		4732	17	17	0	1.02	92.75	17	0	0.28	92.75
		5048	16	16	0	1.05	99.99	16	0	0.08	99.99
		5408	15	15	0	1.08	94.81	15	0	0.12	94.81
		5824	14	14	0	1.10	100.00	14	0	0.04	100.00
		5853	13	14	7.69	1.05	79.01	13	0	0.06	89.63
		6309	13	13	0	1.07	99.81	13	0	0.09	99.81
		6842	12	12	0	1.03	92.57	12	0	0.19	92.57
		6883	12	12	0	1.03	92.31	12	0	0.38	92.31
		7571	11	11	0	1.04	98.18	11	0	0.35	98.18
		8412	10	10	0	1.05	98.18	10	0	0.22	98.18
		8898	9	9	0	1.03	90.00	9	0	0.45	90.00
		10816	7	8	14.29	1.04	98.78	8	14.29	0.71	98.78
Scholl	297	1394	50	51	2.00	12.08	100.00	51	2.00	12.28	100.00
		1452	48	49	2.08	16.10	86.84	49	2.08	11.08	86.84
		1483	47	48	2.13	11.30	96.84	48	2.13	11.30	96.84
		1515	46	47	2.17	11.55	99.82	47	2.17	11.55	99.82
		1548	45	46	2.22	11.92	89.82	46	2.22	11.92	89.82
		1584	44	45	2.27	11.63	88.89	45	2.27	11.63	88.89
		1620	43	44	2.33	11.91	92.75	44	2.33	11.91	92.75
		1659	42	43	2.38	11.97	99.99	43	2.38	11.97	99.99
		1699	41	42	2.44	12.28	94.81	42	2.44	12.28	94.81
		1742	40	41	2.50	11.08	100.00	41	2.50	11.08	100.00
		1787	39	40	2.56	11.96	79.01	40	2.56	11.96	79.01
		1834	38	39	2.63	11.45	99.81	39	2.63	11.55	99.81
		1883	37	38	2.70	12.22	92.57	38	2.70	12.30	92.57
		1935	36	37	2.77	12.30	92.31	37	2.77	12.10	92.31
		>1991	>35	>36	>2.86	>12.10	>98.18	>36	>2.86	>11.55	>98.18
		2049	34	35	2.94	11.55	98.18	35	2.94	12.03	98.18
		2111	33	34	3.03	12.30	90.00	34	3.03	12.85	90.00
Scholl	297	2177	32	33	3.13	12.10	98.78	33	3.13	11.55	98.78
		2247	31	32	3.23	11.55	100.00	32	3.23	11.92	100.00
		2322	30	31	3.33	12.03	86.84	31	3.33	11.63	86.84
		2402	29	30	3.45	12.85	96.84	30	3.45	11.91	96.84
		2488	28	29	3.57	12.84	99.82	29	3.57	11.97	99.82
		2580	27	28	3.70	12.81	92.57	28	3.7	12.28	92.57
		2680	26	27	3.85	11.77	92.31	27	3.85	11.08	92.31
		2787	25	26	4.00	12.63	98.18	26	4.00	11.99	98.18
Total Optimal Solution (or Lower Bound) Found from 62 Problem Instances			35	1.49/inst.		5.70/inst.	94.61/inst.	36	0.94/inst.	4.88/inst.	95.51/inst.

Notes: $E = \sum \frac{t}{mc} \times 100$ E = Efficiency of Balance. m^* is the optimal solution (data set). MMAS** (no local search). DE1*** = DE/best/2 to Exponential 1 position. m is the number of stations. % is the average relative deviation from the best-known solution.

7. Conclusions and Suggestions

Recently, the U-shaped line has been utilized in many production lines in place of the traditional straight-line configuration due to the use of the just-in-time principle. The shape of U-lines improves visibility and allows for the construction of stations containing tasks on both sides of the line. This arrangement, combined with cross-trained operators, provides greater flexibility in station construction than is available with a comparable straight production line. The UALBP and the DE algorithm of the metaheuristic for assigning tasks to stations are presented in this paper. The performance of the metaheuristic was applied to solve a large number of benchmark problems obtained from previously published research. The computational results indicate that one of the metaheuristic rules (DE algorithm) can be satisfied by the proposed algorithm, and the computational requirements are not high. This study has taken a step in the direction of finding good metaheuristic rules for solving the UALBP-1. For further research, it would be interesting to use other metaheuristics, (e.g., bee algorithm, particle swarm optimization, simulated annealing, etc.) and find more flexible solutions of the larger UALBP.

The improved DE algorithm DE/Best/2 to Exponential Crossover 1 was the most effective method with the minimum search time for finding optimal solutions, and the basic DE algorithm was the worst because it spent the maximum amount of time searching for optimal solutions when compared to the other DE methods. In the optimal solution search for the number of workstations, it was found that every method received the same answer when compared to the other DE methods.

The comparison of the method in this paper, UALBP-1 by using the DE algorithm, with other metaheuristic methods for the medium-sized UALBP (21–45 tasks) and large-scale UALBP (75–297 tasks) leads to the conclusion that the basic DE algorithm and the improved DE algorithm is better at generating optimal solutions in the search for workstations and spends less time searching for optimal solutions than MMAS (no local search).

Further studies should develop the DE algorithm methods with more difficult problems, such as the mixed-/multi-model line, stochastic task time, and U-lines with other characteristics, as well as develop other metaheuristic principles for solving the large-scale UALBP and develop the principles of other metaheuristic methods or other types of methods for solving the UALBP-2, i.e., given the number of stations (m), minimize the cycle time (Min. c, given m), and the UALBP-E, i.e., maximize the line efficiency (E) for c and m being variable (Max. E, given c, m).

In addition, the proposed DE algorithms in this study [37] can be applied to solve more realistic assembly line balancing problem in many industries; garment, automobile, electronical appliance; etc. for productivity improvement by minimizing the workstations and labor costs.

Author Contributions: P.S. designed the algorithm and validated of algorithm; N.K. and P.K. gathered data and validated of algorithm; K.C. summarized the computation and conclusion.

Acknowledgments: We would like to thank Industrial Engineering Department, Faculty of Engineering, Ubon Ratchathani University and Manufacturing Engineering Department, Faculty of Engineering, Mahasarakham University, for funding this research.

Conflicts of Interest: The authors declare no conflicts of interest.

References

1. Chen, S. Just-In-Time U-Shaped Assembly Line Balancing. Ph.D. Thesis, Lehigh University, Bethlehem, PA, USA, 2003.
2. McMullen, P.; Frazier, G.V. Using Simulated Annealing to Solve a Multiobjective Assembly Line Balancing Problem with parallel workstations. *Int. J. Prod. Res.* **1998**, *10*, 2717–2741. [CrossRef]
3. Suresh, G.; Sahu, S. Stochastic Assembly Line Balancing Using Simulated Annealing. *Int. J. Prod. Res.* **1994**, *8*, 1801–1810. [CrossRef]
4. Kriengkorakot, N.; Pianthong, N. The Assembly Line Balancing Problem: Review Articles. *KKU Eng. J.* **2007**, 34133–34140.

5. Garg, H. A hybrid GSA-GA algorithm for constrained optimization problem. *Int. Sci.* **2019**, *478*, 499–523. [CrossRef]
6. Garg, H. A hybrid PSO-GA algorithm for constrained optimization problem. *Appl. Math. Comput.* **2016**, *274*, 292–305. [CrossRef]
7. Garg, H. An efficient biogeography based optimization algorithm for solving reliability optimization problems. *Swarm Evolut. Comput.* **2015**, *24*, 1–10. [CrossRef]
8. Garg, H. An approach for solving constrained reliability-redundancy allocation problems using Cuckoo search algorithm. *J. Appl. Sci.* **2015**, *4*, 14–25. [CrossRef]
9. Sethanan, K.; Pitakaso, R. Improved differential evolution algorithms for solving generalized assignment problem. *Expert Syst. Appl.* **2016**, *45*, 450–459. [CrossRef]
10. Miltenburg, J.; Wijingaard, J. The U-line Line Balancing Problem. *Memet. Sci.* **1994**, *40*, 1378–1388. [CrossRef]
11. Miltenburg, J.; Sparling, D. Optimal solution algorithms for the U-line balancing problem. Ph.D. Thesis, McMaster University, Hamilton, ON, Canada, 1995.
12. Sparling, D.; Miltenburg, J. The mixed-model U-line balancing problem. *Int. J. Prod. Res.* **1998**, *36*, 485–501. [CrossRef]
13. Urban, T.L. Optimal Balancing of U-Shaped Assembly Lines. *Memet. Sci.* **1998**, *44*, 738–741. [CrossRef]
14. Miltenburg, J. Balancing U-lines in a multiple U-line facility. *Eur. J. Oper. Res.* **1998**, *15*, 1–23. [CrossRef]
15. Veres, P.; Banyai, T.; Illes, B. Optimization of In-Plant Production Supply with Black Hole Algorithm. *Solid State Phenom.* **2017**, *261*, 503–508. [CrossRef]
16. Banyai, T.; Illes, B.; Banyai, A. Smart Scheduling: An Integrated First Mile and Last Mile Supply Approach. *Complexity* **2018**, *2018*, 1–15. [CrossRef]
17. Nearchou, A.C. Multi-objective balancing of assembly lines by population heuristics. *Int. J. Prod. Res.* **2008**, *46*, 466–481. [CrossRef]
18. Vila, M.; Pereira, J. A branch-and-bound algorithm for assembly line worker assignment and balancing problem. *Comput. Oper. Res.* **2013**, *40*, 3045–3055. [CrossRef]
19. Scholl, A.; Klein, R. ULINO: optimally balancing U-Shaped JIT assembly lines. *Int. J. Prod. Res.* **1999**, *37*, 721–736. [CrossRef]
20. Martinez, U.; Duff, W.S. Heuristic approaches to solve the U-shaped line balancing problem augmented by Genetic Algorithms. *Inf. Eng. Des. Sym.* **2004**, *11*, 246–259.
21. Gokcen, H.; Agpak, K.; Gencer, C.; Kizilkaya, E. A shortest route formulation of simple U-type assembly line balancing problem. *Appl. Math. Mode.* **2005**, *29*, 373–380. [CrossRef]
22. Gutjahr, A.L.; Nemhauser, G.L. An algorithm for the line balancing problem. *Memet. Sci.* **1964**, *11*, 308–315. [CrossRef]
23. Gokcen, H.; Agpak, K. A goal programming approach to simple U-line balancing problem. *Eur. J. Oper. Res.* **2006**, *171*, 577–585. [CrossRef]
24. Chiang, W.C.; Urban, T.L. The Stochastic U-line Balancing Problem: A Heuristic Procedure. *Eur. J. Oper. Res.* **2006**, *175*, 1767–1781. [CrossRef]
25. Erel, E.; Sabuncuoglu, I.; Aksu, B.A. Balancing of U-type Assembly Systems Using Simulated Annealing. *Int. J. Prod. Res.* **2001**, *39*, 3003–3015. [CrossRef]
26. Sotskov, Y.N.; Dolgui, A.; Portmann, M.C. Stability analysis of an optimal balance for an assembly line with fixed cycle time. *Eur. J. Oper. Res.* **2006**, *168*, 783–797. [CrossRef]
27. Hwang, R.; Katayama, H.; Gen, M. U-shaped assembly line balancing problem with genetic algorithm. *Eur. J. Oper. Res.* **2008**, *46*, 4637–4649. [CrossRef]
28. Andreas, C.N. A Differential Evolution Algorithm for Simple Assembly Line Balancing. *IFAC. Int. Fed. Auto. Cont.* **2005**, *16*, 1462–1467.
29. Pitakaso, R. Differential evolution algorithm for simple assembly line balancing type 1 (SALBP-1). *J. Ind. Prod. Eng.* **2015**, *32*, 104–114. [CrossRef]
30. Qin, A.K.; Hang, V.L.; Suganthan, P.N. Differential Evolution Algorithm with Strategy Adaptation for Global Numerical Optimization. *IEEE Trans. Evolut. Comput.* **2009**, *13*, 398–417. [CrossRef]
31. Zou, D.; Liu, H.; Gao, L.; Li, S. An improved differential evolution algorithm for the task assignment problem. *Eng. Appl. Artif. Intell.* **2011**, *24*, 616–624. [CrossRef]

32. Zeng, X.; Wong, W.; Sun Leung, S. An operator allocation optimization model for balancing control of the hybrid assembly lines using Pareto utility discrete differential evolution algorithm. *Comput. Oper. Res.* **2012**, *39*, 1145–1159. [CrossRef]
33. Jackson, J.R. A computing procedure for a line balancing problem. *Memet. Sci.* **1956**, *2*, 261–271. [CrossRef]
34. Bowman, E.H. Assembly Line Balancing by Linear Programming. *Oper. Res.* **1960**, *8*, 385–389. [CrossRef]
35. Ronkkonen, J.; Kukkonen, S.; Price, K.V. Real-Parameter Optimization with Differential Evolution. *IEEE Congr. Evolut. Comput.* **2005**, *8*, 506–513.
36. Kriengkorakot, N. Metaheuristic approach for assembly line balancing problem. Ph.D. Thesis, Ubon Ratchathani University, Ubon Ratchathani, Thailand, 2008.
37. Pitakaso, R.; Sethanan, K. Modified differential evolution algorithm for simple assembly line balancing with a limit on the number of machine types. *Eng. Optim.* **2015**, *48*, 253–271. [CrossRef]

© 2018 by the authors. Licensee MDPI, Basel, Switzerland. This article is an open access article distributed under the terms and conditions of the Creative Commons Attribution (CC BY) license (http://creativecommons.org/licenses/by/4.0/).

Mathematical and Computational Applications

MDPI

Article

Improved Differential Evolution Algorithm for Flexible Job Shop Scheduling Problems

Prasert Sriboonchandr *, Nuchsara Kriengkorakot and Preecha Kriengkorakot

Industrial Engineering, Department, Faculty of Engineering, Ubon Ratchathani University,
Ubon Ratchathani 34190, Thailand
* Correspondence: prasert.sr.58@ubu.ac.th; Tel.: +66-81-697-1020

Received: 20 July 2019; Accepted: 5 September 2019; Published: 6 September 2019

Abstract: This research project aims to study and develop the differential evolution (DE) for use in solving the flexible job shop scheduling problem (FJSP). The development of algorithms were evaluated to find the solution and the best answer, and this was subsequently compared to the meta-heuristics from the literature review. For FJSP, by comparing the problem group with the makespan and the mean relative errors (MREs), it was found that for small-sized Kacem problems, value adjusting with "DE/rand/1" and exponential crossover at position 2. Moreover, value adjusting with "DE/best/2" and exponential crossover at position 2 gave an MRE of 3.25. For medium-sized Brandimarte problems, value adjusting with "DE/best/2" and exponential crossover at position 2 gave a mean relative error of 7.11. For large-sized Dauzere-Peres and Paulli problems, value adjusting with "DE/best/2" and exponential crossover at position 2 gave an MRE of 4.20. From the comparison of the DE results with other methods, it was found that the MRE was lower than that found by Girish and Jawahar with the particle swarm optimization (PSO) method (7.75), which the improved DE was 7.11. For large-sized problems, it was found that the MRE was lower than that found by Warisa (1ST-DE) method (5.08), for which the improved DE was 4.20. The results further showed that basic DE and improved DE with jump search are effective methods compared to the other meta-heuristic methods. Hence, they can be used to solve the FJSP.

Keywords: improved differential evolution algorithm; flexible job shop scheduling problem; local search and jump search

1. Introduction

Nowadays, the goal of businesses and industry is to reduce costs, and this is affected by production scheduling. Efficient production scheduling can reduce production expenses and time, resulting in on-schedule delivery of goods to customers and a competitive advantage for the firm. The issues of production scheduling concern the sequencing and machine assignment for each order. Owing to the requirement of the modern manufacturing processes for greater flexibility, the job shop scheduling problem (JSP) is an important type of production scheduling; furthermore, the flexible job shop scheduling problem (FJSP) was developed from the classical JSP. Job operations are allocated to every given machine in the production process. It is possible to assign any job to more than one machine as per the machine's capability and, consequently, constructing an environment that is similar to the actual industry [1]. FJSP is an NP-Hard problem of the combinatorial optimization type, which has a complex solution. Applying a mathematical method to determine the exact algorithms for an optimal solution takes a moderate amount of time and wastes money; in particular, there is a big problem with the great number of variables and the limitations of the method. Metaheuristics are developed to solve FJSP to find a near optimal production schedule and to shorten the time required to solve the problem. It includes Tabu Search, the genetic algorithm (GA), particle swarm optimization (PSO), and ant colony optimization (ACO).

Differential evolution was proposed by Storn and Price [2]. It involves the optimization algorithm by using a population of each generation to search for the solution. Differential evolution calls the member of any generation vector and component of the vector point. Furthermore, it calls a number of points for each vector dimension. Each point can be compared to a gene from the GA method. The method of differential evolution is widely popular owing to its simplicity, various types of solution, and appropriate solutions. In some cases, DE results represent the global optimum.

Hence, in this research, we develop the differential evolution algorithm for solving in the flexible job shop scheduling problem with the target of minimizing the makespan.

This paper is structured as follows: In Section 2, the literature review is presented; Section 3 presents the mathematical model of the FJSP; Section 4 describes the general structure of the DE; Section 5 presents the results of the experiment on DE to solve the FJSP; Section 6 presents the comparison of DE with other metaheuristics; and Section 7 presents the conclusions and suggestions.

2. Literature Review

2.1. Flexible Job Shop Scheduling Problems by Using Other Metaheuristic Methods

To summarize the relevant literature and research on the solution to the flexible job shop scheduling problem by using metaheuristics, Xia and Wu [3] studied a hybrid of the PSO and SA methods for multiple purposes. Kanate [4] researched the development of a metaheuristic called the makespan tree for sequencing jobs on machines. Subsequently, two metaheuristics, the genetic algorithm and particle swarm optimization, were developed. Both metaheuristics use the makespan tree as a part of their method to solve the flexible job shop problems with the objective of minimizing the makespan. The findings for the job scheduling problems showed that the makespan tree outperformed the non-delay by 11.80%, improved the earliest finish time by 13.60%, and reduced the shortest processing time by 17.41%. In comparison, PSO had better results than GA by 0.97%. Tang et al. [5] researched the use of a hybrid algorithm for the flexible job-shop scheduling problem by combining chaos particle swarm optimization with the genetic algorithm in order to minimize the makespan. Wannaporn and Arit [6] applied the modified genetic algorithm to the flexible job-shop scheduling problem. This included the following processes: (1) Selecting chromosomes by the fuzzy roulette wheel selection method; (2) operating the crossover by the cluster-crossover operator to calculate the similarity between chromosomes for the crossover; (3) processing mutations using the mutation-local search operator to determine the diversity of the population, resulting in an optimal solution. The objective of this research was to minimize the makespan. Thanyaporn et al. [7] developed and improved the mixed integer programming (MIP) for an advanced planning and scheduling (APS) problem as a technique for production planning and scheduling. Its various considered constraints were the machine capacity, operation sequence, multi-machine due dates, multi-order, and product structure, comprising multiple steps and items. Each item could be processed by any given set of machines, and there was an extension to involve the considered constraints as a preventive maintenance time window. The objective was to minimize the total costs from idle production time, earliness, and tardiness, and furthermore, to optimize production scheduling. Examples of solutions were presented in four models which were (1) APS1, APS without alternative machines, (2) APS1PM, APS without alternative machines and PM, being similar to model 1 but with preventive maintenance included, (3) APS2, APS with alternative machines, and (4) APS2PM, APS with alternative machines and PM, being similar to model 3 but with preventive maintenance included. Hamid et al. [8] studied the machine scheduling in production—a content analysis. Which found that there were 132 surveyed papers regarding machine scheduling problems in production. The results were applied as an approach for proposing future research. It was found that, generally, the objective was to minimize the makespan. For the problem-solving approach, simple heuristics, as the shortest processing time first (SPT), and meta-heuristics, as the genetic algorithm, were employed. Li and Gao [9] studied an effective hybrid genetic algorithm and Tabu Search for flexible job shop scheduling problem. Luan et al. [10] studied improved whale algorithm for solving the flexible job shop scheduling problem and Li et al. [11] studied the hybrid artificial bee colony algorithm with a rescheduling strategy for solving the flexible job shop scheduling problems.

2.2. Flexible Job Shop Scheduling Problems by Using Differential Evolution Algorithm

Based on a study of the literature and research regarding the solutions to flexible job shop scheduling problems by using the differential evolution algorithm, Wisittipanich [12] proposed the application of adapted differential evolution algorithms to minimize the makespan in the FJSP. The modification of algorithms aimed to improve the efficiency of original DE by balancing its exploration and exploitation abilities to avoid the common problem of premature convergence. The first adapted DE was called the DE with subgroup strategy. In this algorithm, the population is divided into groups, and the population in each group employs different strategies to search for the new solution simultaneously in order to extract the strength of any strategy and compensate for the weakness of each strategy. This led to an increase in the overall performance efficiency. The second adapted DE was the DE with the switching strategy. This algorithm allowed the entire population to change searching strategies when there was no improvement to the solution. Thus, the chance of being trapped at a local optimum was decreased. The efficiency of two modified DEs was examined in solving an experimental problem and compared with the result of the original DE. The solutions provided by both modified DE algorithms were comparable or of a higher quality than the solutions from the original DE. Yuan and Xu [13] studied flexible job shop scheduling using hybrid differential evolution algorithms. Hybridization comprised the development of a mechanism to use the discrete differential evolution algorithm to solve the flexible job shop scheduling problem and second, enhancement of the local search ability in the DE framework. The objective was to find minimize the makespan. Bhaskara et al. [14] on the use of the differential evolution algorithm for the flexible job shop scheduling problem, applied the local search algorithm with the objective of minimizing the makespan.

2.3. Differential Evolution Algorithm for Solving Other Problems

The differential evolution algorithm is perceived as a modern method. It has become a favorable method to employ in various areas including operational research problems, such as application in the blocking flow shop scheduling problems to reduce production time [15]. Furthermore, in was applied in job shop production. Zhang et al. [16] adapted this method to deal with the job shop problem in order to minimize the total tardiness. For applications in the supply chain management problem, in particular, for the vehicle routing problem in the supply chain, the relevant studies have been carried out with different objectives and constraints under various conditions. For example, Cao and Lai [17] adapted the method to the open vehicle routing problem with fuzzy demands to reduce the total traveled distance. Lai and Cao [18] applied the method to solve the VRP with pickups and deliveries and time windows aimed at minimizing the total traveled distance. Xu and Wen [19] employed the DE in the unidirectional logistics distribution vehicle routing problem with no time windows and achieved the shortest total distance. By adapting DE with the agricultural management problem, Cruz et al. [20] applied the optimal control problem to determine the optimal control of nitrogen gas in lettuce. Moreover, DE was employed in a proposal on crop planning using the multi-objective differential evolution algorithm. This study had the objective of minimizing irrigation water usage and maximizing the total yield and net profit planting under various conditions [21]. DE was further applied to other problems, for example, in the electric power system, it was used to solve the optimal power flow, resulting in voltage stability enhancement and cost minimization [22]. In the wireless sensor network system, DE was proposed to prolong the lifetime of the system by preventing it from overloading [23]. An adaptation of DE, moreover, was used in the chemical industry in order to determine the optimal criteria for a chemical process [24].

3. Flexible Job Shop Scheduling Problem Pattern and Mathematical Model

We only focused on the FJSP in this study, and the details of the problem are as follows:

3.1. Flexible Job Shop Scheduling Problem (FJSP)

This production system is similar to the job shop scheduling problem but with more flexibility. As an explanation, any job includes a specific operation that can be processed by 1 or more machines owing to the various capabilities of the individual machine. As shown in Table 1, there are 3 jobs, each with a different set of operations. Every operation is allowed to select any machine from a given set, for example, in job 2, operation 1 can be performed on 2 machines, with M2 processing 10 time units and M3 processing 7 time units, while M1 shows "-", referring to an inability to operate.

Table 1. Example of flexible job shop scheduling problem pattern.

Job	Operation	Machines		
		M_1	M_2	M_3
J_1	$O_{1,1}$	5	-	3
	$O_{1,2}$	-	5	10
	$O_{1,3}$	5	9	-
J_2	$O_{2,1}$	-	10	7
	$O_{2,2}$	20	6	-
	$O_{3,3}$	2	-	11
J_3	$O_{3,1}$	2	5	4
	$O_{3,3}$	2	5	10

Figure 1 shows the general pattern of flexible job shop scheduling problem systems, which consist of a system with C work stations. In each work station, there will be a number of identical parallel machines. Each work station has its own specific route and can choose to perform the tasks assigned to one of the parallel machines and can be at the same work station. Considering the complexity of the model, the model of the flexible job shop scheduling problem systems that allows for recursive operation is the most redundant model.

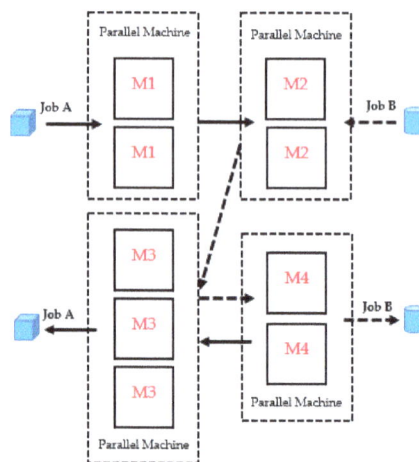

Figure 1. Flexible job shop scheduling problem pattern.

To measure the production scheduling efficiency, there are various effective evaluations based on the production characteristics such as determining the minimization of the makespan (C_{max}), the number of tardy jobs, and the maximum lateness. This paper employed the minimized makespan

(C_{max}) for evaluation in accordance with Equation (1). While C1, C2, ... , C_p are possible solutions to produce scheduling, the solution resulting in the longest processing time was selected. Moreover, Z refers to the objective of production scheduling—to achieve the lowest value—as described in Equation (1). Figure 2 shows a sample of solutions from the total processing time of entire jobs:

$$\text{Minimize } Z = f\,(C_1, C_2, \dots, C_p).\tag{1}$$

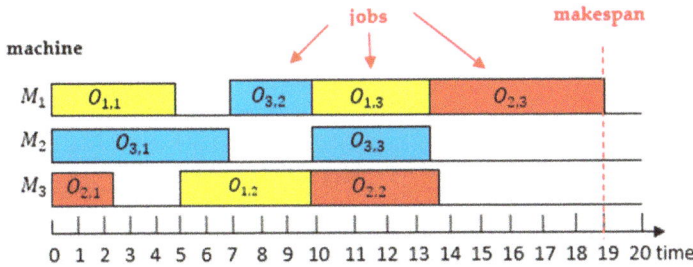

Figure 2. Gantt chart of solutions.

The relative error is the difference between the measured and actual values, and it is generally expressed as a percentage (%). When the measured value is proximate to the actual value, this indicates high correctness or accuracy. The relative error is calculated using Equation (2), as follows;

$$\%\text{Relative Error}: \text{RE} = \frac{C_{min} - \text{BKS}}{\text{BKS}} \times 100\tag{2}$$

where C_{min} is the optimal solution to the algorithm, and BKS is the best known solution.

3.2. Mathematical Model of the Flexible Job Shop Scheduling Problem

The mathematical model of the FJSP, proposed by Kanate [4], includes many relevant binary variables, which are as follows:

3.2.1. Indices

i	Machine
j, k	Job
h, l	Operation

3.2.2. Parameter

M	Mathematically large real number
$P_{i,j,h}$	Processing time for operation h of job j on machine i
$O_{i,j,h}$	Operation h of job j on machine i
$a_{i,j,h}$	Constant to define if job j at operation h is able to be processed by i—equal to 1 for the ability to process and 0 for the inability to process

3.2.3. Decision Variables

C_{max}	Makespan
$t_{j,h}$	Start time of the processing operation h of job j on any machine
$f_{j,h}$	Finish time of the processing operation h of job j on any machine
$y_{i,j,h}$	Binary variable equal to 1 while processing operation h of job j on machine i

$X_{i,j,h,k,l}$ Binary decision variable equal to 1 while processing operation h of job j on machine $\left(O_{i,j,h}\right)$, which comes before operation l of job k on machine i $\left(O_{i,k,l}\right)$.

The mathematical model of the flexible job shop scheduling problem can be written as:

$$\text{Minimize } C_{max} = \max \{C_i\}, i = 1, 2, 3, \ldots, n \tag{3}$$
$$\text{st.}$$

$$t_{j,h} + y_{i,j,h} \times P_{i,j,h} \le f_{i,h}; \quad \forall(i,j,h) \tag{4}$$

$$f_{j,h} \le t_{j,h+1}; \quad \forall(j,h) \tag{5}$$

$$f_{j,h} \le C_{max}; \quad \forall(j,h) \tag{6}$$

$$y_{i,j,h} \le a_{i,j,h}; \quad \forall(i,j,h) \tag{7}$$

$$t_{j,h} + P_{i,j,h} \le t_{k,l} + \left(1 - x_{i,j,h,k,l}\right)M; \quad \forall(i,j,h,k,l) \tag{8}$$

$$\sum_i y_{i,j,h} = 1; \quad \forall(h,j) \tag{9}$$

$$\sum_j \sum_h x_{i,j,h,k,l} = y_{i,k,l}; \quad \forall(i,k,l) \tag{10}$$

$$\sum_k \sum_j x_{i,j,h,l,k} = y_{i,j,h}; \quad \forall(i,j,h) \tag{11}$$

$$x_{i,j,h,j,h} = 0; \quad \forall(i,j,h) \tag{12}$$

$$t_{j,h} \ge 0; \quad \forall(j,h) \tag{13}$$

$$f_{j,h} \ge 0; \quad \forall(j,h) \tag{14}$$

$$y_{i,j,h} \in \{0,1\} \tag{15}$$

$$x_{i,j,h,k,l} \in \{0,1\} \tag{16}$$

Equation (3) is a target equation to produce the minimum makespan, in order to reduce the production time, contributing to lower costs and product delivery time. Next, Equations (4) and (5) restrict the order based on the priority order in which each job is processed (precedence constraint). Equation (6) imposes a constraint on the makespan, whereby all job operations must be finished in a time less than or equal to the makespan. For Equation (7), only machines available for processing can be selected. Equation (8) is used to ensure that each machine can process, at most, one job at a time. Constrained by Equation (9), any job operation can be assigned to process on one machine only. Equations (10) and (11) create the sequence of job priority on machine i—Equation (10) selects the predecessor job and Equation (11) selects the next job. Equation (12) is the constraint that ensures that each job operation cannot be processed before its release time on machine i. Subsequently, Equations (13) and (14) restrict the start and finish times of every job to positive real numbers. Furthermore, Equations (15) and (16) specify $y_{i,j,h}$ and $x_{i,j,h,k,l}$ as binary variables.

4. General Differential Evolution Algorithm

The general differential evolution algorithm has general procedures [25], as shown in Figure 3.

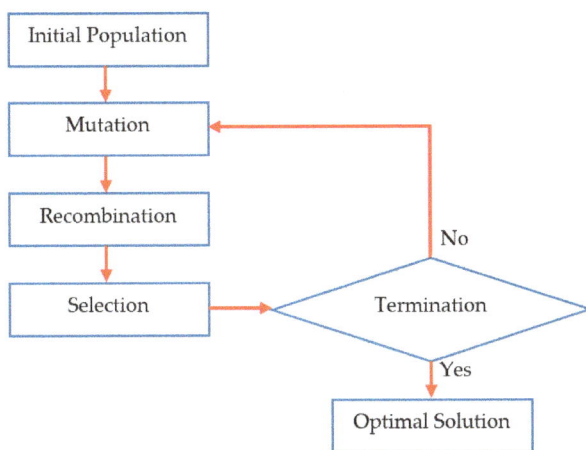

Figure 3. General differential evolution algorithm procedure.

The procedures of the differential evolution algorithm consist of (1) the initial population, and (2) mutation to generate mutant vector by differentiating a dimension of the vector. The calculation for mutating a vector's dimensions is shown in Equation (17):

$$V_{m,n,G} = X_{r1,n,G} + F(X_{r2,n,G} - X_{r3,n,G}) \tag{17}$$

where m is the number of vectors of each generation, n is vector dimension, G is the iteration round, such as round 1, 2, or 3, and $r1$, $r2$, and $r3$ are 3 random vectors. F is the scaling factor, where $V_{m,n,G}$ is a mutant vector for vector m at vector dimension n in iteration round G. Moreover, $X_{r1,n,G}$ is firstly the random target vector at dimension n in iteration round G. $X_{r2,n,G}$ and $X_{r3,n,G}$ are, respectively, the second and third random target vectors. Thus, the mutant vector of vector m at dimension n in round G equals the value of target vector $r1$ at dimension n round G plus scaling factor F times the difference between target vectors $r2$ and $r3$. (3) Recombination of the trial vector is generated by exchanging vector dimensions. The equation employed for generating a trial vector is Equation (18):

$$U_{m,n,G} = \begin{cases} V_{m,n,G} \text{ if } rand_{mn} \leq CR \text{ or } D_m = D_{m_{rand}} \\ X_{m,n,G}, \text{ else.} \end{cases} \tag{18}$$

where $U_{m,n,G}$ is the trial vector of vector m at dimension n in round G, $rand_{mn}$ is a random real number between [0, 1] of vector m at dimension n, CR refers to the crossover rate, D_m is the dimension of vector m. Furthermore, $D_{m_{rand}}$ denotes a random integer number of vector m in the range [1, N], where N is the vector size. Consequently, the value of trial vector m in dimension n in iteration round G equals the mutant vector when there is a random number of vector m at dimension n that is less than CR; in other words, $D_{m_{rand}}$ equals D_m. (4) The formula for selecting a target vector for the next round described in Equation (19). To solve the equation for the minimum target value (minimization), the less than or equal to sign is used:

$$X_{m,n,G+1} = \begin{cases} U_{m,n,G} \text{ if } f(U_{m,n,G}) \leq f(X_{m,n,G}) \\ X_{m,n,G}, \text{ else.} \end{cases} \tag{19}$$

where, $X_{m,n,G+1}$ is the target vector m at dimension n in round $G + 1$, where the vector with a better fitness function value is selected in comparison with the value of the target vector and trial vector in round G.

4.1. Procedure of FJSP by Using Differential Evolution Algorithm

The procedure of FJSP uses the general differential evolution algorithm. The values used in the calculation must be set. The variables are as follows: *Iterate* = round, *NP* = number in population, *F* = scaling factor, and *CR* = crossover rate. In this problem's calculation, these suitable variables were set from the experiment as *Iterate* = 1, *NP* = 5, *F* = 0.8, and *CR* = 0.8. The general differential evolution algorithm "DE/rand/1" and binomial crossover were used in the calculation.

4.1.1. Calculation Using the General Differential Evolution Algorithm DE/rand/1 and Binomial Crossover

The flexible job shop scheduling problem in Table 2 provides the details of the processing time on each machine. By explanation,

Row 1 indicates the numbers of jobs and machines, including 4 jobs and 5 machines.

Row 2 shows the data on job 1, comprising 3 operations. In operation 1, there are 5 available machines, which are machine 1 with a processing time of 2, machine 2 with a processing time of 5, machine 3 with a processing time of 4, machine 4 with a processing time of 1, and machine 5 with a processing time of 2. Operation 2 includes machine 1 with a processing time of 5, machine 2 with a processing time of 4, machine 3 with a processing time of 5, machine 4 with a processing time of 7, and machine 5 with a processing time of 5. Furthermore, operation 3 has machine 1 with a processing time of 4, machine 2 with a processing time of 5, machine 3 with a processing time of 5, machine 4 with a processing time of 4, and machine 5 with a processing time of 5.

Rows 3–5 show the data on jobs 2–4. The processing time of each operation can be likewise described, as shown in row 2, and put into categories, as shown in Table 3.

Table 2. Sample problems of Kacem et al. [26,27].

45.
35. 1 2 2 5 3 4 4 1 5 2 5 1 5 2 4 3 5 4 7 5 5 5 1 4 2 5 3 5 4 4 5 5
35. 1 2 2 5 3 4 4 7 5 8 5 1 5 2 6 3 9 4 8 5 5 5 1 4 2 5 3 4 4 5 4 5 5
45. 1 9 2 8 3 6 4 7 5 9 5 1 6 2 1 3 2 4 5 5 4 5 1 2 2 5 3 4 4 2 5 4 5 1 4 2 5 3 2 4 1 5 5
25. 1 1 2 5 3 2 4 4 5 1 2 5 1 5 2 1 3 2 4 1 5 2

Table 3. Sample problem with four jobs and five machines.

Jobs	Operations	Machines				
		M1	M2	M3	M4	M5
J1	$O_{1,1}$	2	5	4	1	2
	$O_{1,2}$	5	4	5	7	5
	$O_{1,3}$	4	5	5	4	5
J2	$O_{2,1}$	2	5	4	7	8
	$O_{2,2}$	5	6	9	8	5
	$O_{2,3}$	4	5	4	54	5
J3	$O_{3,1}$	9	8	6	7	9
	$O_{3,2}$	6	1	2	5	4
	$O_{3,3}$	2	5	4	2	4
	$O_{3,4}$	4	5	2	1	5
J4	$O_{4,1}$	1	5	2	4	12
	$O_{4,2}$	5	1	2	1	2

To improve the solution to round 1, the following procedure is completed:

Step 1: Generate the initial population

The initial population is randomized from a number between [0, 1], where the dimension or position (D) equals the number of operations (12), and the number of populations (P; 5) results in the target vector of a sample, as shown in Table 4.

Table 4. Target vectors of sample problems.

NP	Dimensions, D											
	1	2	3	4	5	6	7	8	9	10	11	12
1	0.55	0.32	0.70	0.12	0.64	0.89	0.96	0.81	0.38	0.55	0.27	0.71
2	0.17	0.80	0.94	0.93	0.44	0.36	0.77	0.35	0.13	0.42	0.17	0.11
3	0.42	0.35	0.15	0.61	0.10	0.34	0.93	0.51	0.08	0.59	0.63	0.50
4	0.65	0.72	0.30	0.58	0.02	0.74	0.59	0.17	0.14	0.07	0.73	0.31
5	0.72	0.32	0.04	0.20	0.89	0.28	0.42	0.67	0.15	0.49	0.09	0.81

Step 2: Differentiation of a dimension or mutation

Three vectors, r_1, r_2, and r_3, are randomly picked from the total population equal to 5 in order to generate a mutant vector of the entire 5 target vectors. As indicated by the sample problems, the randomized vectors (r_1, r_2, r_3) from target vector 1 consist of vectors 2, 5, and 3. The other concerned target vectors are demonstrated in Table 5.

Table 5. Randomized vectors r_1, r_2, and r_3.

Random Vector	r1	r2	r3
1	2	5	3
2	3	3	5
3	4	2	2
4	5	1	1
5	1	4	4

Calculation of the mutant vector ($V_{m,n,G}$) can be performed by substituting the randomized vector into Equation (20). Thus, the mutant vectors of sample problems are presented in Table 6.

$$V_{m,n,G} = X_{r1,n,G} + F(X_{r2,n,G} - X_{r3,n,G}) \tag{20}$$

Table 6. Mutant vectors of sample problems in round 1.

Mutation	1	2	3	4	5	6	7	8	9	10	11	12
1	0.77	0.74	0.42	0.11	2.02	0.24	−0.25	0.67	0.27	0.22	−0.91	0.73
2	0.80	0.13	0.74	0.24	1.78	0.50	−0.47	0.04	1.81	0.38	−0.43	0.37
3	0.16	0.70	0.56	0.61	1.45	0.43	−0.74	0.24	1.30	−0.08	0.02	0.40
4	0.81	0.43	0.24	−0.36	0.03	0.09	0.97	0.69	−0.70	1.04	1.19	1.04
5	0.87	1.30	0.65	0.69	1.05	1.39	0.23	−0.86	1.31	0.60	0.31	0.91

Step 3: Crossover

The number in the range [0, 1] at the target vector position is randomly picked and is comparative to the crossover rate (CR) predefined for the crossover. Table 7 shows the random numbers for the crossover.

Table 7. Random numbers for the crossover.

Vector	1	2	3	4	5	6	7	8	9	10	11	12
1	0.83	0.05	0.75	0.80	0.95	0.39	0.29	0.60	0.30	0.44	0.59	0.65
2	0.68	0.36	0.48	0.47	0.70	0.96	0.04	0.76	0.64	0.42	0.16	0.44
3	0.32	0.40	0.97	0.38	0.63	0.69	0.71	0.92	0.65	0.83	0.92	0.49
4	0.56	0.18	0.06	0.38	0.47	0.23	0.11	0.85	0.80	0.30	0.65	0.02
5	0.81	0.35	0.70	0.50	0.89	0.89	0.84	0.29	0.01	0.21	0.41	0.83

Subsequently, the trial vector is calculated with Equation (21). In comparison, if a random number is less than or equal to CR = 0.8, the mutant vector in the same position will be selected as the obtained trial vector of that position. For differential cases, the target vector in the same position is the answer to the trial vector of the concerned position. Table 8 shows the obtained trial vectors.

$$U_{ji,G+1} = \begin{cases} V_{ji,G+1} \ if\,(randb(j) \leq CR) \ or \ j = rnbr\,(i) \\ X_{ji,G} \quad if\,(randb(j) > CR) \ or \ j \neq rnbr\,(i) \end{cases} \tag{21}$$

Table 8. Trial vectors from round 1.

Trial Vector	1	2	3	4	5	6	7	8	9	10	11	12
1	0.55	0.74	0.42	0.12	0.64	0.24	−0.25	0.67	0.27	0.22	−0.91	0.73
2	0.80	0.13	0.74	0.24	1.78	0.50	−0.47	0.04	1.81	0.38	−0.43	0.37
3	0.16	0.70	0.56	0.61	1.45	0.43	−0.74	0.24	1.30	−0.08	0.02	0.40
4	0.65	0.72	0.30	0.58	0.02	0.74	0.59	0.17	0.14	0.07	0.73	0.31
5	0.72	0.32	0.04	0.20	0.89	0.28	0.42	0.67	0.15	0.49	0.09	0.81

Step 4: Fitness Evaluation

To decode the value of each dimension, the values are sorted based on the rank order value (ROV) in ascending order. Furthermore, the values are positioned in accordance with the vector order, as described in Table 9.

Table 9. Results of decoding.

Vector	11	7	4	10	6	9	3	1	5	8	12	2
1	−0.91	−0.25	0.12	0.22	0.24	0.27	0.42	0.55	0.64	0.67	0.73	0.74
$O_{i,j}$	1, 1	1, 2	1, 3	2, 1	2, 2	2, 3	3, 1	3, 2	3, 3	3, 4	4, 1	4, 2

As a consequence of the sequencing machine operation, the machine with the earliest completion time is selected. If the machine operation consumes time equally, selection becomes random. This is described in Table 10 and illustrated by the Gantt chart in Figure 4.

Table 10. Results of the sequencing machine operation.

Vector	1	2	3	4	5	6	7	8	9	10	11	12
1	0.55	0.74	0.42	0.12	0.64	0.24	−0.25	0.67	0.27	0.22	−0.91	0.73
$O_{i,j}$	3,2	4,2	3,1	1,3	3,3	2,2	1,2	3,4	2,3	2,1	1,1	4,1
M	2	2	4	1	1	5	2	4	3	1	4	1
PT	1	1	6	4	2	5	4	1	4	2	1	1

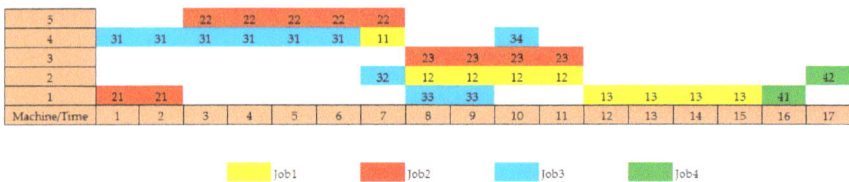

Figure 4. Gantt chart of sample problems.

Figure 4 presents a Gantt chart for solving the problem of a sample with 4 jobs, 5 machines, 12 operations, and a makespan of 17. The machines on the critical part are machines 1, 5, and 2 which consequentially interfere with solutions using the local search method.

Step 5: Selection

By comparing the fitness factor values of target vectors (Table 11) with the fitness factor values of the trial vectors (Table 12), the better vectors are selected as the target values for the next round. This is shown in Table 13.

Table 11. Target vectors of round 1.

Target Vector	1	2	3	4	5	6	7	8	9	10	11	12	Target
1	0.55	0.32	0.70	0.12	0.64	0.89	0.96	0.81	0.38	0.55	0.27	0.71	21
2	0.17	0.80	0.94	0.93	0.44	0.36	0.77	0.35	0.13	0.42	0.17	0.11	20
3	0.42	0.35	0.15	0.61	0.10	0.34	0.93	0.51	0.08	0.59	0.63	0.50	22
4	0.65	0.72	0.30	0.58	0.02	0.74	0.59	0.17	0.14	0.07	0.73	0.31	24
5	0.72	0.32	0.04	0.20	0.89	0.28	0.42	0.67	0.15	0.49	0.09	0.81	19

Table 12. Trial vectors of round 1.

Trial Vector	1	2	3	4	5	6	7	8	9	10	11	12	Target
1	0.55	0.74	0.42	0.12	0.64	0.24	−0.25	0.67	0.27	0.22	−0.91	0.73	18
2	0.80	0.13	0.74	0.24	1.78	0.50	−0.47	0.04	1.81	0.38	−0.43	0.37	25
3	0.16	0.70	0.56	0.61	1.45	0.43	−0.74	0.24	1.30	−0.08	0.02	0.40	16
4	0.65	0.72	0.30	0.58	0.02	0.74	0.59	0.17	0.14	0.07	0.73	0.31	17
5	0.72	0.32	0.04	0.20	0.89	0.28	0.42	0.67	0.15	0.49	0.09	0.81	26

Table 13. Target vectors for the next round.

Vector	1	2	3	4	5	6	7	8	9	10	11	12	Target
1	0.55	0.74	0.42	0.12	0.64	0.24	−0.25	0.67	0.27	0.22	−0.91	0.73	18
2	0.17	0.80	0.94	0.93	0.44	0.36	0.77	0.35	0.13	0.42	0.17	0.11	20
3	0.16	0.70	0.56	0.61	1.45	0.43	−0.74	0.24	1.30	−0.08	0.02	0.40	16
4	0.65	0.72	0.30	0.58	0.02	0.74	0.59	0.17	0.14	0.07	0.73	0.31	17
5	0.72	0.32	0.04	0.20	0.89	0.28	0.42	0.67	0.15	0.49	0.09	0.81	19

4.1.2. Procedure of FJSP by Using the Improved Differential Evolution Algorithm

(1) The improved DE was developed by applying four mutation equations [28], DE/rand-to-best/1/bin, DE/rand/2/bin, DE/rand/1/exp 2 position, and DE/best/2/exp 2 position, as seen in Equations (22), (23), (24), and (25), as follows:

$$V_{m,n,G} = X_{r1,n,G} + F1(X_{r2,n,G} - X_{r3,n,G}) + F2\left(X_{best,n,G} - X_{r1,n,G}\right) \tag{22}$$

$$V_{m,n,G} = X_{r1,n,G} + F(X_{r2,n,G} - X_{r3,n,G} + X_{r4,n,G} - X_{r5,n,G}) \tag{23}$$

$$V_{m,n,G} = X_{r1,n,G} + F(X_{r2,n,G} - X_{r3,n,G} + X_{r4,n,G} - X_{r5,n,G}) \tag{24}$$

$$V_{m,n,G} = X_{best,n,G} + F(X_{r2,n,G} - X_{r3,n,G} + X_{r4,n,G} - X_{r5,n,G}) \tag{25}$$

Let r1, r2, r3, r4, and r5 denote the vectors which are randomly selected from a set of target vectors j, which represents the best vector found so far in the algorithm. F is a predefined integer parameter (scaling factor). In the proposed heuristics, F is set to 2; i is the vector number which ranges from 1 to NP, and j is the position of a vector which runs from 1 to D.

(2) The improved DE was developed by applying one crossover or recombination equation at exponential crossover position 2, as seen in Equation (26) [29], as follows:

$$U_{i,j,G} = \begin{cases} V_{i,j,G} \text{ when } j \leq rand_{i,1} \text{ and } j \geq rand_{i,2} \\ X_{i,j,G} \text{ when } rand_{i,1} < j < rand_{i,2} \end{cases} \tag{26}$$

As the predefined parameters in the proposed heuristics, let $randb_i$ be a random number between 0 and 1 and let CR be the recombination probability. $randb_i$, $randb_{i,1}$, and $randb_{i,2}$ are random integer numbers which represent the position of a vector; these random numbers range from 1 to D.

On the basis of the explanations in steps 1–4, the improved DE is shown in Algorithm 1.

Algorithm 1. Pseudo-code of the improved DE for the FJSP

Setup the initial DE parameter
Do while from first iteration to final iteration
   Do while from first DE to final DE
      Setup the initial parameters: job, operation, machine, processing time,
      operation sequence, machine assignment.
      Do while from first task to final task
         Find the start/following task where the fitness is the makespan of the data instances
         Input the scaling factor, crossover rate, NP, job assignment, machine assignment, and
         local search to data list
         Produce the four mutation equations:

$$V_{m,n,G} = X_{r1,n,G} + F1\left(X_{r2,n,G} - X_{r3,n,G}\right) + F2\left(X_{best,n,G} - X_{r1,n,G}\right)$$

$$V_{m,n,G} = X_{r1,n,G} + F\left(X_{r2,n,G} - X_{r3,n,G} + X_{r4,n,G} - X_{r5,n,G}\right)$$

$$V_{m,n,G} = X_{r1,n,G} + F\left(X_{r2,n,G} - X_{r3,n,G} + X_{r4,n,G} - X_{r5,n,G}\right)$$

$$V_{m,n,G} = X_{best,n,G} + F\left(X_{r2,n,G} - X_{r3,n,G} + X_{r4,n,G} - X_{r5,n,G}\right)$$

         Developed by applying the two crossover or recombination equations:

$$U_{i,j,G} = \begin{cases} V_{i,j,G} \text{ when } j \le rand_{i,1} \text{ and } j \ge rand_{i,2} \\ X_{i,j,G} \text{ when } rand_{i,1} < j < rand_{i,2} \end{cases}.$$

         Produce the new target vector (selection/process):

$$X_{m,n,G+1} = \begin{cases} U_{m,n,G} \text{ if } f\left(U_{m,n,G}\right) \le f\left(X_{m,n,G}\right) \\ X_{m,n,G}, \text{else} \end{cases}.$$

      End do
      End do
      Select the best solution from all DEs in the iteration
   End do
   Show/select the best solution from all DEs in all iterations

---

4.1.3. Procedure of FJSP by the Using Local Search with the Jump Search

The flexible job shop scheduling problem points to the optimal target of the minimum makespan. Considering the pathway of any latest complete operation, it denotes the critical pathway for flexible job shop scheduling. As shown in Figures 5 and 6, the critical pathway is $S \rightarrow O_{2,1} \rightarrow O_{4,1} \rightarrow O_{4,2} \rightarrow O_{4,3} \rightarrow O_{2,2} \rightarrow O_{2,3} \rightarrow O_{2,4} \rightarrow T$.

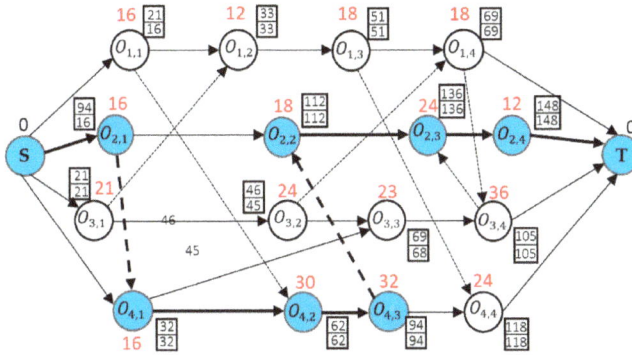

Figure 5. Disjunction graph of flexible job shop scheduling.

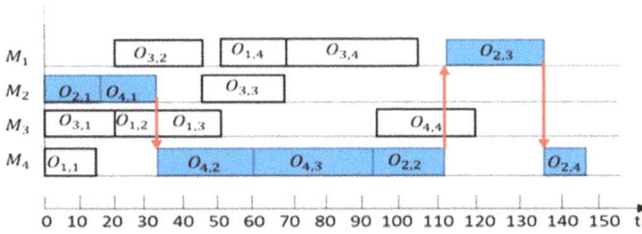

Figure 6. Critical pathway for flexible job shop scheduling.

Figure 6 reveal the critical pathway for flexible job shop scheduling. This pathway is able to determine solutions using the local search and jump search methods through an algorithm that identifies target values from the sorted critical pathway by thoroughly checking any operation on the critical pathway. In Figure 6, operation $O_{2,1}$ is checked for possible intervention by a predecessor under these circumstances and priorities with a lower processing time and compatibility with an operating machine. Accordingly, operations are checked in the following consecutive order $O_{4,1} \to O_{4,2} \to O_{4,3} \to O_{2,2} \to O_{2,3} \to O_{2,4}$. Then, the fitness value is calculated in each round until completion of a set number of iterations.

5. Analysis of the Results from the Experiment on DE for Solving FJSP

To solve the flexible job shop scheduling problem, the Matlab program running on a personal computer (Core i5, 2.5 GHz, 8.00 GB RAM, Windows 7 operating system) was applied. It was developed by metaheuristic algorithms for the solutions focusing on the makespan.

Calculation factors were derived from the experiment based on optimization and relevant research. The findings were as follows: NP = 150, Iterate = 200, F = 2, CR = 0.8, number of iterated local search = 500 rounds. Moreover, the operation sequence that gave priority to operating the most remaining operations as well as machine assignment by choosing a machine with the minimum workload (MWL) were determined.

5.1. Results of Solving the Flexible Job Shop Scheduling Problem with Sample Problems from Kacem et al.

The results of solving the flexible job shop scheduling problem with sample problems from Kacem et al. [26,27] are shown in Table 14, and a Gantt chart of solutions to problem K01 is shown in Figure 7.

Table 14. Summary of solving the flexible job shop scheduling problem with sample problems from Kacem et al. [26,27].

Problem	BKS	Mutation Strategy			
		DE *	DE **	DE ***	DE ****
K01	11	12	12	11	11
		(9.09)	(9.09)	(0.00)	(0.00)
K02	14	15	15	15	15
		(7.14)	(7.14)	(7.14)	(7.14)
K03	11	11	11	11	11
		(0.00)	(0.00)	(0.00)	(0.00)
K04	7	7	7	7	7
		(0.00)	(0.00)	(0.00)	(0.00)
K05	11	12	12	12	12
		(9.09)	(9.09)	(9.09)	(9.09)
MRE		5.06	5.06	3.25	3.25

* DE/rand to best/1/Bin; ** DE/rand/2/Bin; *** DE/rand/1/Exp Crossover Position 2/Local Search with Jump Search; **** DE/best/2/Exp Crossover Position 2/Local Search with Jump Search; Mean relative error (MRE); Best known solution (BKS).

In Table 14, the result of solving the flexible job shop scheduling problem with small size problems [26,27] is revealed. The solutions optimizing differential evolution algorithms that provide the most optimal solutions are a combination of DE/rand/1 and exponential position 2 as well as a combination of DE/best/2 and exponential position 2, resulting in a makespan of 11. These solutions are illustrated in the Gantt chart in Figure 7 for problem K01. A further finding is that the lowest MRE is 3.25.

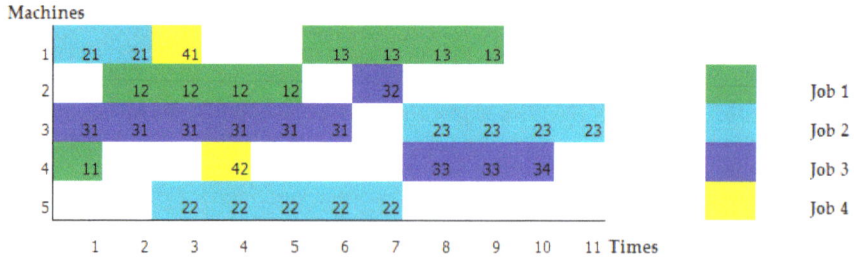

Figure 7. Gantt chart of solutions to flexible job shop scheduling problem K01 presented by Kacem et al. [26,27].

5.2. Results of Solving the Flexible Job Shop Scheduling Problem with Sample Problems of Brandimarte

The results of solving the flexible job shop scheduling problem with the sample problems of Brandimarte [30] can be found in Table 15 and the Gantt chart of solutions for problem Mk1 (Figure 8).

Table 15. Summary of solving the flexible job shop scheduling problem with sample problems Brandimarte [30].

Problem	BKS	Mutation Strategy			
		DE *	DE **	DE ***	DE ****
Mk1	40	43	43	40	40
		(7.50)	(7.50)	(0.00)	(0.00)
Mk2	27	28	28	28	28
		(7.69)	(7.69)	(7.69)	(7.69)
Mk3	204	204	204	204	204
		(0.00)	(0.00)	(0.00)	(0.00)
Mk4	60	71	71	71	71
		(18.33)	(18.33)	(18.33)	(18.33)
Mk5	174	178	178	179	179
		(2.30)	(2.30)	(2.87)	(2.87)
Mk6	59	73	73	73	73
		(23.73)	(23.73)	(23.73)	(23.73)
Mk7	143	149	149	148	146
		(4.20)	(4.20)	(3.50)	(2.10)
Mk8	523	528	528	528	528
		(0.96)	(0.96)	(0.96)	(0.96)
Mk9	307	324	321	323	321
		(5.54)	(4.56)	(5.21)	(4.56)
Mk10	212	234	233	236	235
		(10.38)	(9.90)	(11.32)	(10.85)
MRE		8.06	7.92	7.36	7.11

* DE/rand to best/1/Bin; ** DE/rand/2/Bin; *** DE/rand/1/Exp Crossover Position 2/Local Search with Jump Search; **** DE/best/2/Exp Crossover Position 2/Local Search with Jump Search; Mean relative error (MRE).

Table 15 shows the results of solving the flexible job shop scheduling problem for medium size problems using the sample problems of Brandimarte [30]. The solution optimizing the differential evolution algorithm that provides the most optimal solution is DE/best/2 combined with exponential position 2 and DE/rand/1 combined with exponential position 2, which results in a makespan of 40 and 204. This value is equal to the BKS value of the data set from sample Mk1 and Mk3. Furthermore, the lowest MRE value obtained is 7.11, as shown in the Gantt chart in Figure 8.

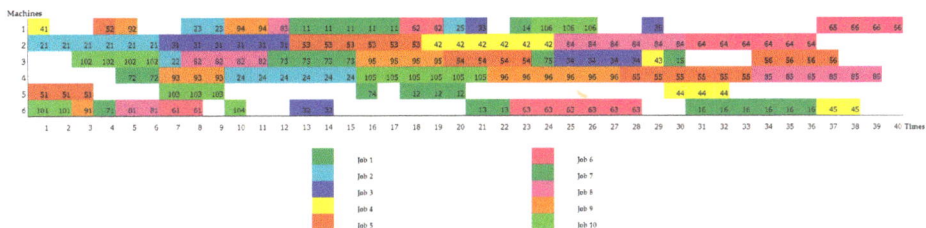

Figure 8. Gantt chart of solutions for flexible job shop scheduling problem Mk1 presented by Brandimarte [30].

5.3. Results of Solving the Flexible Job Shop Scheduling Problem with Sample Problems of Dauzere-Peres and Paulli

The results of solving the flexible job shop scheduling problem with the sample problems of Dauzere-Peres and Paulli [31] can be found in Table 16.

Table 16. Summary of solving the flexible job shop scheduling problem with sample problems Dauzere-Peres and Paulli.

Problem	BKS	Mutation Strategy			
		DE *	DE **	DE ***	DE ****
01a	2530	2895 (14.42)	2750 (8.70)	2615 (3.36)	2645 (4.55)
04a	2555	2859 (11.90)	2770 (8.41)	2650 (3.72)	2610 (2.15)
07a	2396	2759 (15.15)	2650 (10.60)	2650 (10.60)	2510 (4.76)
09a	2074	2281 (9.98)	2269 (9.40)	2210 (6.56)	2150 (3.66)
11a	2078	2378 (14.44)	2366 (13.86)	2221 (6.88)	2200 (5.87)
MRE		13.18	10.19	6.22	4.20

* DE/rand to best/1/Bin; ** DE/rand/2/Bin; *** DE/rand/1/Exp Crossover Position 2/Local Search with Jump Search; **** DE/best/2/Exp Crossover Position 2/Local Search with Jump Search; Mean relative error (MRE).

Table 16 shows the results of solving the flexible job shop scheduling problem for large size problems using the sample problems of Dauzere-Peres and Paulli [31]. The solution optimizing the differential evolution algorithm that provides the best solution is DE/best/2 combined with exponential position 2, which results in a makespan of 2610. When compared with the other DE algorithm, it obtained the lowest value that is 4.20.

6. The Results of the Comparison of the DE Algorithm with Other Metaheuristic Methods

6.1. Results of Solving the Flexible Job Shop Scheduling Problem with Sample Problems of Brandimarte

A comparison of the differential evolution algorithm with other metaheuristic algorithms, GA and PSO, for the 10 comparative sample problems of Brandimarte [30] is given in Table 17.

Table 17. Summary of comparing the differential evolution algorithm with other metaheuristic algorithms.

Problem	$n \times m \times k$ *	BKS **	Chen et al. (GA) [32]	Girish and Jawahar (PSO) [33]	DE-FJSP
		C_{max}	C_{max}	C_{max}	C_{max}
Mk01	$10 \times 6 \times 55$	40	40 (0.00)	40 (0.00)	40 (0.00)
Mk02	$10 \times 6 \times 58$	27	29 (6.89)	27 (0.00)	28 (7.69)
Mk03	$15 \times 8 \times 150$	204	204 (0.00)	204 (0.00)	204 (0.00)
Mk04	$15 \times 8 \times 90$	60	63 (4.76)	62 (3.22)	71 (18.33)
Mk05	$15 \times 4 \times 106$	174	181 (3.86)	178 (2.24)	179 (2.87)
Mk06	$10 \times 15 \times 150$	59	60 (1.66)	78 (24.35)	73 (23.73)
Mk07	$20 \times 5 \times 100$	143	148 (3.38)	147 (2.72)	146 (2.10)
Mk08	$20 \times 10 \times 225$	523	523 (0.00)	523 (0.00)	528 (0.96)
Mk09	$20 \times 10 \times 240$	307	308 (0.32)	341 (9.97)	321 (4.56)
Mk10	$20 \times 15 \times 240$	212	212 (0.00)	252 (15.07)	235 (10.85)
MRE			2.08	7.75	7.11

* n = Job, m = Machine, k = Operation; ** Best known solution (BKS) ; FJSP = Flexible job shop scheduling problem; GA = Genetic algorithm; PSO = Particle swarm optimization.

Table 17 shows the comparative results for solving the flexible job shop scheduling problem using the differential evolution algorithm versus various dimension-optimizing algorithms. The value of mean of relative error (MRE) is lower than the value obtained by Girish and Jawahar [33] with PSO (7.75), while the improved DE has a value of 7.11. The work by Chen et al. [32] showed a value of 2.08 with the, while DE provided a greater value of 7.11. However, some problems, including Mk1 and Mk3, resulted in an equally good makespan in comparison with the BKS value of the data set.

6.2. Results of Solving the Flexible Job Shop Scheduling Problem with Sample Problems of Dauzere-Peres and Paulli

A comparison of the differential evolution algorithm with other metaheuristic algorithms, 1ST-DE, for the 5 comparative sample problems of Dauzere-Peres and Paulli [31] is given in Table 18.

Table 18. Summary of comparing the differential evolution algorithm with other metaheuristic algorithms.

Problem	$n \times m \times k$ *	BKS **	Wisittipanich (1ST-DE) [12]	DE-FJSP
		C_{max}	C_{max}	C_{max}
01a	$10 \times 5 \times 196$	2530	2645 (4.55)	2645 (4.55)
04a	$10 \times 5 \times 196$	2555	2616 (2.39)	2610 (2.15)
07a	$15 \times 8 \times 293$	2396	2582 (7.76)	2510 (4.76)
09a	$15 \times 8 \times 293$	2074	2153 (3.81)	2150 (3.66)
11a	$15 \times 8 \times 293$	2078	2221 (6.88)	2200 (5.87)
MRE			5.08	4.20

* n = Job, m = Machine, k = Operation; ** Best known solution (BKS); FJSP = Flexible job shop scheduling problem.

Table 18 shows the comparative results for solving the flexible job shop scheduling problem using the differential evolution algorithm versus various dimension-optimizing algorithms. The value of mean of relative error (MRE) is lower than the value obtained by Wisittipanich [12] with 1ST-DE (5.08), while the improved DE has a value of 4.20. From the computational results, we can see that the improved DE algorithms with jump search are effective methods when compare with the basic DE and some meta-heuristic method.

7. Conclusions and Suggestions

This section presents the conclusions of this study. The differential evolution algorithm was used to solve the flexible job shop scheduling problem and optimize the dimensions. Among sample problems, the makespan and the BKS value of the data set were compared and the mean of relative error (MRE) was calculated. The sample problems of Kacem et al. were used as examples of small-sized problems. The dimensions were optimized by using "DE/rand/1" combined with exponential crossover position 2, as well as "DE/best/2" combined with exponential crossover position 2, which resulted in the minimum MRE value of 3.25. The sample problems of Brandimarte were used as examples of medium-sized problems. The dimensions were optimized with the combination of "DE/best/2" and exponential crossover position 2, providing a minimized MRE value of 7.11. Furthermore, the sample problems of Dauzere-Peres and Paulli were used as examples of large-sized problems. The dimensions were optimized with the combination of "DE/best/2" and exponential crossover position 2, providing a minimized MRE value of 4.20. Hence, the improved differential evolution in this research was able to solve the flexible job shop scheduling problem.

The DE algorithm proposed in this study can be applied to solve problems in the manufacturing industry in Thailand, such as mold and die manufacturing, the flexible job shop scheduling problem of the work center (FJSSPWC), or flexible job shop scheduling problem with preventive maintenance of the machine.

Author Contributions: P.S. designed the algorithm, summarized the computation, wrote the conclusions, and validated the algorithm; N.K. and P.K. gathered data and validated the algorithm.

Acknowledgments: We would like to thank the Industrial Engineering Department, Faculty of Engineering, Ubon Ratchathani University for funding this research.

Conflicts of Interest: The authors declare no conflicts of interest.

References

1. Karthikeyan, S.; Asokan, P.; Nickolas, S. A hybrid discrete firefly algorithm for multi-objective flexible job shop scheduling problem with limited resource constraints. *Int. J. Adv. Manuf. Technol.* **2014**, *72*, 1567–1579. [CrossRef]

2. Storn, R.; Price, K. Differential evolution—A simple and efficient heuristic for global optimization over continuous space. *J. Glob. Optim.* **1997**, *11*, 341–359. [CrossRef]

3. Xia, W.; Wu, Z. An effective hybrid optimization approach for multi-objective flexible job-shop scheduling problems. *Comput. Ind. Eng.* **2005**, *48*, 409–425. [CrossRef]

4. Kanate, P. Algorithm Development for Solving Flexible Job Shop Scheduling Problem. Ph.D. Thesis, Kasetsart University, Krung Thep Maha Nakhon, Thailand, 2011.

5. Tang, J.; Zhang, G.; Lin, B.; Zhang, B. A Hybrid Algorithm for Flexible Job-shop Scheduling Problem. *Procedia Eng.* **2011**, *15*, 3678–3683. [CrossRef]

6. Wannaporn, T.; Arit, T. Modified Genetic Algorithm for Flexible Job-Shop Scheduling Problems. *Procedia Comput. Sci.* **2012**, *12*, 122–128.

7. Thanyapon, U.; Pupong, P.; Kwanniti, K. Solving an Advanced Planning and Scheduling Problem with Preventive Maintenance Time Window Constraints by Mixed Integer Programming Models. *Thai J. Oper. Res.* **2016**, *4*, 1–15. Available online: https://www.tci-thaijo.org/index.php/TJOR/article/view/60994 (accessed on 22 July 2019).

8. Hamid, A.; Christoph, H.; Michael, D. Machine scheduling in production: A content analysis. *Appl. Math. Model.* **2017**, *50*, 279–299.

9. Li, X.; Gao, L. An Effective Hybrid Genetic Algorithm and Tabu Search for Flexible Job Shop Scheduling Problem. *Int. J. Prod. Econ.* **2016**, *174*, 93–110. [CrossRef]

10. Luan, F.; Cai, Z.; Wu, S.; Jiang, T.; Li, F.; Yang, J. Improved Whale Algorithm for Solving the Flexible Job Shop Scheduling Problem. *Mathematics* **2019**, *7*, 384. [CrossRef]

11. Li, X.; Peng, Z.; Du, B.; Guo, J.; Xu, W.; Zhuang, K. Hybrid artificial bee colony algorithm with a rescheduling strategy for solving flexible job shop scheduling problems. *Comput. Ind. Eng.* **2017**, *113*, 10–26. [CrossRef]

12. Wisittipanich, W. Minimizing Makespan in Flexible Job Shop Problems by Adapting the Differential Evolution. *Thai J. Oper. Res.* **2015**, *3*, 40–50.

13. Yuan, Y.; Xu, H. Flexible job shop scheduling using hybrid differential evolution algorithms. *Comput. Ind. Eng.* **2013**, *65*, 246–260. [CrossRef]

14. Bhaskara, P.; Padmanabhan, G.; Satheesh, K.B. Differential Evolution Algorithm for Flexible Job Shop Scheduling Problem. *Int. J. Adv. Prod. Mech. Eng.* **2015**, *5*, 71–79.

15. Wang, L.; Pan, Q.K.; Tasgetiren, M.F. Minimizing the total flow time in a flow shop with blocking by using hybrid harmony search algorithms. *Expert Syst. Appl.* **2010**, *37*, 7929–7936. [CrossRef]

16. Zhang, R.; Song, S.; Wu, C. A hybrid differential evolution algorithm for job shop scheduling problems with expected total tardiness criterion. *Appl. Soft Comput.* **2013**, *13*, 1448–1458. [CrossRef]

17. Cao, E.; Lai, M. The open vehicle routing problem with fuzzy demands. *Expert Syst. Appl.* **2010**, *37*, 15. [CrossRef]

18. Lai, M.; Cao, E. An improved differential evolution algorithm for vehicle routing problem with simultaneous pickups and deliveries and time windows. *Eng. Appl. Artif. Intell.* **2010**, *23*, 188–195.

19. Xu, H.; Wen, J. Differential Evolution Algorithm for the Optimization of the Vehicle Routing Problem in Logistics. In Proceedings of the 2012 Eighth International Conference on Computational Intelligence and Security, Guangzhou, China, 17–18 November 2012; pp. 48–51.

20. Cruz, I.L.; van Willigenburg, L.G.; van Straten, G. Efficient differential evolution algorithms for multimodal optimal control problems. *Appl. Soft Comput.* **2003**, *3*, 97–122. [CrossRef]
21. Adeyemo, J.; Otieno, F. Differential evolution algorithm for solving multi-objective crop planning model. *Agric. Water Manag.* **2010**, *97*, 848–856. [CrossRef]
22. Abou El Ela, A.A.; Abido, M.A.; Spea, S.R. Optimal power flow using differential Evolution algorithm. *Electr. Power Syst. Res.* **2010**, *80*, 878–885. [CrossRef]
23. Kuila, P.; Jana, P.K. A novel differential evolution based clustering Algorithm for wireless sensor networks. *Appl. Soft Comput.* **2014**, *25*, 414–425. [CrossRef]
24. Sharma, S.; Rangaiah, G.P. An improved multi-objective differential evolution with a termination criterion for optimizing chemical processes. *Comput. Chem. Eng.* **2013**, *56*, 155–173. [CrossRef]
25. Pitakaso, R.; Parawech, P.; Jirasirierd, G. Comparisons of Different Mutation and Recombination Processes of the DEA for SALB-1. In Proceedings of the Institute of Industrial Engineers Asian Conference, Taipei, Taiwan, 18–20 July 2013; pp. 1571–1579.
26. Kacem, I.; Hammadi, S.; Borne, P. Approach by localization and multiobjective evolutionary optimization for flexible job-shop scheduling problems. *IEEE Trans. Syst. Man Cybern.* **2002**, *32*, 1–13. [CrossRef]
27. Kacem, I.; Hammadi, S.; Borne, P. Pareto-optimality approach for flexible job-shop scheduling problems: hybridization of evolutionary algorithms and fuzzy logic. *Math. Comput. Simul.* **2002**, *60*, 245–276. [CrossRef]
28. Nguyen, S.; Kitchitvichynukul, V.; Wisittipanich, W. *ET-Lib User's Guide. Volume 2. Differential Evolution*; Asian Institute of Technology: Khlong Luang, Thailand, 2013.
29. Poontana, S.; Nuchsara, K.; Preecha, K.; Krit, C. U-Shaped Assembly Line Balancing by Using Differential Evolution Algorithm. *Math. Comput. Appl.* **2018**, *23*, 79.
30. Brandimarte, P. Routing and Scheduling in a Flexible Job Shop by Tabu Search. *Ann. Oper. Res.* **1993**, *41*, 157–183. [CrossRef]
31. Dauzere-Peres, S.; Paulli, J. An integrated approach for modeling and solving the general multiprocessor job-shop scheduling problem using tabu search. *Ann. Oper. Res.* **1997**, *70*, 281–306. [CrossRef]
32. Chen, H.; Ihlow, J.; Lehmann, C. A genetic algorithm for flexible job-shop scheduling. In Proceedings of the IEEE International Conference on Robotics and Automation, Detroit, MI, USA, 10–15 May 1999; pp. 1120–1125.
33. Girish, B.S.; Jawahar, N. A particle swarm optimization algorithm for flexible job shop scheduling problem. In Proceedings of the IEEE International Conference on Automation Science and Engineering, Bangalore, India, 22–25 August 2009; pp. 298–303.

© 2019 by the authors. Licensee MDPI, Basel, Switzerland. This article is an open access article distributed under the terms and conditions of the Creative Commons Attribution (CC BY) license (http://creativecommons.org/licenses/by/4.0/).

Mathematical and Computational Applications

MDPI

Article

Pool-Based Genetic Programming Using Evospace, Local Search and Bloat Control

Perla Juárez-Smith [1], **Leonardo Trujillo** [1,*], **Mario García-Valdez** [1], **Francisco Fernández de Vega** [2] **and Francisco Chávez** [3]

[1]   Tecnológico Nacional de México/Instituto Tecnológico de Tijuana, Tijuana BC C.P. 22430, Mexico
[2]   Departamento de Tecnología de los Computadores y de las Comunicaciones, Universidad de Extremadura, 06800 Mérida, Spain
[3]   Departamento de Ingeniería Sistemas Informáticos y Telemáticos, Universidad de Extremadura, 06800 Mérida, Spain
*   Correspondence: leonardo.trujillo@tectijuana.edu.mx; Tel.: +52-(664)-607-8400 (ext. 234)

Received: 29 July 2019; Accepted: 27 August 2019; Published: 29 August 2019

Abstract: This work presents a unique genetic programming (GP) approach that integrates a numerical local search method and a bloat-control mechanism within a distributed model for evolutionary algorithms known as EvoSpace. The first two elements provide a directed search operator and a way to control the growth of evolved models, while the latter is meant to exploit distributed and cloud-based computing architectures. EvoSpace is a Pool-based Evolutionary Algorithm, and this work is the first time that such a computing model has been used to perform a GP-based search. The proposal was extensively evaluated using real-world problems from diverse domains, and the behavior of the search was analyzed from several different perspectives. The results show that the proposed approach compares favorably with a standard approach, identifying promising aspects and limitations of this initial hybrid system.

Keywords: Genetic Programming; Bloat; NEAT; Local Search; EvoSpace

1. Introduction

Within the field of Evolutionary Computation (EC), the Genetic Programming (GP) paradigm includes a variety of algorithms that can be used to evolve computer code or mathematical models, and has had success in a variety of domains. Even the first version of GP, proposed by Koza in the 1990s and commonly referred to as tree-based GP or standard GP [1], is still being used today. This paper focuses on a recent variant of GP called neat-GP-LS [2] that integrates what we consider as fundamental elements of any state-of-the-art GP method, e.g., bloat control and local search (LS) techniques.

However, one discouraging aspect of integrating LS methods into a GP search is the increase in algorithm complexity (execution time might increase if the total number of generations is kept constant, but, since the algorithm converges more quickly, fewer generations are required to reach the same level of performance). One way to minimize this issue is by porting the search process to massively parallel architectures [3]. However, another approach is to move towards distributed EC systems (dEC) [4–6]. There are several possible benefits from this approach. First, it is much simpler to develop and use a distributed system than developing low-level code for GPUs or FPGAs [3,7]. The need for strict synchronization policies, for instance, is greatly reduced in a distributed framework compared to a GPU or FPGA implementation. Second, it is possible to leverage cheaper computing power that is already accessible, rather than investing in specialized hardware [8,9]. Finally, the robustness and asynchronous nature of an evolutionary search can easily deal with unexpected errors or dropped connections in a distributed environment. In this work, we use a distributed platform designed to run using heterogeneous computing resources called EvoSpace, a conceptual model for the development of distributed pool-based algorithms [8–10]. While it has been applied in standard black-box optimization benchmarks and collaborative-interactive evolutionary algorithms [11], it has not been studied in a GP-based search.

To summarize, the present paper proposes a hybrid distributed GP system that integrates a recent bloat control mechanism and a LS operator for parameter optimization of GP trees. Bloat control is performed by neat-GP, which uses speciation and the well-known method of fitness sharing to control the growth of program trees [12]. For the LS process, the method from [13,14] is used, where the individual trees are enhanced with numerical weights in each node, and these are then optimized using a trust region optimizer [15]; this strategy has proven to be beneficial in several recent learning problem [16,17]. This work shows that the EvoSpace model can easily exploit the speciation process performed by neat-GP, maintaining the same level of performance as the sequential version even though evolution is now performed in an asynchronous manner.

The remainder of this work is organized as follows. Section 2 presents relevant background and related research. Section 3 describes how the proposed system is ported to a distributed framework. A summary and conclusions are outlined in Section 4.

2. Background

This section described the neat-GP algorithm and a method to integrate LS in GP. In addition, a brief overview of EvoSpace model is provided.

2.1. neat-GP

The neat-GP algorithm [12] is based on the operator equalization [18] family of bloat control methods, in particular the Flat-OE [19] algorithms and the NeuroEvolution of Augmenting Topologies algorithm (NEAT) [20].

The neat-GP algorithm has the following main features: The initial population only contains shallow trees (3 levels), while most GP algorithms initialize the search with small- and medium-sized trees (depth of 3–6 levels).

Individual trees are grouped into species, using a similarity measure that is based on their size and shape. With the following measure we can group individuals: given a tree T, let n_T represent the size of the tree (number of nodes) and d_T its depth (number of levels). Moreover, let $S_{i,j}$ represent the shared structure between both trees starting from the root node (upper region of the trees), which is also a tree, as seen in Figure 1. Then, the dissimilarity between two trees T_i and T_j is given by

$$\delta_T\left(T_i, T_j\right) = \beta \frac{N_{i,j} - 2n_{s_{i,j}}}{N_{i,j} - 2} + (1 - \beta)\frac{D_{i,j} - 2d_{s_{i,j}}}{D_{i,j} - 2}, \tag{1}$$

where $N_{i,j} = n_{T_i} + n_{T_j}$, $D_{i,j} = d_{T_i} + d_{T_j}$, and $\beta \in [0,1]$; a degenerate case arises when both trees have a single node (only the root node), in this case $\delta_T = 0$.

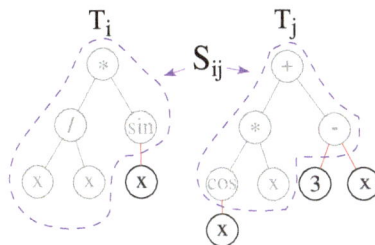

Figure 1. Example of the shared structure $S_{i,j}$ between two trees T_i and T_j ([2], with permission from Springer).

Each time an individual T_i is produced, it is compared to a randomly chosen individual T_j, sequentially from different species. This is done by first randomly shuffling the species, and then if $\delta_T(T_i, T_j) < h$, with threshold h an algorithm parameter, then the tree T_i is assigned to the species of T_j, and no further comparisons are carried out. When the condition described above is never satisfied, a new species is created for the tree T_i.

To promote the formation of several species fitness sharing is used, in this way the individuals in large species (with many trees) are penalized more than individuals from smaller (with fewer trees) species. Assuming a minimization problem, neat-GP penalizes individuals with

$$f'(T_i) = |S_u| f(T_i), \tag{2}$$

where $f(T_i)$ is the raw fitness of the tree, $f'(T_i)$ is the penalized or adjusted fitness, S_u is the species to which T_i belongs, and $|S_u|$ is the number of individuals in species S_u. However, the best individual (with the best fitness) from each species are not penalized, this protects the elite individuals from each species. Moreover, penalization is most important during selection for parents, which considered the computed adjusted value of fitness. Selection is done deterministically, sorting the population based on adjusted fitness. In this way, individuals with very bad adjusted fitness will not produce offspring, but this high selective pressure is offset by protecting the elite individuals from each species, such that the best individual from each species has a good chance of producing offspring.

2.2. Local Search in Genetic Programming

Particularly, we focus on symbolic regression problems, where the goal is to search for the symbolic expression $K^O : \mathbb{R}^p \to \mathbb{R}$ that best fits a particular training set $\mathbb{T} = \{(x_1, y_1), \dots, (x_n, y_n)\}$ of n input/output pairs with $x_i \in \mathbb{R}^p$ and $y_i \in \mathbb{R}$ defined as

$$(K^O, \theta^O) \leftarrow \underset{K \in \mathbb{G}; \theta \in \mathbb{R}^m}{arg\ min}\ f(K(x_i, \theta), y_i)\ \text{with}\ i = 1, \dots, n\ , \tag{3}$$

where $\mathbb{G}$ is the solution or syntactic space defined by the primitive set $\mathbb{P}$ of functions and terminals; f is the fitness function that is based on the difference between a program's output $K(x_i, \theta)$ and the desired output y_i; and θ is a particular parametrization of the symbolic expression K, assuming m real-valued parameters. The goal of the LS method is to optimize the parameters of each GP solution.

Following [13,14], the search includes on additional search operator which is not common in GP, an LS process that is used to optimize the implicit parameters in GP individuals. This allows the search to use subtree mutation and crossover to explore the search space, or syntax space, and uses the LS process to perform fine tuning of the individuals in parameter space.

As suggested in [21], for each individual K in the population, we add a small linear upper tree above the root node, such that $K' = \theta_2 + \theta_1(K)$ where K' represents the new program output, while θ_1 and θ_2 are the first two parameters from θ, as shown in Figure 2.

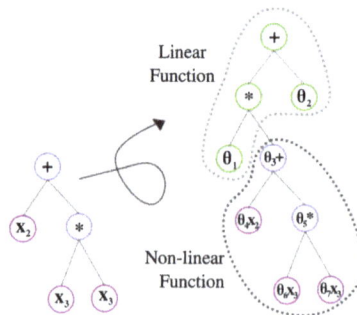

Figure 2. Example of the tree transformation for the LS process ([2], with permission from Springer).

In this way, for all the other nodes n_k in the tree K we add a weight coefficient $\theta_k \in \mathbb{R}$, such that each node is now defined by $n'_k = \theta_k n_k$, where n'_k is the new modified node, $k \in \{1,...,r\}$, $r = |Q|$ and Q is the tree representation. Notice that each node has a unique parameter that can be modified to help meet the overall optimization criteria of the non-linear expression. When the search starts, the parameters are initialized to $\theta_k = 1$. Then, during the evolutionary process, when subtree mutation or crossover exchange genetic material (syntax) between individuals, these also include the corresponding parameter values. In general, each GP individual is considered to be a nonlinear expression that the LS operator must fit to the problem data. This can be done using different methods, but here a trust region optimizer is used [22], following [13,14].

One of the most important things to consider is that the local search optimizer can substantially increase the underlying computational cost of the search, particularly when individual trees are very large. While applying the local search strategy to all trees might produce good results [13], it is preferable to reduce to a minimum the amount of trees to which it is applied.

2.3. Integration LS into neat-GP

The neat-GP-LS algorithm was recently proposed to integrate the neat-GP search with an LS process [2], showing the ability to improve performance and generate compact and simple solutions. Figure 3 shows the main modules in this algorithm. Another interesting result reported in [2] was that neat-GP-LS exhbited very little performance variance on all tested problems, suggesting that the meta-heuristic search is robust.

Given the reliance of neat-GP-LS on the speciation process, as defined for neat-GP, the following observations are of note. First, species tend to grow in size when the individuals in the species have good fitness, and they grow more when they include the best solution in the entire population. Second, while species with bigger trees tend to appear as evolution progresses, diversity is maintained throughout the search. Third, while species are different, in terms of the size and shape of individuals they contain, it is common for all species to include at least some highly fit individuals. Finally, species grow in size when they contain highly fit individuals, and this increased exploitation is beneficial because the LS tends to produce high levels of improvement in those particular species.

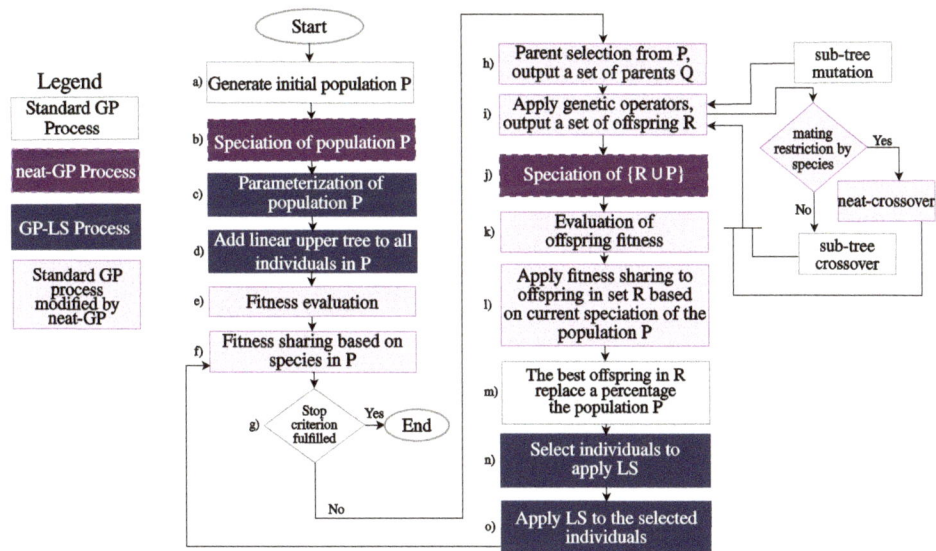

Figure 3. General flow diagram of the neat-GP-LS algorithm ([2], with permission from Springer).

2.4. EvoSpace

The EvoSpace model for evolutionary algorithms (EA) follows a pool-based approach [8,9], where the search process is conducted by a collection of possibly heterogeneous processes that cooperate using a shared memory or population pool. We refer to such algorithms as pool-based EAs (PEAs) and highlight the fact that such systems are intrinsically parallel, distributed and asynchronous.

In EvoSpace, distributed nodes (called EvoWorkers) asynchronously interact with the pool; their job is to take a subset of individuals from the central pool, which is called a sample, and evolve them for a certain number of generations (or until a given termination criterion is met), and return the new population of offspring back to the pool. The general scheme is depicted in Figure 4.

This means that EvoSpace has two main components, a set of EvoWorkers and a single instance of an EvoStore. The EvoStore container manages a set of objects representing individuals in a EA population. EvoWorkers pull a subset of individuals from the EvoStore making them unavailable to other workers. Moreover, individuals are removed from the EvoStore as a random subset or sample of the population. Once a EvoWorker has a sample to work on, it can perform a partial evolutionary process, and then return the newly evolved subpopulation to the EvoStore where the new individuals replace those found in the original sample; at this point, replaced or reinserted individuals can be taken by others clients. Figure 5 shows the distributed architecture of the EvoSpace model with GP. The figure shows that on the Server the EvoSpace manager and HTTP communication framework are performed, while different samples of individuals from the population are sent to EvoWorkers where evolution takes place.

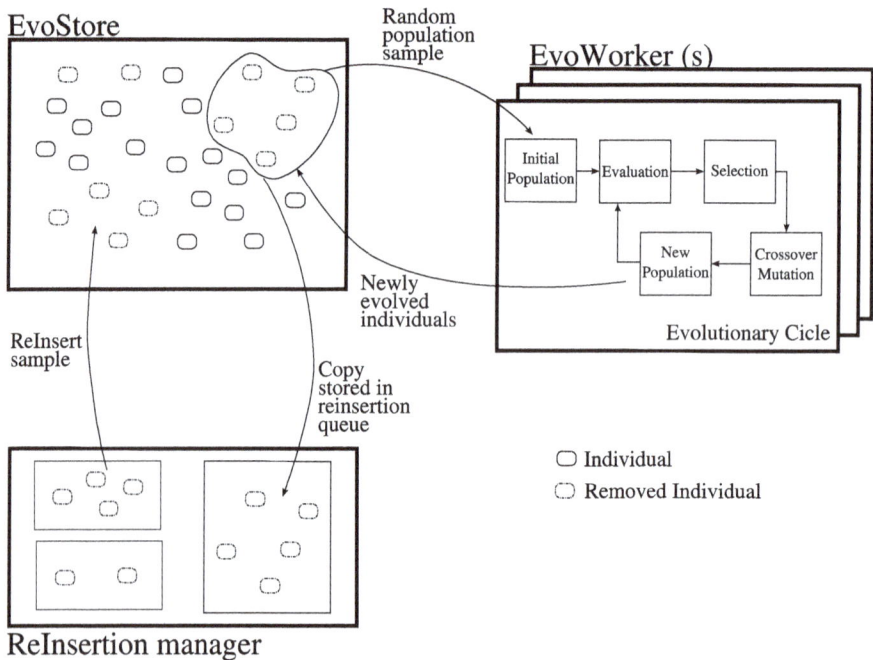

Figure 4. Main components and data flow within the EvoSpace model.

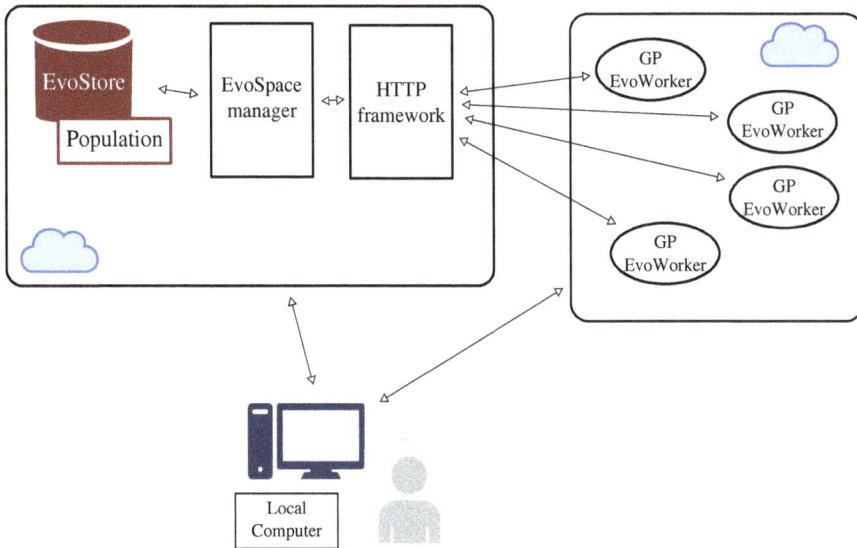

Figure 5. EvoSpace distributed architecture.

EvoSpace was conceived as a model for cloud-based evolutionary algorithms and is general enough to be amenable to any type of population-based algorithm. Several works have shown that this general approach can solve standard black-box optimization problems [9] and even interactive evolution tasks [11]. It has been shown, as expected, that distributing costly fitness function evaluations will help reduce the total run-time of the algorithm [9].

3. Distributing neat-GP-LS into the EvoSpace Model

In this work, we present the first implementation of a GP algorithm on EvoSpace.

Since neat-GP-LS already divides the population into species, it seems straightforward to exploit this structure and distribute individuals to EvoWorkers by sending complete species to each.

3.1. The Intra-Species Distance and Re-Speciation

One aspect of neat-GP-LS that is not asynchronous is the speciation process. In the sequential and synchronous versions, speciation occurs at specific moments during the search, as shown in Figure 3. However, since EvoSpace is asynchronous, EvoWorkers return samples to the population pool at different moments in time. When an EvoWorker returns a sample, it is not correct to assume that all of the new individuals actually belong in the same species. It is possible that the species diverged during the local evolution carried out on the EvoWorker.

To solve this issue, we track the level of homogeneity within each species, which is measured before a species leaves the pool and when the new species returns from the EvoWorker. If a significant change is detected, then a flag is raised that tells EvoSpace that the population should go through a new speciation process or re-speciation. This is done by computing what is referred to as the intra-species distance. Basically, in each species, we compute the dissimilarity measure using Equation (1), between each tree T_i and its nearest neighbor T_j (the individual with which Equation (1) is minimum within the species), calling this value nn_{T_i}. Then, the intra-species distance D_{S_l} for species S_l is the average of all nn_{T_i} considering all $T_i \in S$.

The D_{S_l} values could be used in different ways to trigger a re-speciation process. In this work, we can say that D_{S_l} is the intra-species distance before S_l is taken as a sample by an EvoWorker, and we can define $\hat{D}_{S_l}$ as the intra-species distance of species S_l computed with the population returned by the EvoWorker. If $\hat{D}_{S_l} > D_{S_l}$ for any species in the population, then a re-speciation event is triggered. Basically, this causes a synchronization event, where the EvoStore waits for all species to return and the population goes through the speciation process once more. Figure 6 shows the basic scheme of the proposes implementation. Compared to Figure 5, the new implementation in Figure 6 accounts for specific elements of the neat-GP algorithm. In particular, the speciation process is carried out on the server, such that instead of sending random samples of individuals to the EvoWorkers, complete species are sent and a local evolutionary process is carried out. In this case, the number of EvoWorkers used depends on the number of species in the population.

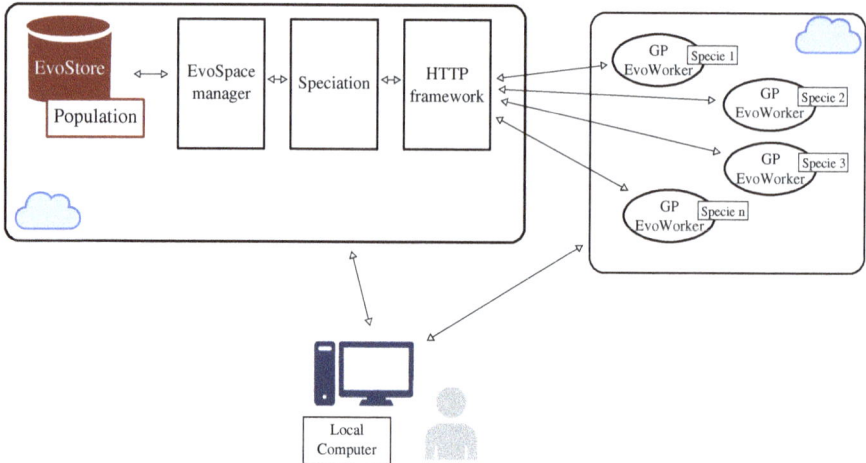

Figure 6. Implementation of the neat-GP-LS algorithm in EvoSpace, where the samples taken by each EvoWorker correspond to a complete species.

3.2. Experiments and Results

We analyzed and evaluated the integration of the neat-GP-LS algorithm in a PEA known as the EvoSpace model. EvoSpace was designed for problems where fitness computation might be expensive; in this work, we were only interested in studying the effects of implementing neat-GP-LS as a PEA. In particular, we wanted to determine if there are any significant and substantial effects on the convergence of the algorithm, the solutions qualities on all the population and the behavior of the bloating phenomena.

For simplicity, the distributed framework was simulated using multiple CPU threads, such that each EvoWorker was assigned to a specific thread. When the number of EvoWorkers exceeded the number of threads, then several workers could share a single thread.

All experiments were carried out using real world symbolic regression problems, where the objective is to minimize the fitness function. All problems are summarized in Table 1.

When a species was sent to an EvoWorker, we performed a short local evolutionary search, basically a standard GP search using the parameters specified in Table 2. The number of EvoWorkers depended on the number of species in the EvoStore, and we assumed that an EvoWorker was always available for any species in the EvoStore. In addition, the local evolution performed in an EvoWorker iterated for 10 generations, applying the LS operator with probability of 0.50.

Table 1. Symbolic regression real world problems.

Problems	No. Instances	No. Features	Description
Housing [23]	506	14	Concerns housing values in suburbs of Boston.
Concrete [24]	1030	9	The concrete compressive strength is a highly nonlinear function of age and ingredients.
Energy Heating [25]	768	9	This study looked into assessing the heating load requirements of buildings as a function of building parameters.
Energy Cooling [25]	768	9	This study looked into assessing the cooling load requirements of buildings as a function of building parameters.
Tower [26]	5000	26	An industrial data set of a gas chromatography measurement of the composition of a distillation tower.
Yacht [27]	308	7	Delft data set, used to predict the hydodynamic performance of sailing yachts from dimensions and velocity.

Table 2. Parameters used in real world problems.

Parameter	neat-GP-LS
Runs	30
Population	100
Generations	10
Training set	70%
Testing set	30%
Operators Crossover (p_c), Mutation (p_m)	p_c=0.9, p_m=0.1
Tree initialization	Ramped Half-and-Half, maximum depth 6.
Function set	+,-,x,sin,cos,log,sqrt,tan,tanh, constants
Terminal set	Input variables and constants as indicated in each real-world problem.
Selection for reproduction	Eliminate the $p_{worst} = 50\%$ worst individuals of each species.
Elitism	Do not penalize the best individual of each species.
Species threshold value	$h = 0.15$ with $\beta = 0.5$
Local optimization probability	$P_s = 0.5$

Figure 7 shows a single run of the PEA version of neat-GP-LS on the Housing, Concrete and Energy Cooling problems. The plots show the convergence of the training and testing RMSE, as well as the average size of the population given in number of tree nodes. The horizontal axis represents the number of samples taken from the EvoStore. Note that the number of samples over different problems and over different runs l varied due to the randomness of the individual population and the speciation process, and due to the asynchronous nature of the EvoSpace model, which makes it unfeasible to aggregate the behavior of multiple runs into a single plot. Therefore, these plots only show a single run, but the behavior of the algorithm in these examples is in fact representative of the convergence behavior of most runs. One notable observation is the almost identical behavior of both training and testing MAE in all of the runs, showing that the algorithm generalizes in a consistent manner relative to training performance. The size of the population is also quite informative. Notice that, while the average size fluctuates in all cases, the algorithm is in general producing compact solutions. This is particularly clear when the search process terminates and the final sample is returned to the EvoStore.

The results are summarized in Figures 8 and 9, which show a box plot comparisons between the sequential neat-GP-LS algorithm and the PEA implementation in EvoSpace, respectively, for test RMSE and the average size of the population. Table 3 presents the p-values of the Friedman test, where bold values indicate that the null hypothesis is rejected at the $\alpha = 0.05$ confidence level. The null hypothesis states that the medians of the two groups are the same. Notice that, on three (Concrete, Energy Heating and Tower) out of the six problems, the EvoSpace version performed worse than the sequential algorithm in terms of RMSE, since the null-hypothesis were rejected. Conversely, if we consider the three problems in which the PEA version and the sequential algorithm performed equivalently based on test RMSE (i.e., the null hypothesis is not rejected), the Housing Energy Cooling and Yacht problems, EvoSpace produced smaller trees and thus was more effective at bloat control. Therefore, we can state with some confidence that the modified search dynamics introduced in the distributed version of the algorithm do alter the effectiveness of the search. On the one hand, the quality of the results seemed to depend on the problem. On the other hand, in all cases where the EvoSpace implementation achieved equivalent performance, it was significantly and substantially less affected by bloat, producing more parsimonious and compact solutions.

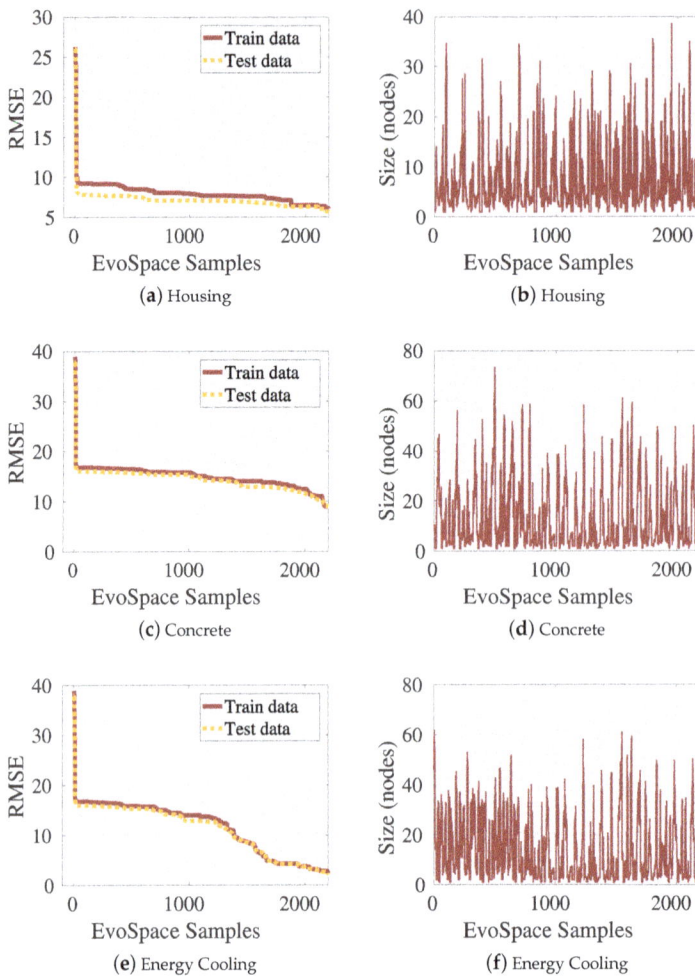

Figure 7. Performance of a single run of the PEA implementation of neat-GP-LS in EvoSpace for: *Housing* (**a**), (**d**); *Concrete* (**b**), (**e**); and *Energy Cooling* (**d**), (**f**). The plots in the left column show the evolution of the training and testing RMSE. The plots in the right column show the evolution of the average program size. All plots are ordered based on the number of samples taken from the EvoStore.

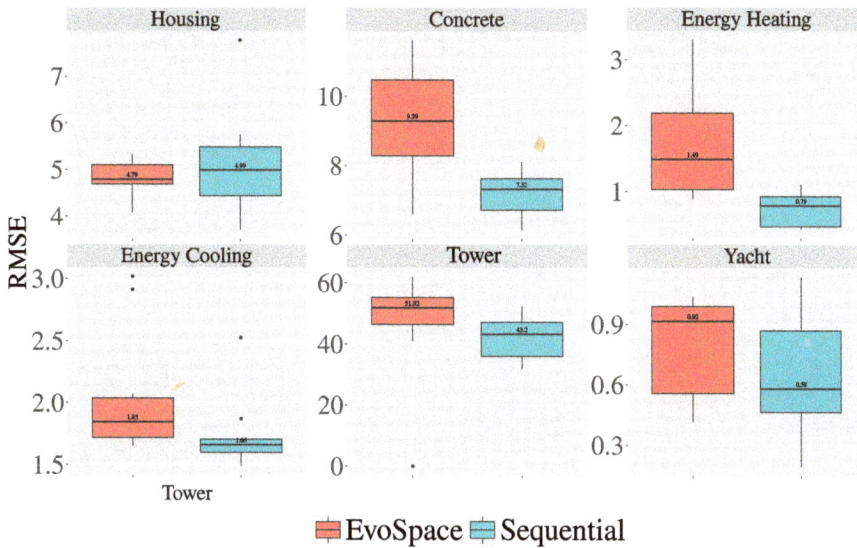

Figure 8. Box plot comparison of the sequential and the EvoSpace implementation of the neat-GP-LS algorithm on the testing RMSE.

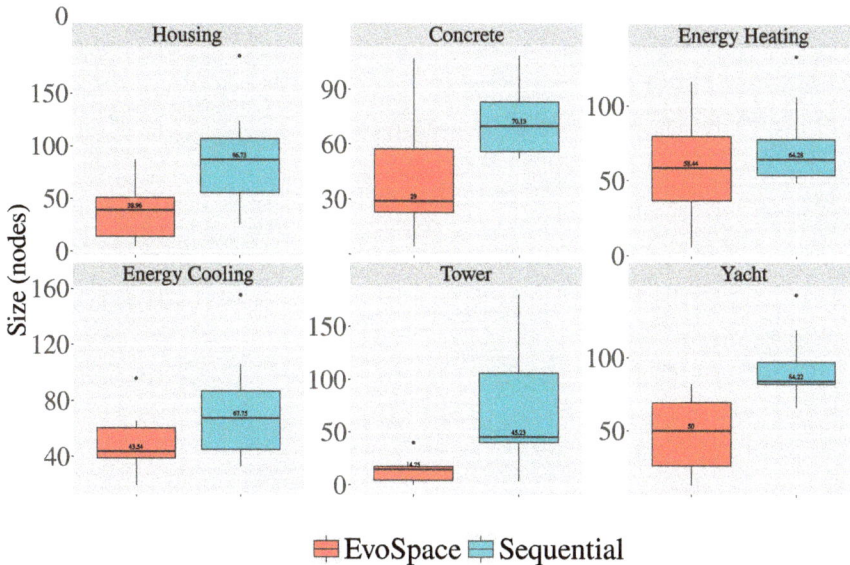

Figure 9. Box plot comparison of the sequential and the EvoSpace implementation of the neat-GP-LS algorithm on the average size of individuals given in number of nodes.

It is reasonable to assume that larger learning problems, in terms of number of instances and features, are in general more difficult to solve. Moreover, difficult problems usually require more complex or larger solutions to effectively model their structure. The three problems where RMSE performance of the EvoSpace implementation was statistically worse (Concrete, Energy Heating and Tower) are also three of the four largest problems used in our experiments, in terms of total number of instances and number of features (see Table 1). Since the EvoSpace search dynamics pushes the search towards smaller program sizes, with statistical significance in five of the six problems (including all problems in which RMSE performance was worse), a plausible explanation of the results can be formulated. The EvoSpace implementation is controlling bloat too aggressively, severely impacting learning in the more difficult test cases. Therefore, future variants of the implementation will need to allow the search to explore large program sizes to evolve more accurate models.

Table 3. Friedman test *p*-values, comparing the sequential neat-GP-LS and the EvoSpace implementation based on test RMSE and average size of the final population. Bold indicates that the null-hypothesis was rejected at the $\alpha = 0.05$ significance level.

	test	size
Problem	*p*-value	
Housing	0.2733	**0.0114**
Concrete	**0.0010**	**0.0010**
Energy Cooling	0.0578	**0.0285**
Energy Heating	**0.0114**	1.000
Tower	**0.0285**	**0.0114**
Yacht	0.2059	**0.0114**

Finally, Figure 10 analyzes the re-speciation process based on the intra-species distance. The plot shows how D_{S_l} changes over for each of the species in the population, using a single run of the algorithm on the Housing problem, zooming in on the first 225 samples taken by the EvoWorkers. Each vertical line represents the difference between D_{S_l} and $\hat{D}_{S_l}$. When a line is black (shorter lines), it means that a re-speciation event was not triggered, and when a line is red (longer lines) this means that a re-speciation event could have been triggered by a sample. We can see that, at the beginning of the run, speciation events are more frequent and, as the search progresses begins to converge, these events become infrequent.

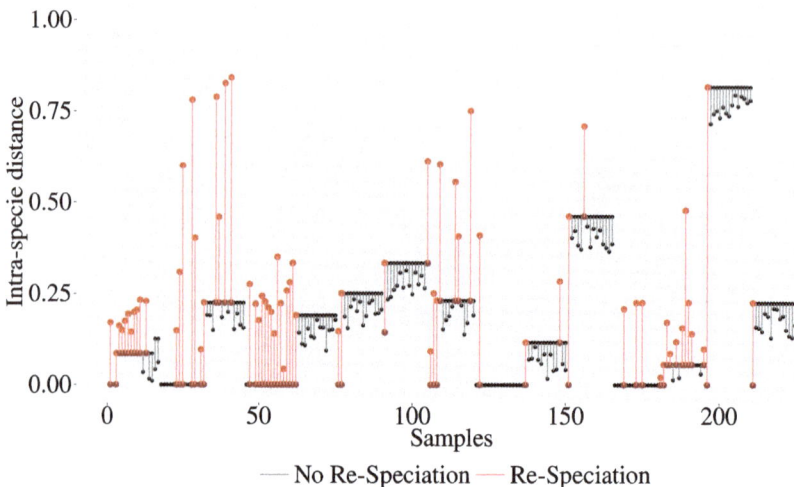

Figure 10. Analysis of the re-speciation process using the intra-species distance.

4. Conclusions and Future Work

This work presents, to the authors' knowledge, the first implementation of a GP system in a Pool-based EA, using the EvoSpace model. The PEA approach is particularly well suited for the speciation-based neat-GP search, allowing for a straightforward strategy to distribute the population over the processing elements of the system (EvoWorkers). It is notable that the performance of the PEA version was not equivalent to the sequential one, in two key respects. On the one hand, it did not reach the same level of performance on some problems. On the other hand, on the problems where it performed equivalently, or better, it was able to reduce solution size significantly.

Future work will center around eliminating the synchronization required by the speciation process in the EvoSpace implementation. Another interesting extension is to consider other elements in the speciation process besides program size and shape, such as program semantics, program behavior or solution novelty. Moreover, we would like to integrate a wider range of parameter local search methods, particularly gradient free methods, and to combine them with other forms of local optimizers that work at the level of syntax or semantics. Finally, it will be important to deploy the proposed algorithms in high-performance computing platforms, to tackle large scale big data problems, where distributing the computational load becomes a requirement.

Author Contributions: L.T., P.J.-S. and M.G.-V. conceived and designed the experiments; P.J.-S. performed the experiments; P.J.-S., L.T., F.F.d.V. and F.C. analyzed the data; F.F.d.V. and F.C. contributed analysis tools; P.J.-S. and L.T. wrote the paper; and M.G.-V., F.F.d.V. and F.C. provided feedback and improved the manuscript.

Acknowledgments: This work was funded by CONACYT (Mexico) project No. FC-2015-2/944 Aprendizaje evolutivo a gran escala, and TecNM (Mexico) project no. 6826-18-p. The first author was supported by CONACYT doctoral scholarship 332554. The authors would like to thank Spanish Ministry of Economy, Industry and Competitiveness and European Regional Development Fund (FEDER) under projects TIN2014-56494-C4-4-P (Ephemec) and TIN2017-85727-C4-4-P (DeepBio); and Junta de Extremadura Project IB16035 Regional Government of Extremadura, Consejeria of Economy and Infrastructure, FEDER.

Conflicts of Interest: The authors declare no conflict of interest.

References

1. Koza, J.R. *Genetic Programming: On the Programming of Computers by Means of Natural Selection*; MIT Press: Cambridge, MA, USA, 1992.
2. Juárez-Smith, P.; Trujillo, L.; García-Valdez, M.; Fernández de Vega, F.; Chávez, F. Local search in speciation-based bloat control for genetic programming. *Genet. Program. Evolvable Mach.* **2019**. [CrossRef]
3. Langdon, W.B. A Many Threaded CUDA Interpreter for Genetic Programming. In Proceedings of the 13th European Conference on Genetic Programming (EuroGP 2010), 7–9 April 2010, Istanbul, Turkey; pp. 146–158.
4. Gong, Y.J.; Chen, W.N.; Zhan, Z.H.; Zhang, J.; Li, Y.; Zhang, Q.; Li, J.J. Distributed Evolutionary Algorithms and Their Models. *Appl. Soft Comput.* **2015**, *34*, 286–300. [CrossRef]
5. Kshemkalyani, A.; Singhal, M. *Distributed Computing: Principles, Algorithms, and Systems*, 1st ed.; Cambridge University Press: New York, NY, USA, 2008.
6. Gebali, F. *Algorithms and Parallel Computing*, 1st ed.; Wiley Publishing: Hoboken, NJ, USA, 2011.
7. Goribar-Jimenez, C.; Maldonado, Y.; Trujillo, L.; Castelli, M.; Gonçalves, I.; Vanneschi, L. Towards the development of a complete GP system on an FPGA using geometric semantic operators. In Proceedings of the 2017 IEEE Congress on Evolutionary Computation (CEC), San Sebastian, Spain, 5–8 June 2017; pp. 1932–1939.
8. García-Valdez, M.; Trujillo, L.; Fernández de Vega, F.; Merelo Guervós, J.J.; Olague, G. EvoSpace: A Distributed Evolutionary Platform Based on the Tuple Space Model. In Proceedings of the 16th European Conference on Applications of Evolutionary Computation, Vienna, Austria, 3–5 April 2013; Springer: Berlin/Heidelberg, Germany, 2013; pp. 499–508.
9. García-Valdez, M.; Trujillo, L.; Merelo, J.J.; Fernández de Vega, F.; Olague, G. The EvoSpace Model for Pool-Based Evolutionary Algorithms. *J. Grid Comput.* **2015**, *13*, 329–349. [CrossRef]

10. García-Valdez, M.; Mancilla, A.; Trujillo, L.; Merelo, J.J.; de Vega, F.F. Is there a free lunch for cloud-based evolutionary algorithms? In Proceedings of the 2013 IEEE Congress on Evolutionary Computation, Cancun, Mexico, 20–23 June 2013; pp. 1255–1262.
11. Trujillo, L.; García-Valdez, M.; de Vega, F.F.; Merelo, J.J. Fireworks: Evolutionary art project based on EvoSpace-interactive. In Proceedings of the 2013 IEEE Congress on Evolutionary Computation, Cancun, Mexico, 20–23 June 2013; pp. 2871–2878.
12. Trujillo, L.; Muñoz, L.; Galván-López, E.; Silva, S. neat Genetic Programming: Controlling bloat naturally. *Inf. Sci.* **2016**, *333*, 21–43. [CrossRef]
13. Z-Flores, E.; Trujillo, L.; Schütze, O.; Legrand, P. Evaluating the Effects of Local Search in Genetic Programming. In *EVOLVE—A Bridge between Probability, Set Oriented Numerics, and Evolutionary Computation V*; Springer International Publishing: Cham, Switzerland, 2014; pp. 213–228.
14. Z-Flores, E.; Trujillo, L.; Schütze, O.; Legrand, P. A Local Search Approach to Genetic Programming for Binary Classification. In Proceedings of the 2015 Annual Conference on Genetic and Evolutionary Computation (GECCO '15), 11–15 July 2015, Madrid, Spain; pp. 1151–1158.
15. Sorensen, D. Newton's Method with a Model Trust Region Modification. *SIAM J. Numer. Anal.* **1982**, *16*. [CrossRef]
16. Z-Flores, E.; Abatal, M.; Bassam, A.; Trujillo, L.; Juárez-Smith, P.; Hamzaoui, Y.E. Modeling the adsorption of phenols and nitrophenols by activated carbon using genetic programming. *J. Clean. Prod.* **2017**, *161*, 860–870. [CrossRef]
17. Enríquez-Zárate, J.; Trujillo, L.; de Lara, S.; Castelli, M.; Z-Flores, E.; Muñoz, L.; Popovič, A. Automatic modeling of a gas turbine using genetic programming: An experimental study. *Appl. Soft Comput.* **2017**, *50*, 212–222. [CrossRef]
18. Dignum, S.; Poli, R. Operator Equalisation and Bloat Free GP. In Proceedings of the 11th European Conference on Genetic Programming (EuroGP 2008), Naples, Italy, 26–28 March 2008; pp. 110–121.
19. Silva, S. Reassembling Operator Equalisation: A Secret Revealed. In Proceedings of the 13th Annual Conference on Genetic and Evolutionary Computation (GECCO '11), 12–16 July 2011, Dublin, Ireland; pp. 1395–1402.
20. Stanley, K.O.; Miikkulainen, R. Evolving Neural Networks Through Augmenting Topologies. *Evol. Comput.* **2002**, *10*, 99–127. [CrossRef] [PubMed]
21. Kommenda, M.; Kronberger, G.; Winkler, S.M.; Affenzeller, M.; Wagner, S. Effects of constant optimization by nonlinear least squares minimization in symbolic regression. In Proceedings of the Genetic and Evolutionary Computation Conference (GECCO '13), Amsterdam, The Netherlands, 6–10 July 2013; pp. 1121–1128.
22. Byrd, R.H.; Schnabel, R.B.; Shultz, G.A. A trust region algorithm for nonlinearly constrained optimization. *SIAM J. Numer. Anal.* **1987**, *24*, 1152–1170. [CrossRef]
23. Quinlan, J.R. Combining Instance-Based and Model-Based Learning. In Proceedings of the Tenth International Conference on Machine Learning, Amherst, MA, USA, 27–29 June 1993; pp. 236–243.
24. Yeh, I.C. Modeling of strength of high-performance concrete using artificial neural networks. *Cem. Concr. Res.* **1998**, *28*, 1797–1808. [CrossRef]
25. Tsanas, A.; Xifara, A. Accurate quantitative estimation of energy performance of residential buildings using statistical machine learning tools. *Energy Build.* **2012**, *49*, 560–567. [CrossRef]
26. Vladislavleva, E.J.; Smits, G.F.; den Hertog, D. Order of nonlinearity as a complexity measure for models generated by symbolic regression via pareto genetic programming. *IEEE Trans. Evol. Comput.* **2009**, *13*, 333–349. [CrossRef]
27. Ortigosa, I.; Lopez, R.; Garcia, J. A neural networks approach to residuary resistance of sailing yachts prediction. *Proc. Int. Conf. Mar. Eng.* **2007**, *2007*, 250.

© 2019 by the authors. Licensee MDPI, Basel, Switzerland. This article is an open access article distributed under the terms and conditions of the Creative Commons Attribution (CC BY) license (http://creativecommons.org/licenses/by/4.0/).

Mathematical and Computational Applications

MDPI

Article

How Am I Driving? Using Genetic Programming to Generate Scoring Functions for Urban Driving Behavior

Roberto López [1], **Luis Carlos González Gurrola** [1,*], **Leonardo Trujillo** [2], **Olanda Prieto** [1], **Graciela Ramírez** [1], **Antonio Posada** [2], **Perla Juárez-Smith** [2] and **Leticia Méndez** [1]

[1] Facultad de Ingeniería, Universidad Autónoma de Chihuahua, Circuito No. 1, Nuevo Campus Universitario, Apdo. postal 1552, Chihuahua 31240, Mexico; jrlopez@uach.mx (R.L.); oordaz@uach.mx (O.P.); galonso@uach.mx (G.R.); nmendez@uach.mx (L.M.)

[2] Departamento de Ingeniería en Electrónica y Eléctrica, Instituto Tecnológico de Tijuana, Calzada Tecnológico SN, Tomas Aquino, Tijuana 22414, Mexico; leonardo.trujillo@tectijuana.edu.mx (L.T.); antonio.posada.gonzalez@gmail.com (A.P.); pjuarez@tectijuana.edu.mx (P.J.-S.)

* Correspondence: lcgonzalez@uach.mx

Received: 7 March 2018; Accepted: 30 March 2018; Published: 3 April 2018

Abstract: Road traffic injuries are a serious concern in emerging economies. Their death toll and economic impact are shocking, with 9 out of 10 deaths occurring in low or middle-income countries; and road traffic crashes representing 3% of their gross domestic product. One way to mitigate these issues is to develop technology to effectively assist the driver, perhaps making him more aware about how her (his) decisions influence safety. Following this idea, in this paper we evaluate computational models that can score the behavior of a driver based on a risky-safety scale. Potential applications of these models include car rental agencies, insurance companies or transportation service providers. In a previous work, we showed that Genetic Programming (GP) was a successful methodology to evolve mathematical functions with the ability to learn how people subjectively score a road trip. The input to this model was a vector of frequencies of risky maneuvers, which were supposed to be detected in a sensor layer. Moreover, GP was shown, even with statistical significance, to be better than six other Machine Learning strategies, including Neural Networks, Support Vector Regression and a Fuzzy Inference system, among others. A pending task, since then, was to evaluate if a more detailed comparison of different strategies based on GP could improve upon the best GP model. In this work, we evaluate, side by side, scoring functions evolved by three different variants of GP. In the end, the results suggest that two of these strategies are very competitive in terms of accuracy and simplicity, both generating models that could be implemented in current technology that seeks to assist the driver in real-world scenarios.

Keywords: genetic programming; driving scoring functions; driving events; risky driving; intelligent transportation systems

---

1. Introduction

It is a well known problem that reckless driving affects society in a variety of ways, with noteworthy impacts on health, economic and social issues. Government offices in charge of planning and deploying urban programs could largely benefit by technological tools which can provide decision-support information to develop appropriate public policies. On their part, taxicab companies or on-line transportation services could also benefit from such tools to optimize resources or reduce economic risks related to insurance or maintenance. At the same time, insurance companies could work on coverage plans designed according to a driver's profile which could, at least partially, be based on automatic tools that exploit useful information obtained from regular driving trips [1,2].

For these reasons, there has been a growing amount of interest in the development of approaches that can extract information from sensors embedded in almost every modern mobile device, such as GPS, gyroscope, magnetometer and accelerometer. This opens the possibility for a broad spectrum of new applications that exploit different ways of collecting and analyzing sensor data. In particular several works have proposed to transform sensor readings to deliver a driving safety score, with a small but varied set of techniques, such as simple penalty functions [3], fuzzy classifiers [4] or Bayesian models [5]. Moreover, other authors have suggested to expand the amount of sensors used to score driving behaviors, such as using real time data about weather, road conditions and traffic density, among others [6].

The problem of calculating a driving score based on the performance of the driver could be seen as a computational learning task, where given a feature vector that contains the frequency of risky maneuvers the goal is to assign a score to represent the driver's performance in a risk-safety scale (this problem could also be seen as a machine learning problem for human-rating). Recently, in [7] the authors presented the evaluation of seven different Machine Learning (ML) approaches to learn how individuals assigned a driving score. From this comparison a clear winner emerged, this being the Genetic Programming (GP) approach. Since then, a question that remained open was to evaluate if other variants of GP could attain better results. In this manuscript we approach this question by evaluating three different GP variants that have proved to be successful for a variety of problems, namely GPTIPS [8], neatGP [9] and neat-GP-LS [10]. The first method generates models linear in parameters, the second builds non-linear models using a bloat control mechanism, and the last one extends neat-GP by including a local search operator. Results suggest that GPTIPS as well as neatGP-LS are very competitive at evolving mathematical scoring functions, and given their accuracy and simplicity could be integrated in current car technology to assist the driver.

The rest of the paper is organized as follows. In Section 2 we overview related work on the problem domain. Next, in Section 3 we describe our proposed approach, including building the dataset, posing the learning problem and applying GP to solve it. In Section 4 we detail the results obtained, and finally in Section 5 we present the main conclusions.

2. Related Work

The exponential growth of the world population has increased traffic flow in all cities worldwide. Therefore, new and improved urban policies must be implemented, and the use of current technology is becoming a mandatory requirement to alleviate the social impact of road accidents [11]. Recent works deal with detecting and classifying sensor data from mobile devices, as well as sensors embedded in automobiles [3]. Due to the increase availability of such devices, there is a surge of possible applications for the data gathered by them. Opportunely determining the mental and physical state of a driver through an analysis of their behavior might help mitigate the number of car accidents. For instance the work in [6] details the social impact of traffic accidents in the UK, and approaches the problem via a context-aware architecture that links data from sensors, driver behavior and road infrastructure to jointly help decrease the possibility of road accidents. While most works attempt to distinguish between aggressive from calm driving, drunk and careless driving might also be identified. For example, Dai et al. [12] tried to detect drunk driving through an inexpensive platform using mobile phone sensors, sending notifications to the proper authorities and to the driver.

However, modeling the behavior of a driver is a very complex task, it involves a large number of variables and subjective data. Self perception of driving skills is very biased, some drivers might believe that their driving style is safe, whilst a driving companion may have an entirely different opinion. Detecting aggressive behavior while driving is a problem that has been addressed through several ML techniques, that exploit specific types of sensor data. In a large study presented in [1], a mobile application was designed using a crowd-sourcing approach. Data from sensors embedded in mobile devices was recorded by the application and used in conjunction with automobile characteristics and road conditions. This distributed telematics platform allowed the authors to successfully gather data regarding unsafe behaviors whilst driving. Their approach used a multivariate Gaussian model to determine the likelihood of normal versus abnormal driving maneuvers. The work detailed in [13,14] also deals with classification of unsafe driving practices, from sensor data within mobile phones. The use of such sensors is extremely prevalent due to the deep penetration of the aforementioned devices among current drivers. Most of the works that attempt to detect aggressive or reckless behaviors while driving, focus on accurately detecting specific driving events, like sudden stops, intrusive lane changes or speeding. This is usually done by posing a classification problem. ML approaches that have been used to solve this problem include Fuzzy Logic Systems and Time Series analysis [4,13,15–17], to mention two prominent techniques. However, we argue that classifying these events is only the first step towards properly determining unsafe driving behaviors. To properly score a driving trip, one should account for the total frequency of the events with regards to the total length of the trip. The nature of these events are vastly studied under the concept of Insurance Telematics [2], where they are called *Figure of Merits* (FoMs). Once these FoMs or risky maneuvers are detected, a new regression problem appears, which is related to calculating a driving score. For instance, in the work of [5] the authors attempt to score driving trips using a Bayesian classifier to differentiate between risky and safe maneuvers. In [7], the authors compared a Fuzzy Inference System [1], a Safety Index [13], a Bayesian regressor [5], a Multi-layer perceptron, a Random Forrest, a Support Vector regressor and a GP approach to learn how individuals score 200 virtual road trips, where each trip was represented by a feature vector containing the frequency of risky maneuvers. Results showed that the GP strategy was superior than competitors, even in some cases with statistical significance. We depart from this point, where a main comparison of approaches was already done, positioning GP as the best performer. On top of these results this manuscript compares the performance of three flavors of GP, with the aim to know if a better performance can be found.

3. Methodology

In this section we briefly describe the datasets that will be used. We also present the different flavors of GP that are compared. Finally, we present the performance metric and the statistical test that is used to validate the analysis.

3.1. The Dataset

Given that our aim is to compared GP-strategies on the first place, but also to contextualize our results given prior results, we will use the same dataset employed in [7]. This dataset is composed of 200 virtual trips, where each trip is characterized by a vector of frequencies, with each position of the vector associated to a particular risky maneuver. Each trip (vector) is also associated to a Driving score (target value) that was computed as the mean value of all scores given to it by a group of drivers. The risky maneuvers considered in the vector are analogous to those found in [18], which suggests those types of incidents are closely related to safety on the road. As an example, Table 1 shows how a single road trip was characterized by a human observer based on the frequency of each risky maneuver. To establish the ground truth for the driving score, the descriptive vector of each trip was shown to several human observers (a total of 40) who were then asked to provide a subjective score for the trip on a scale of 1 (very unsafe) to 10 (very safe). This means that each observer imposed subjective criteria when deriving their score, based on what events they considered to be the most correlated with driving safety.

Table 1. A sample of the survey used to evaluate each driving event as safe or unsafe, based on the frequency of each DE.

Driving Event (id Number in the Feature Vector)	Value (Frequency)	Score for the Travel
Distance (x_1)	7	
Avg. Velocity (x_2)	6	
# of acceleration events (x_3)	5	
# of sudden starts (x_4)	3	8
# of abrupt lane changes (x_5)	2	
# of intense brakes (x_6)	7	
# of sudden stops (x_7)	0	
# of abrupt steerings (x_8)	1	

To compare approaches, two experiments were envisioned. In the first experiment, now called 1–10 scale, the dataset was used the way it is, since the target values for all the feature vectors contains a score in the range [1, 10]. For the second experiment, the targets were fitted to a 1–4 scale following the criteria proposed in [3]. Both experiments were planned to analyzed the performance of the approaches in a fine and coarse version of the dataset. Note that, for both scales, the minimum value corresponds to a totally reckless trip, and the highest value to a totally safe trip.

3.2. Genetic Programming and Tested Flavors

GP uses an evolutionary search to derive small programs, operators or models. In most cases, GP is used to solve different kinds of ML tasks, with the most common being symbolic regression, producing Symbolic Regression Models (SRMs). These models represent the relation between the input variables and the dependent output variable. Therefore, in this work the problem is posed as a symbolic regression one, where GP is used to evolve the scoring functions, taking as input each of the frequency features described above. In standard GP, and all the variants used here, the SRMs are represented using a tree structure, where internal nodes contain elements from a set of basic mathematical operations called the function set. Tree leaves contain input values, in this case each of the x_i features and random constants. The fitness function used in all cases is the Root Mean Squared Error (RMSE) ($RMSE = \sqrt{\frac{1}{n}\sum_{i=1}^{n}\left(pred_i - target_i\right)^2}$, where $target_i$ is the score in the survey for the $i - th$ road trip, $pred_i$ is the predicted score given by a particular scoring function.) between the estimated score given by an individual SRM and the ground truth score on all of the samples in a training set. As for any other evolutionary algorithm, in GP, special genetic operators are used to build new SRMs (offspring) from previous ones (parents) that were chosen stochastically, with a bias that is based on the fitness of the solution. In order to be a self-contained manuscript, the general pseudocode of a GP search is summarized in Algorithm 1. The specific strategies of GP evaluated in this manuscript are described next.

Algorithm 1 Genetic Programming pseudocode.

1: **for** $i = 1$ to *NumOfGenerations* (*or until an acceptable solution is found*) **do**
2:    **if** 1st generation **then**
3:       generate initial population with primitives (variables, constants and elements from the function set)
4:    **end if**
5:    Calculate fitness (minimize RMSE) of population members
6:    Select n parents from population (based on fitness)
7:    Stochastically apply Genetic operators to generate n offspring
8: **end for**
9: Return best individual (based on fitness) found during search

---

3.2.1. GPTIPS V2

GPTIPS implements a multi-gene strategy to represent individuals in the population. Basically, the representation of a single individual includes a collection of trees, and the final SRM is constructed as a linear combination of all the trees. This is a linear in parameter model, with the model weights fitted with linear regression. GPTIPS also uses a multi-objective selection pressure, where model complexity (size) is also included as a selection criterion. A good aspect of the GPTIPS tool is that it offers a variety of post execution reports, such as summary information about the latest generation, the best solutions on the Pareto front, considering both the training and testing partitions, the complexity of the proposed solutions in terms of number of nodes, RMSE and R^2 of the best programs, among other useful stats [8] (http://gptips.sourceforge.net/).

3.2.2. neatGP

The neatGP method [9] combines GP with the NeuroEvolution of Augmenting Topologies algorithm (NEAT) [19]. It was proposed as a bloat-free GP search based on the results obtained by Flat Operator Equalization (Flat-OE) [20]. In general, neatGP preserves a diverse population of individuals of different shapes and sizes, by using speciation techniques and standard fitness sharing. Results on both regression and classification tasks have show that neatGP can produce very accurate models that are orders of magnitude smaller than the ones produced by standard GP [9]. The implementation used in this paper was developed on top of the DEAP Library [21].

3.2.3. neatGP-LS

One way to enhance a global search algorithm or metaheuristic is by embedding additional search operators that can improve the overall exploitation ability of the search process. In particular, GP utilizes search operators that operate at the level of syntax, i.e., they modify the syntactic representation of the programs or SRMs. This can make the search more inefficient when what is required are small steps in the solution space. That is one of the reasons that GPTIPS, for instance, uses linear regression to fit the linear parameters of the evolved models. Similarly, neatGP-LS integrates a local search (LS) mechanism within neatGP, by first parameterizing each solution by adding a real-valued weight θ to each node and fitting those weights using a trust region optimizer [10]. In standard GP these weights are usually ignored and considered to be $\theta = 1$, in neat-GP this represents their initial value but the numerical optimizer can tune the weights as required. One effect of embedding a LS method within GP is that smaller solutions in the population have a better chance of surviving after the LS is applied, since they are more susceptible to suboptimal parameter values.

3.3. Statistical Analysis: Friedman Test and Critical Difference Diagram

To validate the differences in the output generated by each GP strategy, we will conduct the Friedman test [22], which is a non-parametric test that helps identify if there are statistical differences in the variances of multiple treatments. For all the experiments, we will use a significance level of $\alpha = 0.05$. To present this analysis, we will make use of a Critical Difference Diagram (CDD), proposed by [23]. CDD shows in an intuitively manner how different approaches are ranked, being favored those that are located rightmost in the horizontal bar. It also shows when no statistical difference was found between a group of treatments, joining them with a thick bar.

4. Results

In this section we present the results of the execution of each GP strategy applied to the problem of finding suitable scoring functions for driving trips. A 10-fold cross validation scheme was applied in all the experiments, where the same data partitions were used by all algorithms. All of the common parameters were set to the same values, summarized in Table 2. All other parameters were set to their default values as reported in the original references.

Table 2. Parameters for GPTIPS, neatGP and neatGP-LS.

Parameter	Value	Units
Population size	100	items
Max. # of generations	100	items
Input variables	8	items
Range of Constants	$[-10, 10]$	items
Training instances	180	road trips
Testing instances	20	road trips
Crossover probability	85	percentage (%)
Mutation probability	15	percentage (%)
Function set	$\times, -, +, \div, \sqrt{x},$	functions
	$\tanh, \exp, \log, x^3,$	
	$MULT3, ADD3,$	
	$negexp, neg, |x|$	

Experiment 1: Targets in the 1–10 scale. The performance of each GP strategy is given in Tables 3–5, for GPTIPS, neatGP and neatGP-LS, respectively. The format for all comparison tables present the performance on each fold, showing the best training RMSE, the RMSE of the best solution found, the size of the best individual given in number of nodes, and the average size of the population. Summary statistics are also given, showing the minimum, maximum, mean, standard deviation, median and linear correlation coefficient of the best solutions, between the ground truth and the model output.

Figure 1 presents a scatter plot analysis of the best model found, plotting the ground truth score and the model output for each training and testing sample of that particular fold, respectively for GPTIPS (a) and neatGP-LS (b), neatGP was omitted in this comparison since it achieved the poorest results. The corresponding linear correlation coefficient, for the training and testing data, are given in the previous tables.

As a concluding remark for this subsection let us examine the distribution of the RMSE scores for all three approaches. Figure 2 shows through boxplots how GP-Tips and neatGP-LS generate models with less error than neatGP.

Table 3. Root Mean Squared Error (RMSE) for the 10-fold cross validation performance of GPTIPS on the 1–10 scale dataset.

GPTIPS				
Fold	Best Train	Best Test	Size Best Ind	Avg. Pop. Size
1	1.1219	1.2211	22	21.1200
2	1.1168	1.2188	25	22.6067
3	1.1329	1.0482	22	20.0633
4	1.0978	1.4509	26	23.6000
5	1.1479	1.0034	25	22.7067
6	1.1296	1.1546	24	24.3533
7	1.1367	1.1857	24	23.2800
8	1.0650	1.7854	25	23.0167
9	1.1356	1.3130	23	22.2233
10	1.0595	1.5560	22	21.1667
Minimum	1.0595	1.0034	22	20.0633
Maximum	1.1479	1.7854	26	24.3533
mean	**1.1144**	**1.2937**	**23.8000**	**22.4137**
SD	0.0306	0.2408	1.4757	1.2994
median	1.1257	1.2199	24.0000	22.6567
Correlation Coefficient	0.8030	0.2868		

Table 4. RMSE for the 10-fold cross validation performance of neatGP on the 1–10 scale dataset.

	neatGP			
Fold	Best Train	Best Test	Size Best Ind	Avg. Pop. Size
1	1.2093	1.5795	46	39.1850
2	1.2584	1.1488	64	45.9650
3	1.8291	1.9927	183	124.3400
4	1.3928	1.3015	132	100.0250
5	1.3064	1.5171	44	27.9500
6	1.2969	1.9753	61	44.3100
7	0.8505	1.5701	625	211.0900
8	1.2261	1.6290	69	54.6200
9	1.2496	1.9254	66	47.0100
10	1.2546	1.4381	44	29.5650
Minimum	0.8505	1.1488	44	27.9500
Maximum	1.8291	1.9927	625	211.0900
mean	**1.2874**	**1.6077**	**133.4000**	**72.4060**
SD	0.2378	0.2845	178.4204	57.7911
median	1.2565	1.5748	65.0000	46.4875
Correlation Coefficient	0.6432	0.6337		

Table 5. RMSE for the 10-fold cross validation performance of neatGP-LS on the 1–10 scale dataset.

	neatGP-LS			
Fold	Best Train	Best Test	Size Best Ind	Avg. Pop. Size
1	1.1690	1.6912	30	11.5900
2	1.2195	1.1623	40	11.9700
3	1.1622	1.2626	22	16.0000
4	1.2188	1.0513	31	18.8400
5	1.1756	1.5657	29	16.1750
6	1.1813	1.0310	36	20.1050
7	1.3758	1.3739	1	3.4200
8	1.1822	1.1456	18	18.5950
9	1.1326	1.4899	31	13.2200
10	1.1917	1.2202	35	16.5700
Minimum	1.1326	1.0310	1	3.4200
Maximum	1.3758	1.6912	40	20.1050
mean	**1.2009**	**1.2994**	**27.3000**	**14.6485**
SD	0.0666	0.2236	11.2551	4.8923
median	1.1818	1.2414	30.5000	16.0875
Correlation Coefficient	0.7735	0.5012		

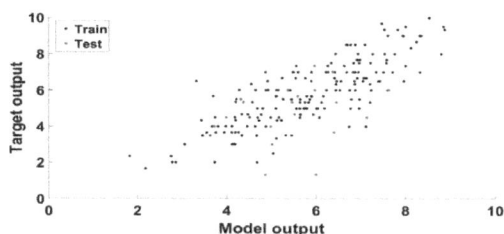

(a) GPTIPS

Figure 1. *Cont.*

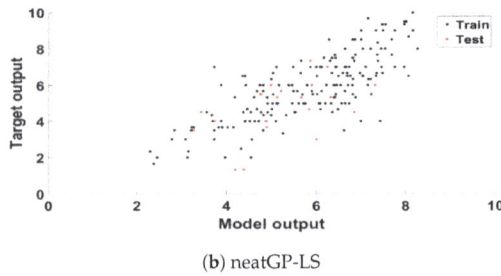

(**b**) neatGP-LS

Figure 1. Scatter plot of the ground truth scores and the model outputs for the training and testing partitions of the best solution found computed for the dataset with a 1–10 scale.

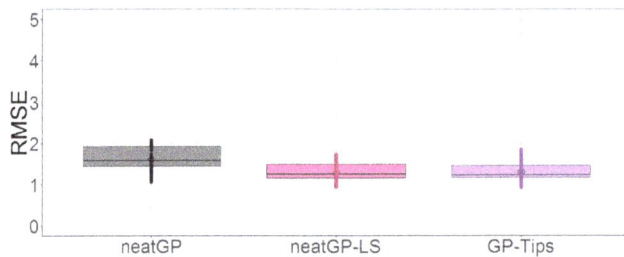

Figure 2. Box plot comparison of all methods based on testing fitness on the dataset with a 1–10 scale.

Experiment 2: Targets in scale 1–4. Tables 6–8 summarize the performance of GPTIPS, neatGP and neatGP-LS, respectively.

The performance over all cross-validation folds is shown in Figure 3, which is a box plot for the testing RMSE. Notice that in the scale 1–4 GPTIPS does not only presents the best performance, but also it shows the less variance.

Table 6. RMSE for the 10-fold cross validation performance of GPTIPS on the 1–4 scale dataset.

	GPTIPS			
Fold	Best Train	Best Test	Size Best Ind	Avg. Pop. Size
1	0.6445	0.7347	22	22.0300
2	0.6567	0.6591	18	17.7833
3	0.6474	0.7408	21	21.8767
4	0.6215	0.9530	23	23.9733
5	0.6604	0.6833	29	24.4567
6	0.6491	0.7931	20	18.0233
7	0.6630	0.5913	22	20.8233
8	0.6524	0.6400	29	22.3667
9	0.6402	0.7241	21	19.9933
10	0.6176	0.9286	23	21.7733
Minimum	0.6176	0.5913	18	17.7833
Maximum	0.6630	0.9530	29	24.4567
mean	**0.6453**	**0.7448**	**22.8000**	**21.3100**
SD	0.0153	0.1182	3.5839	2.2205
median	0.6482	0.7294	22.0000	21.8250
Correlation Coefficient	0.7505	0.6219		

Table 7. RMSE for the 10-fold cross validation performance of neatGP on the 1–4 scale dataset.

	neatGP			
Fold	Best Train	Best Test	Size Best Ind	Avg. Pop. Size
1	0.6070	2.6656	234	174.0650
2	0.5770	0.9034	276	169.5550
3	0.8056	0.8421	464	266.4800
4	0.6513	1.1010	176	118.1200
5	0.4705	0.5493	236	151.3600
6	0.5038	1.1317	245	160.8900
7	0.6748	0.7746	159	110.3400
8	0.7093	0.7463	55	42.1850
9	0.5284	0.9176	205	146.1850
10	0.5491	1.4260	139	65.5550
Minimum	0.4705	0.5493	55	42.1850
Maximum	0.8056	2.6656	464	266.4800
mean	**0.6077**	**1.1058**	**218.9000**	**140.4735**
SD	0.1034	0.5991	107.1888	62.4507
median	0.5920	0.9105	219.5000	148.7725
Correlation Coefficient	0.8805	0.5857		

Table 8. RMSE for the 10-fold cross validation performance of neatGP-LS on the 1–4 scale dataset.

	neatGP-LS			
Fold	Best Train	Best Test	Size Best Ind	Avg. Pop. Size
1	0.6387	0.9852	30	18.9650
2	0.7240	0.7477	29	16.7000
3	0.6675	0.7854	28	14.3150
4	0.6291	0.9384	37	20.5550
5	0.6321	0.7810	21	16.1550
6	0.6611	0.7851	41	24.1900
7	0.7217	0.4519	29	19.7950
8	0.6337	0.6387	34	19.5900
9	0.6622	0.9408	29	15.8750
10	0.6323	1.0198	17	14.0850
Minimum	0.6291	0.4519	17	14.0850
Maximum	0.7240	1.0198	41	24.1900
mean	**0.6603**	**0.8074**	**29.5000**	**18.0225**
SD	0.0359	0.1738	6.9960	3.1629
median	0.6499	0.7852	29.0000	17.8325
Correlation Coefficient	0.7598	0.3352		

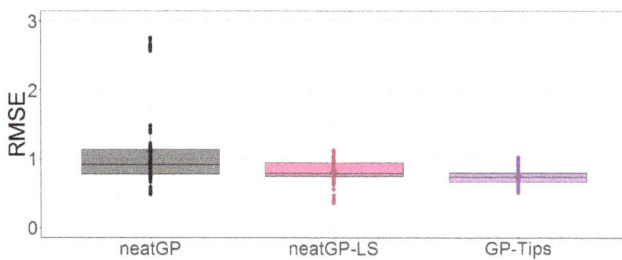

Figure 3. Box plot comparison of all methods based on testing fitness on the dataset with a 1–4 scale.

To ilustrate a resulting evolved model of GP, Table 9 presents one of the best models found by GPTIPS from the perspective of accuracy and complexity.

Table 9. *i*th Best model in the Pareto front, from the stand point of accuracy and complexity (1–4 scale) (GPTIPS).

RMSE Testing	Complexity (# of Nodes)	Model
0.919	37	$y = \dfrac{0.4486\,e^{-1.0\,x_1}\,(x_1 - 1.0\,x_1\,(x_2 - 9.9))}{x_2} - 0.05036\,x_8$ $-\,0.02518\,\tanh(\tanh(x_4))\,(\tanh(\tanh(x_4)) - 1.0\,x_1\,(x_2 - 9.9))$ $-\,0.2356\,x_2 + 3.443$

Statistical analysis. To formally compare the results from each algorithm in terms of test performance (RMSE), we applied a Friedman test [22] and present this results using a Critical Difference Diagram [23], as previously stated.

Figure 4 shows a summary of the results for the 1–10 scale, showing the average rank of each method, indicating that there is no statistically significant difference between the test performance results of all GP variants.

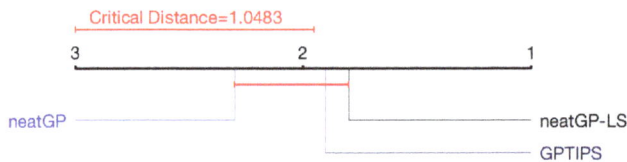

Figure 4. Critical Difference Diagram (CDD) of the Friedman test for the 1–10 scale problem.

Figure 5 presents the same analysis for the 1–4 scale. For the 1–10 scale, neatGP-LS is the top performer, although with no statistical difference. In the 1–4 scale, GPTIPS seems to have advantage, with no statistical difference with neatGP-LS but with statistical difference with neatGP.

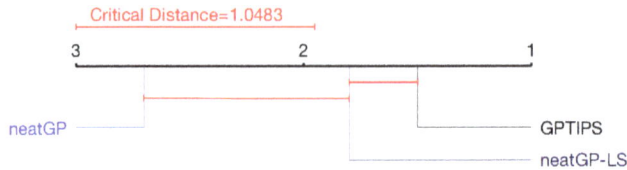

Figure 5. CDD of the Friedman test for the 1–4 scale problem.

Analysis of Features Frequency. To gain some insight about how each method is deriving the predictions of the driving score, Figure 6 plots the frequency with which each input feature appears in all of the best solutions found (a total of 10, one for each fold). Such an analysis can be useful, particularly since the models are quite compact. In the case of neatGP, where the models are larger, the frequency is notably heftier. However, probably the most useful way in which to read this plot is to consider the relative importance of each feature for each method. To associate each feature with its corresponding index variable see Figure 1. In the case of neatGP, feature x_1 is the most used by the models, followed by x_2, and then x_6 and x_8. On the other hand, neatGP-LS prefers x_2, x_4 and

x_6. Additionally GPTIPS uses a smaller number of features, focusing on x_1 and x_2. Notice that some features are practically ignored by neat-GP or GPTIPS, namely x_5 by both and x_5, x_3 and x_4 by GPTIPS. This suggest that these features are not required and may be omitted by a real-time system that must first detect these features before scoring a driving trip.

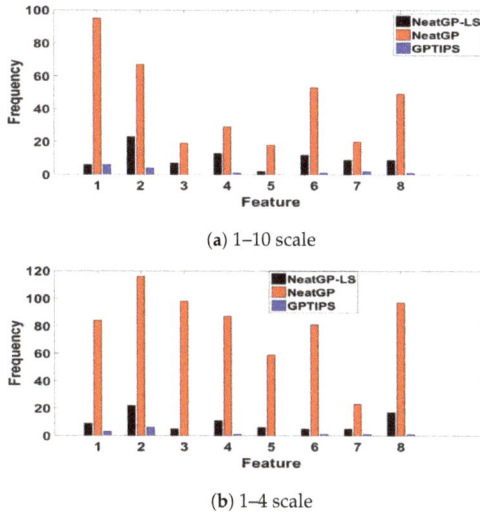

(a) 1–10 scale

(b) 1–4 scale

Figure 6. Frequency of appearance of each DE (feature variables x_1 to x_8) in the best solutions found by each method on the dataset.

5. Conclusions

As cities and urban areas continue to grow, the need for smart technologies that can improve safety is always of great importance. One source of great health and economic impact are traffic accidents, that can range from minor fender benders to life threatening situations. One way to help mitigate these type of accidents is to be able to automatically detect when a person is driving in an unsafe manner, to help bring about the proper corrective measures. This problem can be solved in a two-step process. First, to detect specific types of driving events that are highly correlated with safety. Second, to use the frequency of detected events to derive a safety score for a given driving trip. This paper deals with the latter, building upon previous results were the former has been solved using sensors from mobile devices [24].

This work presents an experimental evaluation of three GP algorithms to solve a difficult real-world problem. GP search is used to evolve scoring functions that take as input the frequency with which a set of driving events are detected during a trip. The goal is to have models that can accurately predict how a human observer would grade a particular trip, based on features such as average speed, distance of a trip, number of lane changes, abrupt steering, sudden stops, among others. A dataset was built, comprised by a total of 200 road trips, each one assembled as a collection of DE, and several human observers graded the trips, with the final score being an average of all scores received.

Using GP, we generated models, expressed as mathematical equations that are able to predict the scores given by humans. Three variants were tested, namely neatGP, neatGP-LS and GPTIPS. In terms of predictive accuracy no statistically significant difference was detected among the methods. However, neatGP did show the largest variance and larger number of outliers from a 10-fold cross validation evaluation. On the other hand, in terms of model size (or complexity) a clear trend was apparent, neatGP produced the largest models, while neatGP-LS and GPTIPS produced similarly concise and compact models. All three methods use a form of bloat control, neatGP and neatGP-LS use speciation and fitness sharing, while GPTIPS uses a multi-objective selection pressure. However, neatGP does not include a numerical local search optimizer, while the other two methods include one, GPTIPS uses linear regression and neatGP-LS a trust region optimizer. This indicates that in real-world problems, traditional bloat control is not sufficient. Local search methods help the evolutionary process tune smaller solutions and improve their chances of survival, curtailing bloat in favor of simpler solutions with properly tuned parameter values. A further analysis about what events are important to construct the scoring functions offers interesting insights. For GPTIPS, the distance (x_1) and average velocity of the trip (x_2) are important features. On its part, neatGP-LS considers the same features, but also the number of abrupt steerings (x_8) and frequency of intense brakes (x_6). In summary, it is clear that GP is a viable approach to solve this problem, and the evolved models are candidates for real-world testing given their accuracy and compact (efficient) form.

Acknowledgments: This project was financed by Google via a 2017 Latin American Research Award and by Proyecto Fronteras de la Ciencia CONACYT FC-2015-2/944 "Aprendizaje evolutivo a gran escala".

Author Contributions: R.L. and L.C.G.G. designed the experiments, implemented the GPTIPS strategy and co-author the paper. O.P., G.R. and L.M. developed the dataset and implemented a Fuzzy strategy as baseline. L.T. co-author the paper and supervised the neatGP and neatGP-LS strategies. A.P. and P.J.-S. implemented the neatGP and neatGP-LS strategies and experiments

Conflicts of Interest: The authors declare no conflict of interest.

Abbreviations

The following abbreviations are used in this manuscript:

GP	Genetic Programming
ML	Machine Learning
SRM	Symbolic Regression Model
ANN	Artificial Neural Networks
SVM	Support Vector Machines
RF	Random Forest
BN	Bayesian Networks
FIS	Fuzzy Inference Systems
BRR	Bayesian Ridge Regression
SVR	Support Vector Regression
LS	Local Search
NEAT	NeuroEvolution of Augmenting Topologies algorithm
Flat-OE	Flat Operator Equalization
DE	Driving Event
RMSE	Root Mean Squared Error

References

1. Castignani, G.; Thierry, D.; Frank, R.; Engel, T. Smartphone-Based Adaptive Driving Maneuver Detection: A Large-Scale Evaluation Study. *IEEE Trans. Intell. Transp. Syst.* **2017**, *18*, 2330–2339.
2. Handel, P.; Skog, I.; Wahlström, J.; Welch, R.; Ohlsson, J.; Ohlsson, M. Insurance telematics: Opportunities and challenges with the smartphone solution. *IEEE Intell. Transp. Syst. Mag.* **2014**, *6*, 57–70.
3. Saiprasert, C.; Thajchayapong, S.; Pholprasit, T.; Tanprasert, C. Driver Behaviour Profiling using Smartphone Sensory Data in a V2I Environment. In Proceedings of the International Conference on Connected Vehicles and Expo (ICCVE), Vienna, Austria, 3–7 November 2014.
4. Aljaafreh, A.; Alshabatat, N.; Al-Din, N.S.M. Driving Style Recognition Using Fuzzy Logic. In Proceedings of the IEEE International Conference on Vehicular Electronics and Safety, Istanbul, Turkey, 24–27 July 2012.
5. Eren, H.; Makinist, S.; Akin, E.; Yilmaz, A. Estimating Driving Behavior by a Smartphone. In Proceedings of the Intelligent Vehicles Symposium, Alcala de Henares, Spain, 3–7 June 2012.
6. Al-Sultan, S.; Al-Bayatti, A.H.; Zedan, H. Context-aware driver behavior detection system in intelligent transportation systems. *IEEE Trans. Veh. Technol.* **2013**, *62*, 4264–4275.

7. López, J.R.; Gonzalez, L.C.; Gómez, M.M.; Trujillo, L.; Ramirez, G. A Genetic Programming Approach for Driving Score Calculation in the Context of Intelligent Transportation Systems. 2017, under review.

8. Searson, D.P. *Handbook of Genetic Programming Applications: Chapter 22-GPTIPS 2: An Open Source Software Platform for Symbolic Data Mining*; Springer: Newcastle, UK, 2015.

9. Trujillo, L.; Munoz, L.; Galvan-Lopez, E.; Silva, S. Neat Genetic Programming: Controlling Bloat Naturally. *Inf. Sci.* **2016**, *333*, 21–43.

10. Juárez-Smith, P.; Trujillo, L. Integrating Local Search Within neat-GP. In Proceedings of the 2016 on Genetic and Evolutionary Computation Conference Companion, Denver, CO, USA, 20–24 July 2016; pp. 993–996.

11. Ali, N.A.; Abou-zeid, H. Driver Behavior Modeling: Developments and Future Directions. *Int. J. Veh. Technol.* **2016**, *2016*, 6952791.

12. Dai, J.; Teng, J.; Bai, X.; Shen, Z.; Xuan, D. Mobile phone based drunk driving detection. In Proceedings of the 2010 4th International Conference on Pervasive Computing Technologies for Healthcare, Munich, Germany, 22–25 March 2010; pp. 1–8.

13. Chaovalit, P.; Saiprasert, C.; Pholprasit, T. A method for driving event detection using SAX with resource usage exploration on smartphone platform. *EURASIP J. Wirel. Commun. Netw.* **2014**, *2014*, 135.

14. Zylius, G. Investigation of Route-Independent Aggressive and Safe Driving Features Obtained from Accelerometer Signals. *IEEE Intell. Transp. Syst. Mag.* **2017**, *9*, 103–113.

15. Davarynejad, M.; Vrancken, J. A survey of fuzzy set theory in intelligent transportation: State of the art and future trends. In Proceedings of the 2009 IEEE International Conference on Systems, Man and Cybernetics, San Antonio, TX, USA, 11–14 October 2009; pp. 3952–3958.

16. Engelbrecht, J.; Booysen, M.; van Rooyen, G.; Bruwer, F. Performance comparison of dynamic time warping (DTW) and a maximum likelihood (ML) classifier in measuring driver behavior with smartphones. In Proceedings of the 2015 IEEE Symposium Series on Computational Intelligence, Cape Town, South Africa, 7–10 December 2015; pp. 427–433.

17. Johnson, D.; Trivedi, M.M. Driving style recognition using a smartphone as a sensor platform. In Proceedings of the 2011 14th International IEEE Conference on Intelligent Transportation Systems (ITSC), Washington, DC, USA, 5–7 October 2011; pp. 1609–1615.

18. Wahlström, J.; Skog, I.; Händel, P. Smartphone-Based Vehicle Telematics: A Ten-Year Anniversary. *IEEE Trans. Intell. Transp. Syst.* **2017**, *18*, 2802–2825.

19. Stanley, K.O.; Miikkulainen, R. Evolving Neural Networks through Augmenting Topologies. *Evolut. Comput.* **2002**, *10*, 99–127, doi:10.1162/106365602320169811.

20. Silva, S.; Dignum, S.; Vanneschi, L. Operator Equalisation for Bloat Free Genetic Programming and a Survey of Bloat Control Methods. *Genet. Program. Evolvable Mach.* **2012**, *13*, 197–238.

21. De Rainville, F.M.; Fortin, F.A.; Gardner, M.A.; Parizeau, M.; Gagné, C. DEAP: A Python Framework for Evolutionary Algorithms. In Proceedings of the 14th Annual Conference Companion on Genetic and Evolutionary Computation GECCO'12, Philadelphia, PA, USA, 7–11 July 2012; pp. 85–92.

22. Conover, W.J. *Practical Nonparametric Statistics*, 3rd ed.; John Wiley: New York, NY, USA, 1999.

23. Demšar, J. Statistical Comparisons of Classifiers over Multiple Data Sets. *J. Mach. Learn. Res.* **2006**, *7*, 1–30.

24. Carlos, M.R.; Gonzalez, L.C.; Ramirez, G. Unsupervised Feature Extraction for the Classification of Risky Driving Maneuvers. 2017, under review.

© 2018 by the authors. Licensee MDPI, Basel, Switzerland. This article is an open access article distributed under the terms and conditions of the Creative Commons Attribution (CC BY) license (http://creativecommons.org/licenses/by/4.0/).

Mathematical and Computational Applications

MDPI

Article

The Construction of a Model-Robust IV-Optimal Mixture Designs Using a Genetic Algorithm

Wanida Limmun [1],*, Boonorm Chomtee [2] and John J. Borkowski [3]

[1] Department of Mathematics and Statistics, Walailak University, Thasala,
     Nakhon Si Thammarat 80160,Thailand
[2] Department of Statistics, Kasetsart University, Chatuchak, Bangkok 10903, Thailand; fsciboc@ku.ac.th
[3] Department of Mathematical Sciences, Montana State University, Bozeman, MT 59717, USA;
     john.borkowski@montana.edu
*   Correspondence: lwanida@mail.wu.ac.th; Tel.: +66-756-3000 (ext. 2035)

Received: 10 April 2018; Accepted: 16 May 2018; Published: 17 May 2018

Abstract: Among the numerous alphabetical optimality criteria is the IV-criterion that is focused on prediction variance. We propose a new criterion, called the weighted IV-optimality. It is similar to IV-optimality, because the researcher must first specify a model. However, unlike IV-optimality, a suite of "reduced" models is also proposed if the original model is misspecified via over-parameterization. In this research, weighted IV-optimality is applied to mixture experiments with a set of prior weights assigned to the potential mixture models of interest. To address the issue of implementation, a genetic algorithm was developed to generate weighted IV-optimal mixture designs that are robust across multiple models. In our examples, we assign models with p parameters to have equal weights, but weights will vary based on varying p. Fraction-of-design-space (FDS) plots are used to compare the performance of an experimental design in terms of the prediction variance properties. An illustrating example is presented. The result shows that the GA-generated designs studied are robust across a set of potential mixture models.

Keywords: mixture experiments; single component constraints; genetic algorithm; IV-optimality criterion

1. Introduction

Industrial product formulations (e.g., food processing, chemical formulations, textile fibers, and pharmaceutical drugs) frequently involve the blending of multiple mixture components. Mixture experiments form a special class of response surface experiments in which the product under investigation is comprised of several components. In this research, we assume that the response of interest is a function only of the proportions of the components that are present in the mixture. The levels of the q experimental factors (x_i; $i = 1, 2, \ldots, q$) in a mixture experiment are component proportions. Thus, each x_i is between zero and one, and the sum of the q component proportions is one. Under these conditions, the experimental region involving q proportions is a regular $q - 1$-dimensional simplex. Typically, there are also single-component constraints (SCCs) defined by lower L_i and upper U_i bounds on the component proportions:

$$0 \leq L_i \leq x_i \leq U_i \leq 1 \quad \text{for } i = 1, 2, \ldots, q. \tag{1}$$

If SCCs are imposed on the component proportions, the experimental region will now be an irregularly shaped polyhedron within the simplex. For a general review of the design and analysis of mixture experiments, see [1,2].

When the mixture region is a simplex, the standard mixture designs such as simplex lattice and simplex centroid designs are often suitable for implementation. However, they are not applicable in situations involving SCCs. When SCCs exist, an extreme vertices design (McLean and Anderson [3]) is one possibility. However, these designs may be very inefficient with respect to prediction variance properties. To improve the prediction variance properties, an algorithmically generated design focused on optimizing an optimality design criterion is commonly used. Optimal designs, however, are, in general, optimal only for a specified model. Numerous approaches have been developed for constructing optimal designs, such as ACED [4], XVERT [5], CONSIM [6], Fedorov [7], DETMAX [8], exchange algorithms [9,10], and genetic algorithms (GAs) [11–15]. GAs have the advantage of being adaptive search algorithms [14]. Recent applications of GAs provide alternative approaches to classic exchange-point algorithms to generate designs. Examples of using GAs to generate designs can be found in Borkowski [11], Heredia-Langner et al. [12,13], Park et al. [15], and Limmun et al. [16].

In this research, we develop and employ a GA that extends the work in Limmun et al. [16] that focused on a single model when constructing an IV-optimal design where the experimental region was an irregularly shaped polyhedral region that is a subspace of a simplex. Their procedure offered a degree of flexibility in its way of constructing designs that allowed it to overcome restrictions that may limit the applicability of other algorithms. The first aim of our research is to consider model misspecification and introduce a weighted IV-optimality criterion determined over a set of potential mixture models, and it will serve to address our second aim: develop a design-generating GA using weighted IV-optimality as its objective function.

The rest of this paper is organized as follows. In Section 2, the Scheffé mixture models and relevant theory relating to mixture experiments of interest are reviewed, and weighted IV-efficiency is defined. Section 3 includes a brief introduction to GAs for constructing designs and an illustration of the steps used in our GA. In Section 4, the proposed scheme is demonstrated with examples, and with concluding statements in Section 5.

2. The Mixture Model and Design Optimality

2.1. Notation and Models

The most common forms of mixture models are the Scheffé (canonical) polynomials. For example, the Scheffé linear model is given by

$$y = \sum_{i=1}^{q} \beta_i x_i + \varepsilon \tag{2}$$

and the Scheffé quadratic model is given by

$$y = \sum_{i=1}^{q} \beta_i x_i + \sum_{i=1}^{q-1} \sum_{j=i+1}^{q} \beta_{ij} x_i x_j + \varepsilon. \tag{3}$$

In these models, y is the response variable; each β_i coefficient represents the expected response when the proportion of the i-th component equals unity; each β_{ij} coefficient represents the nonlinear blending properties of the i-th and the j-th component proportions, and ε is a random error. In this paper, we focus on the Scheffé quadratic model mixture model and possible model misspecification.

The Scheffé mixture models are linear models and can be expressed in the matrix form as

$$y = X\beta + \varepsilon \tag{4}$$

where $y = (y_1, y_2, \ldots, y_n)'$ is the response vector, X is the $N \times p$ model matrix with columns associated with the p model terms (such as linear and cross-products terms), β is the $p \times 1$ vector of model parameters, and $\varepsilon = (\varepsilon_1, \varepsilon_2, \ldots, \varepsilon_n)'$ is the vector of random errors associated with natural variation of y around the underlying surface assumed to be independent and identically normally distributed with zero mean and variance $\sigma^2 I_n$. The prediction properties of a design, specifically the scaled prediction variance, is dependent on the chosen model through model matrix X. The scaled prediction variance $v(x_0)$ is defined as

$$v(x_0) = \frac{N\text{Var}(\hat{y}(x))}{\sigma^2} = Nx_0'\left(X'X\right)^{-1}x_0 \tag{5}$$

where x_0' is the expansion of a mixture $x = (x_1, x_2, \ldots, x_q)$ to vector form corresponding to the p columns of X; N is the design size, which penalizes prediction variances for larger designs, and $\text{Var}(\hat{y}(x))$ is the variance of the estimated response at x. For example, if there are three components and the model is the Scheffé quadratic model, then $x = (x_1, x_2, x_3)$, $x_0' = (x_1, x_2, x_3, x_1x_2, x_1x_3, x_2x_3)$, and X is an $N \times 6$ model matrix.

2.2. Optimality Criteria

Design optimality criteria are single valued criteria that represent different variance or parameter estimation properties of a design. Several criteria have been advanced with the purpose of constructing and comparing and designs. Four commonly used optimality criteria are the D, A, G, and IV criteria. These four criteria are functions of the information matrix $(X'X)$ for a given design. The D and A criteria are focused on parameter estimation, while G and IV criteria are focused on the prediction variance. When a design is being considered for implementation, several of its properties can be evaluated by computing measures of design efficiency. Common D, A, G, and IV design optimality measures are

$$D - \text{efficiency} = \frac{100|X'X|^{1/p}}{N}, \quad A - \text{efficiency} = \frac{100p}{\text{trace}[N(X'X)^{-1}]}$$

$$G - \text{efficiency} = \frac{100p}{\max\limits_{x \in \chi}[v(x_0)]}, \quad IV - \text{efficiency} = \frac{V}{\int_\chi v(x_0)dx_1dx_2\ldots dx_q}$$

where χ is the design space and V is the volume of χ. The D and A criteria are the simplest to handle computationally, and they were the criteria considered in the earliest design generating algorithms (e.g., exchange algorithms). Because designs with high IV or G efficiencies also tend to have good D- and A-efficiencies, this paper will focus on designs that minimize the average prediction variance (i.e., minimize the denominator in the IV-efficiency over the entire experimental region). Designs that minimize the IV-optimality criterion include the IV-optimal design, the Q-optimal design [17], the V-optimal design [18], and the I-optimal design [19,20].

In a review of software approaches to evaluating average prediction variances (APVs), Borkowski [21] showed that it is common for software packages to estimate the APV $= \int_\chi v(x_0)dx_1dx_2\ldots dx_q$ by taking the sample mean of $v(x_0)$ over a fixed set of points (e.g., the candidate set for an exchange algorithm). He demonstrated that estimation of the APV based instead on a random set of evaluation points is unbiased and superior to estimation based on a fixed set of evaluation points. This is one flaw in using software using exchange algorithms to generate IV-optimal designs: the estimated APV is an overestimation of the integral. In our proposed GA, the APV measure is calculated by averaging $v(x_0)$ over a random set of points and will provide an unbiased estimate of the IV-optimality criterion. Additionally, the variability of the estimator will decrease as the size of the random evaluation set increases. In this paper, we use 5000 random points in the evaluation set. Several authors have provided results for the IV-optimal designs, including Borkowski [11], Syafitri et al. [20], and Coetzer and Haines [22]. Further details on the motivation and uses of the optimality criteria can be found in Box and Draper [23], Atkinson et al. [18], and Fedorov [7].

2.3. Weighted IV-Optimality

In this paper, we develop and propose the weighted IV-efficiency, which is defined as

$$\text{Weighted IV} - \text{efficiency} = \sum_{i=1}^{r} w_i IV_i \tag{6}$$

where $IV_i = \frac{V}{\int_\chi v(\mathbf{x}_0)\ dx_1\ dx_2...dx_q}$ is the IV-efficiency for model i, χ is the design space, V is the volume of χ, N is the number of design points, r is the number of reduced models for a given full model, and w_i is the weight for model i.

In terms of design generation, the goal is to use weighted IV-optimality to find a set of points that will minimize the weighted average of the average of the scaled prediction variance over the design region across a set of reduced models. Practically speaking, the goal is to generate a design that protects against having a final model with poor prediction variance properties. Similar to finding an IV-optimal design, the researcher must specify a model. This serves as the "full" model when we consider the weighted IV-optimality. However, unlike IV-optimality, a suite of "reduced" models are also proposed if the full model is misspecified via over-parameterization. Examining Equation (6), note that weighted IV-optimality is not restricted to a particular full model nor is it restricted to mixture models. It easily generalizes to other response surface designs and the associated models that can be fitted by those designs.

The experimenter has the freedom to choose the weighting factors for the full model, the most parsimonious or smallest model, as well as all other intermediate models. To exemplify the use of weighted IV-optimality, a "full" quadratic mixture model and a "smallest" linear mixture model will be considered.

Although there are numerous ways to assign weights in the weighted optimality criterion, in this paper, we assume (i) that not all models should be weighted equally and (ii) that only models having an equal number of model parameters receive equal weight. Here are the reasons for assuming (i) and (ii). Before running the experiment, the experimenter believes that the full model may be the most appropriate, so he/she chooses the maximum weighting factor for the full model. In the analysis phase, however, the full model might be inappropriate because of misspecification due to overparameterization. Thus, it seems reasonable that the weights be reduced as we move further away from the full (or the experimenter's most likely a priori) model via model reduction, and stop with the smallest weight assigned to the model with the fewest number of parameters. We also treat each parameter as equally important under the assumption that the researcher has no prior knowledge or make an educated guess regarding which terms would most likely be removed if a model reduction occurs. Therefore, we assign uniform weights among all models resulting after removal of one term, and then assign a decreasing but uniform weight to all models resulting when two terms are moved, and so on. Therefore, models with more parameters have weights that reflect their greater importance relative to models with fewer parameters. For the proposed method of calculating a weighted IV-efficiency, it is necessary to have a complete enumeration of the set of subset models (reduced models) of interest. If an experimenter has justification for another weighting scheme, then it should certainly be implemented.

One specific weighting scheme consistent with (i) and (ii) above is to use weights for each reduced model based on the numbers of model parameters. Suppose model i has j parameters. The weight we assigned to model i is then defined as

$$w_i = \frac{\psi_j}{m_j} \tag{7}$$

where ψ_j is the jth weighting factor, and $j = q, q+1, \ldots, s-1, s$, where q is the number of parameters of the linear mixture model in Equation (2); s is the number of parameters of the quadratic model in Equation (3), and m_j is the number of reduced models with j parameters. The model i weight $(w_i, i = 1, 2, \ldots, r)$ is nonnegative and the weights satisfy $\sum_{i=1}^{r} w_i = 1$. The relative weighting factor for computing the weighting factors (supplied by the experimenter) is defined as

$$R = \frac{\psi_s}{\psi_q}. \tag{8}$$

In this paper, we assume that the experimenter weights the quadratic mixture model 100 times the weight to be assigned to the linear mixture model; therefore, we use $R = 100$ as the relative weighting factor. The weighting factors are positive and subject to the restriction

$$\psi_q + \psi_{q+1} + \ldots + \psi_{s-1} + \psi_s = 1. \tag{9}$$

There are k equispaced levels of the weighting factor, and $k = s - q + 1$ is the number of levels for a weighting factor. Therefore, the range of the weighting factor is defined as

$$\psi_k = \psi_s - \psi_q = R\psi_q - \psi_q = (R-1)\psi_q. \tag{10}$$

The increment value of the weighting factor can be expressed as

$$INCR = \frac{\psi_k}{k-1} = \frac{(R-1)\psi_q}{k-1}. \tag{11}$$

The values of the weighting factor can be represented as follows:

$$\psi_{q+c} = \psi_q + cINCR; c = 0, 1, \ldots, k-1. \tag{12}$$

Again, the weighting factor must sum to one. The weighting factor can be rewritten in the form:

$$\sum_{i=q}^{s} \psi_i = k\psi_q + \binom{k}{2} INCR = 1. \tag{13}$$

Therefore, the minimum weighting factor can be expressed as

$$\psi_q = \frac{(k-1)INCR}{R-1}. \tag{14}$$

After simplifying, the increment value of the weighting factor is defined as

$$INCR = \frac{2(R-1)}{k(k-1)(R+1)}. \tag{15}$$

For example, suppose the full model is the quadratic mixture model with three component proportions. This model can be expressed as

$$E(y) = \beta_1 x_1 + \beta_2 x_2 + \beta_3 x_3 + \beta_{12} x_1 x_2 + \beta_{13} x_1 x_3 + \beta_{23} x_2 x_3.$$

There are 8 reduced models (the full model and the seven reduced models as shown in Table 1). A 1 or a 0 in any "Terms in Model" column represents the presence or absence of that term in the model, respectively. The column p is the number of model parameters. Using 100 as the relative weighting factor, the increment value of the weighting factor is 0.1634. The weighting factor value and the weight for each model are shown in the ψ_j and w_i columns, respectively. If the 2nd, 3rd, and 4th models have five parameters, then $m_5 = 3$, and the 5-parameter model weighting factor $\psi_5 = 0.3318$ and the individual model weights $w_2, w_3,$ and $w_4 = \psi_5/3 = 0.1106$.

Table 1. The set of reduced models the Scheffé quadratic model with three components.

Model	Terms in Model						p	ψ_j	w_i
	x_1	x_2	x_3	x_1x_2	x_1x_3	x_2x_3			
1st	1	1	1	1	1	1	6	0.4950	0.4950
2nd	1	1	1	0	1	1	5		0.1106
3rd	1	1	1	1	0	1	5	0.3318	0.1106
4th	1	1	1	1	1	0	5		0.1106
5th	1	1	1	0	0	1	4		0.0561
6th	1	1	1	0	1	0	4	0.1683	0.0561
7th	1	1	1	1	0	0	4		0.0561
8th	1	1	1	0	0	0	3	0.0049	0.0049

3. Genetic Algorithms and Constructing IV-Optimal Mixture Designs

3.1. Development of the Genetic Algorithm

Heuristic optimization has been used to solve a variety of experimental design problems. One popular optimization algorithm that is based on the general principle of local improvement is the genetic algorithm (GA). The genetic algorithm was developed by John Holland in 1975 and was popularized through the work of Goldberg in 1989 [24]. Since then, GAs have been used to solve optimization problems for many applications because it is very efficient over a variety of search spaces. Bäck et al. [25] mention that GAs often yield excellent results when applied to complex optimization problems where other methods are either not applicable or turn out to be unsatisfactory. GAs have been successfully applied to experimental design problems using various optimality criteria. For an introduction to GAs, see Michaelewicz [26] and Haupt and Haupt [27].

A GA is a search and optimization technique developed by mimicking the evolutionary principles and the chromosomal processing in the natural selection. A GA takes an initial population of potential solutions to a problem (parent chromosomes). Through evolutionary reproduction operators, the current parent population then passes some of its properties (genes) to produce offspring chromosomes. A subset of the best parent and offspring is retained for the next generation, and the reproduction process is repeated for many generations until an acceptable chromosome has evolved. A GA uses an objective function as a measure of a chromosome's fitness as a solution to the problem of interest. Although a GA is not guaranteed to find the global optimum in a finite number of generations, it is less likely to get trapped at a local optimum than traditional gradient-based search methods when the objective function is not smooth and generally well behaved.

GAs can be very useful for construction of a design when the optimality criterion (objective function) is difficult to work with and/or where there are constraints on the experimental region such as in mixture experiments. We now describe a GA used to generate optimal designs for a mixture experiment with single component constraints (SCCs) that is based on weighted IV-optimality, which extends and modifies the GA approach of Limmun et al. [16] who generated optimal designs for a specified mixture model. Throughout this research paper, we have encoded GA chromosomes using real-value encoding instead of another encoding (e.g., binary) because real-value encoding is easy to interpret, can be modified for many applications, and is flexible enough to allow for a unique representation for every variable.

A chromosome C will represent a potential design (solution) and is encoded as an $N \times q$ matrix, where N is the number of the design points, and q is the number of mixture components. Each row in C is a gene $x_i = \begin{bmatrix} x_{i1} & x_{i2} \dots x_{iq} \end{bmatrix}$ and represents one experimental mixture. The objective function F measures a chromosome's fitness as a solution to the function that we wish to optimize. That is, F takes a chromosome as input and outputs an objective function value. Our GA's objective function F is a weighted IV-efficiency. The goal using a weighted IV-efficiency is to find a set of points that will minimize the weighted average of the scaled prediction variance throughout the entire experimental region for a set of reduced models representing possible model misspecification. Specifically, the goal is to find a design that maximizes objective function F where

$$F = \sum_{i=1}^{r} w_i IV_i \tag{16}$$

where $IV_i = \frac{V}{\int_\chi v(x_0)dx_1 dx_2 ... dx_q}$ is the IV-efficiency for model i, χ is the design space, V is the volume of χ, r is the number of reduced models for a given full model, and w_i is the weight for model i. The GA that we employed includes several genetic operators. Prior to running a GA, the experimenter must specify the r models of interest and the model weights w_i. The steps and operators in our GA are as follows:

Step 1 Specify the GA parameters: population size (M), number of iterations (I), selection method (elitism with random parent pairing), blending rate (α_b), crossover rates $(\alpha_{bc}$ and $\alpha_{wc})$, and mutation rate (α_m).

Step 2 Generate an initial population of M chromosomes (mixture designs). We use the real-value encoding with four decimal places to encode each chromosome. Assume M is odd. To generate the initial population, a random sample is first taken in a hypercube. Then each sampled point in a hypercube is mapped to the constrained mixture space by applying the function used by Borkowski and Piepel [28]. Each experimental mixture is recorded to four-decimal place accuracy.

Step 3 Calculate the IV-efficiency objective function F for each chromosome in the initial population.

Step 4 Find the elite chromosome which is the chromosome that has the largest weighted IV-efficiency F. The remaining $M - 1$ chromosomes are randomly paired for the reproduction process.

Step 5 Produce offspring of the next generation by using genetic operators: *blending, between-parent crossover, within-parent crossover*, and *mutation*. Larger values of genetic parameters are used for the early iterations and the smaller values of genetic parameters are used for the later.

Step 6 Calculate objective function F for each parent/offspring pair. The fitter chromosome in the pair is retained for the next generation.

Step 7 Repeat Steps 5 and 6 for I generations.

Step 8 Apply a local grid search to the best design to further improve objective function F yielding the IV-optimal design. A local grid search searches designs in a small neighborhood of the best design. This is accomplished by perturbing the component proportions by small increments to search for further improvements in F. This continues until no further improvement is found.

3.2. Illustration: A Genetic Algorithm for Constructing an Optimal Mixture Design

To illustrate our algorithm, consider a three-component mixture $(q = 3)$ with the goal of generating a weighted IV-optimal design having $N = 7$ design points. We create an initial population of $M = 5$ chromosomes $(C_1, C_2, C_3, C_4, C_5)$ each having 7 genes where the genes of each chromosome (mixture design) are drawn randomly from the mixture space. One possible realization of the initial population and the associated objective function values are presented in Table 2. C_1 is the elite chromosome because it has the largest F value. Next, the remaining four chromosomes are randomly paired for the reproduction process. Suppose that (C_5, C_4) and (C_2, C_3) are the randomly formed pairs. Applying the reproduction process to C_5 and C_4 yields offspring C_5^* and C_4^*. Table 3 contains a set of random uniform deviates for C_5 and C_4. Boldfacing indicates a probability test is passed (PTIP) at rates $\alpha_b, \alpha_{bc}, \alpha_{wc}$ and $\alpha_m = 0.02$, and that reproduction operation will be performed where $\alpha_b, \alpha_{bc}, \alpha_{wc}$, and α_m represent the success probabilities of blending, between-parent crossover, within-parent crossover, and mutation, respectively.

Table 2. Initial population of five chromosomes.

	C_1			C_2			C_3			C_4			C_5	
x_1	x_2	x_3	x_1	x_2	x_3	x_1	x_2	x_3	x_1	x_2	x_3	x_1	x_2	x_3
0.6531	0.2958	0.0511	0.5242	0.2016	0.2742	0.5771	0.1605	0.2624	0.5160	0.1979	0.2861	0.6705	0.0536	0.2759
0.3612	0.2765	0.3623	0.5858	0.2007	0.2135	0.5229	0.1136	0.3635	0.6487	0.1443	0.2070	0.5256	0.0173	0.4571
0.5077	0.1156	0.3767	0.5367	0.2877	0.1756	0.7316	0.1828	0.0856	0.4404	0.0682	0.4914	0.4927	0.1614	0.3459
0.4479	0.1197	0.4324	0.5214	0.0236	0.455	0.4780	0.2241	0.2979	0.5488	0.2372	0.2140	0.4754	0.0946	0.4300
0.7274	0.0546	0.2180	0.3942	0.1999	0.4059	0.7514	0.2388	0.0098	0.6744	0.1869	0.1387	0.6723	0.0354	0.2923
0.6519	0.2764	0.0717	0.7271	0.1305	0.1424	0.7969	0.1201	0.0830	0.7693	0.2281	0.0026	0.5367	0.0700	0.3933
0.7612	0.2179	0.0209	0.7569	0.1026	0.1405	0.5422	0.2942	0.1636	0.5322	0.1513	0.3165	0.6342	0.1343	0.2315
	$F = 0.2050$			$F = 0.1344$			$F = 0.0848$			$F = 0.0400$			$F = 0.0285$	

Table 3. Random deviates for probability tests on C_5 and C_4.

Blending	Between-Parent Crossover	Within-Parent Crossover		Mutation	
Parent 1	Parent 1	Parent 1	Parent 2	Parent 1	Parent 2
0.9563	0.7198	0.2807	0.4561	0.8851	**0.0049**
0.5209	0.7448	0.5894	0.5453	0.9218	0.9710
0.0158	0.6830	**0.0179**	0.6338	0.1266	0.2420
0.6306	0.6609	0.1799	0.7857	**0.0155**	0.0400
0.1155	0.4510	0.6985	0.0769	0.3290	0.6497
0.6471	**0.0017**	0.6031	0.5732	0.8324	0.0840
0.8924	0.0509	0.9297	**0.0095**	0.6498	0.3173

The first generation is summarized in Table 4 where boldfacing indicates that a gene was changed. For the blending operator, Row 3 of C_5 is blended with a random row (e.g., Row 4) of C_4 with random blending value $\beta = 0.4438$. For the between-parent crossover, Row 6 of C_5 will crossover with a random row (e.g., Row 3) of C_4. For the within-parent crossover, Row 3 of C_5 and Row 7 of C_4 will have within-row crossovers based on random permutations of $(1, 2, 3)$ (e.g., $(3, 2, 1)$, and $(1, 3, 2)$, respectively). For mutation, Row 4 of C_5 and Row 1 of C_4 will be mutated. Then, Components 1, 2, and 3 are randomly selected from Row 4 of C_5 (e.g., $(3, 1, 2)$) and Row 1 of C_4 (e.g., $(2, 3, 1)$), respectively. Two $N(0, 0.1)$ values are generated (e.g., $\varepsilon_{43} = -0.0175$, and $\varepsilon_{12} = 0.102$, respectively), yielding new components $x_{43}^* = 0.43 - 0.0175 = 0.4125$ and $x_{12}^* = 0.1979 + 0.102 = 0.2999$. Respectively, the final component values for offspring Row 4 of C_5 and Row 1 of C_4 are $(0.5534, 0.2885, 0.1581)$ and $(0.4721, 0.2999, 0.2800)$.

The reproduction process continues for the C_2 and C_3 pair, yielding offspring chromosomes C_2^* and C_3^*. The objective function F values at the end of the first generation are summarized in Table 5. Because parents C_2 have a larger F value than its offspring C_2^*, C_2 will appear again in the next generation. However, offspring C_3^*, C_4^*, and C_5^* will replace their parents C_3, C_4, and C_5 in the next generation because they have larger objective function F values. The elite chromosome C_1 is now replaced with C_3^*, which now has the largest F value. Hence, chromosome C_1 will be a part of the reproduction process in the second generation. This process will continue for I generations.

Table 4. First generation reproduction for chromosomes C_5 and C_4.

After Blending						After between-Parent Crossover						After within-Parent Crossover						After Mutation					
C_5			C_4			C_5			C_4			C_5			C_4			C_5			C_4		
x_1	x_2	x_3	x_1	x_2	x_3	x_1	x_2	x_3	x_1	x_2	x_3	x_1	x_2	x_3	x_1	x_2	x_3	x_1	x_2	x_3	x_1	x_2	x_3
0.6705	0.0536	0.2759	0.5160	0.1979	0.2861	0.6705	0.0536	0.2759	0.5160	0.1979	0.2861	0.6705	0.0536	0.2759	0.5160	0.1979	0.2861	0.6705	0.0536	0.2759	0.4721	0.2999	0.2800
0.5256	0.0173	0.4571	0.6487	0.1443	0.2070	0.5256	0.0173	0.4571	0.6487	0.1443	0.2070	0.5256	0.0173	0.4571	0.6487	0.1443	0.2070	0.5256	0.0173	0.4571	0.6487	0.1443	0.2070
0.5239	0.2036	0.2725	0.4404	0.0682	0.4914	0.5225	0.2036	0.2739	0.4467	0.0600	0.4933	0.5225	0.2036	0.2739	0.4467	0.0600	0.4933	0.5225	0.2036	0.2739	0.4467	0.0600	0.4933
0.4754	0.0946	0.4300	0.5176	0.1950	0.2874	0.4754	0.0946	0.4300	0.5176	0.1950	0.2874	0.4754	0.0946	0.4300	0.5176	0.1950	0.2874	0.5534	0.2885	0.1581	0.5176	0.1950	0.2874
0.6723	0.0354	0.2923	0.6744	0.1869	0.1387	0.6723	0.0354	0.2923	0.6744	0.1869	0.1387	0.6723	0.0354	0.2923	0.6744	0.1869	0.1387	0.6723	0.0354	0.2923	0.6744	0.1869	0.1387
0.5367	0.0700	0.3903	0.7693	0.2281	0.0026	0.5304	0.0782	0.3914	0.7693	0.2281	0.0026	0.5304	0.0782	0.3914	0.7693	0.2281	0.0026	0.5304	0.0782	0.3914	0.7693	0.2281	0.0026
0.6342	0.1343	0.2315	0.5322	0.1513	0.3165	0.6342	0.1343	0.2315	0.5322	0.1513	0.3165	0.6342	0.1343	0.2315	0.5322	0.1565	0.3113	0.6342	0.1343	0.2315	0.5322	0.1565	0.3113

Table 5. First generation summary.

Initial Population		Offspring			Next Generation	
Chromosome	F	Chromosome	Partner	F	Chromosome	F
C_1	0.2050	C_1	elite	0.2050	C_1	0.2050
C_2	0.1344	C_2		0.0285	C_2	0.1344
C_3	0.0848	C_3^*		0.3169	C_3^*	0.3169
C_4	0.0400	C_4^*		0.0827	C_4^*	0.0827
C_5	0.0285	C_5^*		0.0486	C_5^*	0.0486
					elite	

4. A Three-Component Mixture Example

Consider the three-component mixture experiment in Crosier [29]. The three SCCs are

$$0.2 \leq x_1 \leq 0.7; \quad 0.05 \leq x_2 \leq 0.65; \quad 0.1 \leq x_3 \leq 0.3.$$

The boundary is formed by six vertices. The full model under consideration is the quadratic mixture model. Weighted IV optimal designs with $N = 7$, 10, and 14 points were generated by the GA using a relative weighting factor $R = 100$. These designs are labeled GA7, GA10, and GA14, respectively. The set of possible reduced models are those in Table 1. We choose $M = 25$ designs to comprise the GA population of chromosomes. In a preliminary study, we investigated the choice of GA parameter values and the number of generations to observe convergence when running the GA. Based on these results, the number of generations was set to 6000, and the ranges of genetic parameter values were $0.02 \leq \alpha_b \leq 0.2; 0.03 \leq \alpha_{bc}, \alpha_{wc} \leq 0.15; 0.05 \leq \alpha_m \leq 0.2$, where $\alpha_b, \alpha_{bc}, \alpha_{wc}$, and α_m represent the success probabilities of blending, between-parent crossover, within-parent crossover, and mutation, respectively. Initially, the parameter values are set to the largest level, and after every 1500 generations, these parameter values are systematically reset to smaller values. The optimal choice operators and the GA parameters across operators is an open research area.

Because no software package can generate weighted IV-optimal designs, we studied how well our weighted IV-optimal designs would fare with respect to D-, A-, G-, and IV-efficiencies. Therefore, we generated IV-optimal designs using the design-generating statistical software package Design-Expert 11 (Stat-Ease (2017)) to generate their versions of IV-optimal designs. These 7, 10, and 14 point designs are labeled DX7, DX10, and DX14, respectively.

The mixtures from the GA and DX designs in the constrained SCC mixture space are shown in Figure 1. The patterns of points are quite different. The majority of points in the GA designs are distributed around the boundary with replications near the overall centroid, while the DX designs also tend to place design points on the boundary but at different locations and the number of replicates differ from the GA designs. Additionally, the DX designs select multiple mixtures as interior points, while the interior points of the GA designs are concentrated with replicates at only one mixture. The fact that the mixtures selected differ between DX and GA designs is based on the fact the DX designs were constructed only to optimize with respect to one model (the quadratic mixture model), while the GA design considered eight potential models when optimizing. Additionally, DX designs were generated using exchange algorithms that restrict mixture selection from only a finite candidate set of mixtures, while GA designs explore the continuum of points in the entire mixture space.

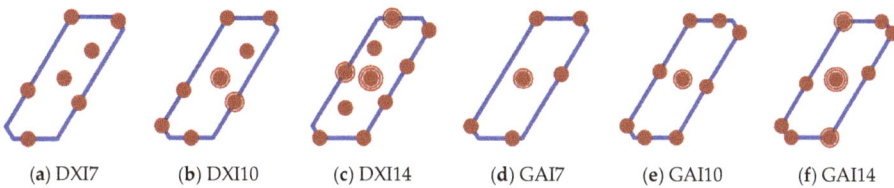

| (a) DXI7 | (b) DXI10 | (c) DXI14 | (d) GAI7 | (e) GAI10 | (f) GAI14 |

Figure 1. The genetic algorithm (GA) designs and Design-Expert (DX) designs.

The next comparison is summarized in Table 6, which contains D-, A-, G- and IV-efficiencies for the six designs. Based on the D-, A-, G-, and IV-efficiency, the GA designs are superior to the DX designs. What is interesting about these results is that, even though Design-Expert generated designs with the goal of optimizing the IV-efficiency for the quadratic mixture model, the GA design was more IV-efficient (despite weighting over eight models) for the quadratic model.

Table 6. The D-, A-, G-, and IV-efficiency.

N	Design	D-Efficiency	A-Efficiency	G-Efficiency	IV-Efficiency
7	DX7	0.2329	0.0036	32.4052	0.2152
	GA7	**0.2729**	**0.0038**	**57.1217**	**0.2503**
10	DX10	0.2443	0.0045	37.2507	0.2291
	GA10	**0.2989**	**0.0057**	**75.7591**	**0.2642**
14	DX14	0.2284	0.0039	36.0426	0.2226
	GA14	**0.2900**	**0.0047**	**68.8525**	**0.2592**

The final comparison uses fraction-of-design-space (FDS) plots to examine model robustness. A model-robust design should perform consistently well with respect to the scaled predictions in the design region for all possible reduced models. The vertical axis represents possible scaled prediction variance $v(x_0)$ values (Equation (5)) and the horizontal axis represents the fraction of the mixture design space that has scaled prediction variance values less than or equal to $v(x_0)$. Thus, a FDS plot is equivalent to a cumulative distribution function plot but with the axes reversed. For additional information of the FDS plots for examining model robustness, see Ozol-Godfrey et al. [30].

Although an FDS plot can be made for each model, we restrict the study to the quadratic (full) mixture model and the most parsimonious linear mixture model because these curves give a manageable summary of that design's prediction variance performance at the extremes in terms of model parameters. These FDS plots are presented in Figure 2.

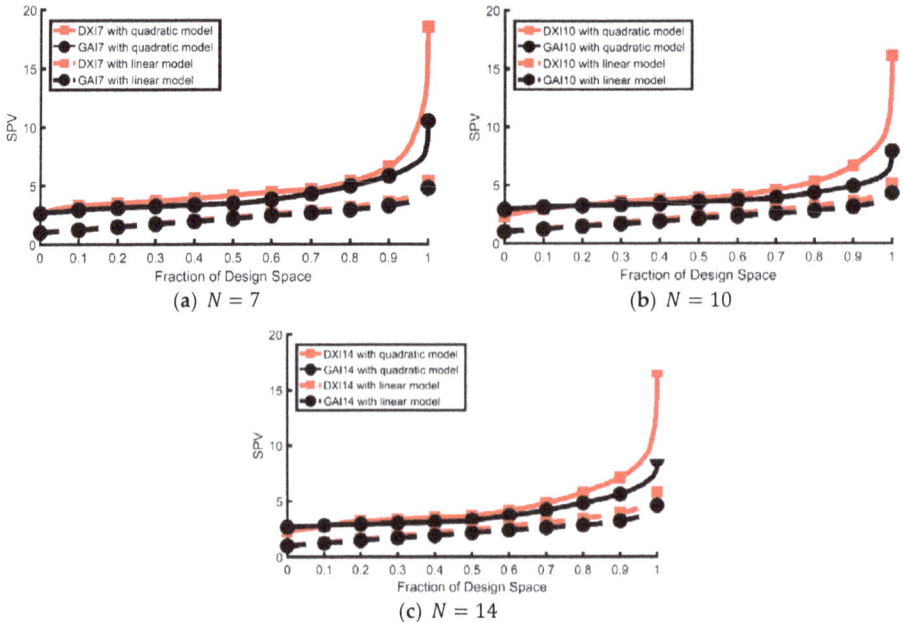

Figure 2. The fraction-of-design-space (FDS) plots for all competing designs.

For the linear mixture model, the GA and the DX designs have similar scaled prediction variance distributions for most of the design space. However, for the quadratic mixture model, the DX designs performs poorly for most of the design space compared to the GA designs. This provides evidence for why the IV-efficiencies are smaller for the GA designs compared to the DX designs. For most of the design space, the GA designs appear to have good robustness properties because it has much flatter and lower FDS curves for both the quadratic and linear mixture models, indicating more desirable prediction variance distributions.

5. Conclusions

In this paper, we introduced the concept of a weighted IV-efficiency, which was then used as the objective function of a GA when generating IV-efficient designs robust to a set of reduced models. The goal of the weighted IV-optimality criterion is to find a set of points that will minimize the weighted average of the average of the scaled prediction variance over the design region across a set of reduced models. The IV-efficiencies are calculated over a set of points randomly selected in the experimental region rather than a fixed set of points to get unbiased estimates of weighted IV-efficiencies. Additional assumptions—(i) that not all models should be weighted equally and (ii) that only models having an equal number of model parameters receive equal weight—were adopted for calculating the weighted IV-efficiency.

We studied the problem of generating weighted IV-optimal designs for mixture experiments where the experimental region in an irregularly shaped polyhedron is formed by single component constraints. The proposed methodology allows movement through a continuous region that includes highly constrained mixture regions and does not require selecting points from a user-defined candidate set of mixtures unlike traditional exchange algorithm approaches.

The GA presented in this research paper is effective for generating model-robust designs. The results from the example show that GA designs performed better than the Design Expert designs when examining the model robustness using FDS plots, as well as the D-, A-, G- and IV-efficiencies. When the experimenter believes that the initial model may not turn out to be the model adopted in the final analysis of the experimental data, GA designs based on weighted IV-efficiency are suggested because they protect against the possibility of having a final model with poor prediction variance properties.

Future research will look for alternatives to using random points to estimate the integrals in the definition of the IV-criterion, additional full and reduced model situations, and alternative schemes for weighting these models when calculating IV-efficiency.

Author Contributions: All authors designed the research. W.L. and J.J.B. performed all simulations and analyses of the results. All authors read and approved the final manuscript.

Conflicts of Interest: The authors declare no conflict of interest.

References

1. Cornell, J.A. *Experiments with Mixtures: Designs, Models, and the Analysis of Mixture Data*, 3rd ed.; John Wiley & Sons, Inc.: New York, NY, USA, 2002.
2. Smith, W.F. *Experimental Design for Formulation*; The American Statistical Association and the Society for Industrial and Applied Mathematics: Philadelphia, PA, USA, 2005.
3. McLean, R.A.; Anderson, V.L. Extreme vertices designs of mixture experiments. *Technometrics* **1966**, *8*, 447–456. [CrossRef]
4. Welch, W.J. ACED: Algorithm for the construction of experimental designs. *Am. Stat.* **1985**, *39*, 146–148. [CrossRef]
5. Snee, R.D.; Marquardt, D.W. Extreme vertices designs for linear mixture model. *Technometrics* **1974**, *16*, 399–408. [CrossRef]
6. Snee, R.D. Experiment design for mixture systems with multicomponent constraints. *Commun. Stat. Theory Methods* **1979**, *17*, 149–159.
7. Fedorov, V.V. *Theory of Optimal Experiments*; Academic Press: New York, NY, USA, 1972.
8. Mitchell, T.J. An algorithm for the construction of D-optimal experimental designs. *Technometrics* **1974**, *16*, 203–210.

9. Huizenga, H.M.; Heslenfeld, D.J.; Molenaar, P.C.M. Optimal measurement conditions for spatiotemporal EEG/MEG source analysis. *Psychometrika* **2002**, *67*, 299–313. [CrossRef]

10. Smucker, B.J.; Del Castillo, E.; Rosenberger, J.L. Exchange Algorithms for Constructing Model-Robust Experimental Designs. *J. Qual. Technol.* **2011**, *43*, 1–15. [CrossRef]

11. Borkowski, J.J. Using Genetic Algorithm to Generate Small Exact Response Surface Designs. *J. Probab. Stat. Sci.* **2003**, *1*, 65–88.

12. Heredia-Langner, A.; Carlyle, W.M.; Montgomery, D.C.; Borror, C.M.; Runger, G.C. Genetic algorithms for the construction of D-optimal designs. *J. Qual. Technol.* **2003**, *35*, 28–46. [CrossRef]

13. Heredia-Langner, A.; Montgomery, D.C.; Carlyle, W.M.; Borror, C.M. Model-robust optimal designs: A genetic algorithm approach. *J. Qual. Technol.* **2004**, *3*, 263–279. [CrossRef]

14. Juang, Y.; Lin, S.; Kao, H. An adaptive scheduling system with genetic algorithms for arranging employee training programs. *Expert Syst. Appl.* **2007**, *33*, 642–651. [CrossRef]

15. Park, Y.; Montgomery, D.C.; Fowler, J.W.; Borror, C.M. Cost-constrained G-efficient response surface designs for cuboidal regions. *Qual. Reliab. Eng. Int.* **2006**, *22*, 121–139. [CrossRef]

16. Limmun, W.; Borkowski, J.J.; Chomtee, B. Using a Genetic Algorithm to Generate D-optimal Designs for Mixture Experiments. *Qual. Reliab. Eng. Int.* **2013**, *29*, 1055–1068. [CrossRef]

17. Myers, R.H.; Montgomery, D.C.; Anderson-Cook, C. *Response Surface Methodology: Process and Product Optimization Using Designed Experiments*, 3rd ed.; John Wiley & Sons Inc.: New York, NY, USA, 2009.

18. Atkinson, A.C.; Donev, A.N.; Tobias, R.D. *Optimal Experimental Design, with SAS*; Oxford University Press: Oxford, UK, 2007.

19. Hardin, R.H.; Sloane, N.J.A. A new approach to construction of optimal designs. *J. Stat. Plan. Inference* **1993**, *37*, 339–369. [CrossRef]

20. Syafitri, U.; Sartono, B.; Goos, P. I-optimal design of mixture experiments in the presence of ingredient availability constraints. *J. Qual. Technol.* **2015**, *47*, 220–234. [CrossRef]

21. Borkowski, J.J. A comparison of Prediction Variance Criteria for Response Surface Designs. *J. Qual. Technol.* **2003**, *35*, 70–77. [CrossRef]

22. Coetzer, R.; Haines, L.M. The construction of D- and I-optimal designs for mixture experiments with linear constraints on the components. *Chemom. Intell. Lab. Syst.* **2017**, *171*, 112–124. [CrossRef]

23. Box, G.E.P.; Draper, N.R. A basis for the selection of a response surface design. *J. Am. Stat. Assoc.* **1959**, *54*, 622–654. [CrossRef]

24. Goldberg, D.E. *Genetic Algorithms in Search, Optimization and Machine Learning*; Addison-Wesley: New York, NY, USA, 1989.

25. Bäck, T.; Hammel, U.; Schwefel, H.P. Evolutionary computation: Comments on the history and current state. *IEEE Trans. Evol. Comput.* **1997**, *1*, 3–17. [CrossRef]

26. Michalewicz, Z. *Genetic Algorithms + Data Structures = Evolution Programs*; Springer-Verlag: New York, NY, USA, 1996.

27. Haupt, R.L.; Haupt, S.E. *Practical Genetic Algorithms*; John Wiley & Sons, Inc.: Hoboken, NJ, USA, 2004.

28. Borkowski, J.J.; Piepel, G.F. Uniform designs for highly constrained mixture experiments. *J. Qual. Technol.* **2009**, *41*, 1–13. [CrossRef]

29. Crosier, R.D. Symmetry in mixture experiments. *Commun. Stat.-Theory Methods* **1991**, *20*, 1911–1935. [CrossRef]

30. Ozol-Godfrey, A.; Anderson-Cook, C.M.; Montgomery, D.C. Fraction of design space plots for examining model robustness. *J. Qual. Technol.* **2005**, *37*, 223–235. [CrossRef]

© 2018 by the authors. Licensee MDPI, Basel, Switzerland. This article is an open access article distributed under the terms and conditions of the Creative Commons Attribution (CC BY) license (http://creativecommons.org/licenses/by/4.0/).

Mathematical and Computational Applications

MDPI

Article

Surrogate-Based Optimization Using an Open-Source Framework: The Bulbous Bow Shape Optimization Case

Joel Guerrero [1,*], Alberto Cominetti [1], Jan Pralits [1] and Diego Villa [2]

[1] DICCA, Dipartimento di Ingegneria Civile, Chimica e Ambientale, Università degli Studi di Genova, 16145 Genova, Italy; alb.cominetti@gmail.com (A.C.); Jan.Pralits@unige.it (J.P.)

[2] DITEN, Dipartimento di Ingegneria Navale, Elettrica, Elettronica e delle Telecomunicazioni, Università degli Studi di Genova, 16145 Genova, Italy; diego.villa@unige.it

* Correspondence: joel.guerrero@unige.it

Received: 25 May 2018; Accepted: 11 October 2018; Published: 13 October 2018

Abstract: Shape optimization is a very time-consuming and expensive task, especially if experimental tests need to be performed. To overcome the challenges of geometry optimization, the industry is increasingly relying on numerical simulations. These kinds of problems typically involve the interaction of three main applications: a solid modeler, a multi-physics solver, and an optimizer. In this manuscript, we present a shape optimization work-flow entirely based on open-source tools; it is fault tolerant and software agnostic, allows for asynchronous simulations, and has a high degree of automation. To demonstrate the usability and flexibility of the proposed methodology, we tested it in a practical case related to the naval industry, where we aimed at optimizing the shape of a bulbous bow in order to minimize the hydrodynamic resistance. As design variables, we considered the protrusion and immersion of the bulbous bow, and we used surrogate-based optimization. From the results presented, a non-negligible resistance reduction is obtainable using the proposed work-flow and optimization strategy.

Keywords: surrogate-based optimization; numerical simulations; shape morphing; bulbous bow; open-source framework

---

1. Introduction

Engineering design and product development using shape optimization can be a very daunting task, especially if the design approach is experimental, where cost and time constraints usually limit the number of prototypes that are possible to construct and test. Moreover, it is required to have an in-depth know-how of the problem being studied to choose which configurations to analyze. In this view, the continuous development of numerical methods and the increase in computing performance have suggested the use of simulation software able to model complex multi-physics problems and optimize the design space, thereby reducing the costs related to prototypes, physical experiments, and field/operational tests.

Proprietary numerical simulation tools allow very detailed analysis, but their use often requires the acquisition of expensive licenses. An alternative to commercial software is the use of free and open-source technology (e.g., GNU General Public License, BSD license, MIT license). This type of software licensing model gives users the freedom to use, read, write, and redistribute the source code, with no price tag attached. More importantly, open-source software has the same capabilities as their commercial counterparts, and it is mature enough so that it can be used in the design and certification process of new products.

With this in mind, we address in this manuscript the implementation of an open-source framework for shape optimization. In particular, we focus our attention on fluid dynamics problems applied to ship design. Nevertheless, the same methodology can be used in any engineering field. In doing so, essentially four tasks need to be addressed:

- How to efficiently and accurately simulate the physics involved.
- How to modify the geometry or mesh.
- How to search the optimal configuration.
- How to interface the applications involved in the optimization loop.

In the context of naval architecture design, the adoption of shape optimization using numerical simulations is not new. The advantages of hull optimization and related devices optimization (such as propellers and hydrofoils) have been demonstrated by several authors. Many design optimization approaches have been used in the design of traditional propellers [1,2], unconventional propulsion systems [3,4], appendages [5], and the optimization of the entire hull shape [6,7]. Recently, a lot of development has been done on the design of unconventional vessels with special mission profiles, from the standard mono-hull shapes [8,9], to unconventional multi-hulls [10,11], or hulls with special shapes, such as the Small Waterplane Area Twin Hull or SWAT [12]. However, even if these design methods have reached a satisfactory maturity level, the hull shape optimization is still confined to the research level due to its high complexity.

One of the main drawbacks of shape optimization for complex industrial applications is the large computational times required to get the outcomes. In naval architecture design (where we are mainly interested in evaluating resistance and seakeeping, among other things), this is usually overcome by the use of highly efficient low-fidelity or medium-fidelity methods, such as the boundary element methods (BEM). However, such approaches are not able to predict the wave pattern generated by the ship, do not include viscous effects, or neglect other nonlinear behaviors (which can be important in order to obtain a proper performance prediction for certain ships). Even though successful applications with low-fidelity methods have been previously reported in the literature [10], the adoption of high-fidelity solvers (such as finite volume method or finite element method solvers) in the design loop can increase the accuracy and reliability of the design process. In this field, many studies have been conducted, for example, Serani et al. [13] successfully used a viscous solver for addressing the shape optimization of a high-speed catamaran with excellent results; however, the computational cost in terms of CPU time (for the whole optimization loop and each single simulation) and the complexity of the optimization loop was too big to be implemented by a manufacturer (particularly small ones, which have limited resources). In light of this, innovative optimization methods to simplify and speed up the entire optimization process are under investigation, such as multi-level optimization [14], surrogate-based optimization [15], proper orthogonal decomposition and dynamic mode decomposition [16], machine learning for interactive design and fluid dynamics simulations [17], data-driven simulations and optimization [18], and generative design [19].

In the shipbuilding industry, the bulbous bow shape is usually designed to reduce the ship wave-making resistance. The hull drag resistance can be split into three main contributions: the wave-making drag, the friction drag, and the wake drag. Depending on the ship speed, the contribution of each component to the total resistance value can be drastically different. Considering the present case, the wave-making component is one of the most important; therefore, its reduction by optimizing the bulbous bow shape can give a not negligible contribution to the whole design (for example, a reduction of some percentage of the fuel consumption). Following the guidelines proposed by Kracht [20], the bulbous bow shape can be defined by six non-dimensional parameters. With the goal of decreasing the computational cost by reducing the design space, and at the same time being able to represent the bulbous bow shape without losing geometrical information, only two design parameters are considered in this study. Nevertheless, the proposed approach can be extended to more complex cases where, for instance, one is interested in having more local control of the bulbous bow shape by using more design variables. Previous work addressing bulbous bow optimization using computational fluid dynamics (CFD) can be found in References [21–25]. The aforementioned sources show the benefits of CFD optimization, but most of them fail in explaining the importance of the application interfacing tool, automation of the loop, and real-time analytics.

Hereafter, we propose a vertical work-flow that could potentially simplify and automate the design and optimization process. All the tools used in this work are open-source, and this translates to cost savings, as no commercial licenses are used. To generate the domain mesh and to compute the hydrodynamic resistance of the hull (including the bulbous bow), we used the multi-physics solver OpenFOAM [26] (version 5.0). To create new geometry candidates, we used the free-form deformation application MiMMO [27]. To reduce the number of optimization iterations and to get an initial screening of the design space, we used surrogate-based optimization. The optimization algorithms and the code coupling interface is provided through the Dakota library [28] (version 6.7). Finally, all the real-time data analytics, quantitative post-processing, and data analytics were performed using Python and bash scripting. Previous attempts in coupling OpenFOAM and Dakota to deal with shape optimization problems are described in References [29–36], but none of them have addressed how to interface different applications in an optimization loop (besides OpenFOAM and Dakota), how to deal with concurrent simulations, work-flow automation, fault-tolerant loops, real-time data analytics, and the importance of knowledge extraction. We aim at coupling all tools needed for shape optimization studies in a streamlined work-flow, with a high degree of automation and flexibility, and using efficient optimization techniques that allow engineers to explore, exploit, and optimize the design space from a limited number of training observations.

2. Description of the Optimization Framework and Optimization Strategy

In a typical shape optimization study, many applications can live together (e.g., solid modeler, shape morpher, mesh generator, multi-physics solver, optimizer, knowledge extraction tools, coupling utilities, and so on), making the optimization loop very difficult to set up, monitor, and control. Moreover, the optimization loop must be fault tolerant, that is, in the event of an unexpected failure of the system (hardware or software), the user should be able to restart from a previously saved state. The optimizer should also be flexible, in the sense that it should provide a variety of optimization methods, design of experiment techniques, and code coupling interfaces to different simulation software and programming languages. Finally, to reduce the long execution times inherent to computational fluid dynamics (CFD) optimization studies, the optimization loop should be able to work in parallel and deploy many simulations at the same time. Furthermore, everything should be done without overwhelming the user. We aim at addressing the findings and recommendations reported in the NASA contractor report *"CFD Vision 2030 Study: A Path to Revolutionary Computational Aerosciences"* [37], where the authors state: *"A single engineer/scientist must be able to conceive, create, analyze, and interpret a large ensemble of related simulations in a time-critical period (e.g., 24 hours), without individually managing each simulation, to a pre-specified level of accuracy"*.

In this work, we successfully coupled many applications used in traditional CFD work-flows to efficiently conduct general CFD optimization studies. We used the library Dakota [28,38] as the optimizer and application interface tool. The Dakota library provides a flexible and extensible interface between simulation codes and iterative analysis methods. This library contains many gradient-based methods and derivative-free methods for design optimization studies, uncertainty quantification, and parameter estimation capabilities. Dakota also contains many design of experiment methods to conduct sensitivity studies. With Dakota, the user is not obliged to use the optimization and space exploration methods implemented on it: one can easily interface Dakota with a third-party optimization library. The library also offers restart capabilities and solution monitoring. Most important, it is software neutral, in the sense that it can interface any application that it is able to parse input/output files via a command line interface.

The optimization loop is illustrated in Figure 1. In this work-flow, the starting point is the initial geometry which is manipulated with a free-form deformation tool to generate new shapes. The MiMMO library [27], which uses radial basis functions (RBF) to shape the geometry [39–42], was used as geometry manipulation tool. With MiMMO, the user is only asked to define a series of control points, these control points are then moved in the design space, and together with the RBF function, they generate a smooth displacement field that represents the new geometry. To have more control over the deformation, the user can define a deformable region using a control box. Then, the continuity between the region to be deformed and the area labeled as undeformed is ensured by using continuous level set functions. In Figure 2, we illustrate the original geometry (left image), where we highlight the control points and the selection box used to deform the base geometry. In this figure (right image), we also show two deformed geometries. In the figure depicted, we only used one control point to deform the geometry, but more control points can be used with no problem. It is worth mentioning that in this framework, we only take one initial geometry. Then, starting from this geometry, we can generate many variations of it. In general, the shape morphing task is computationally inexpensive. We would like to emphasize that, at this point, any geometry manipulation tool can be used; the main reasons in using MiMMO are that the input files used by this library can be easily parameterized and it also offers mesh morphing capabilities (which are not discussed in this manuscript).

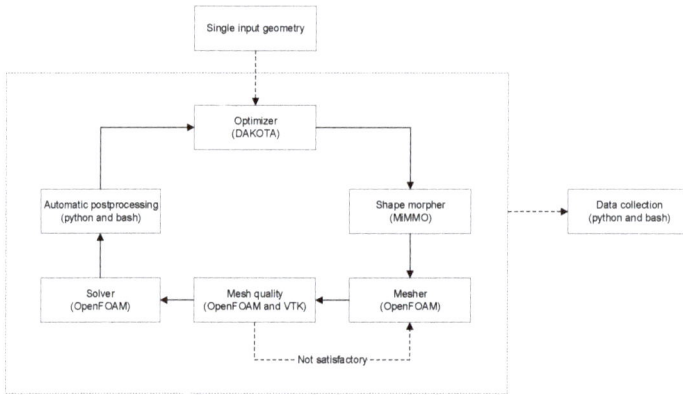

Figure 1. Optimization loop. Notice that it only takes one single geometry. During the whole optimization loop, data are continuously collected.

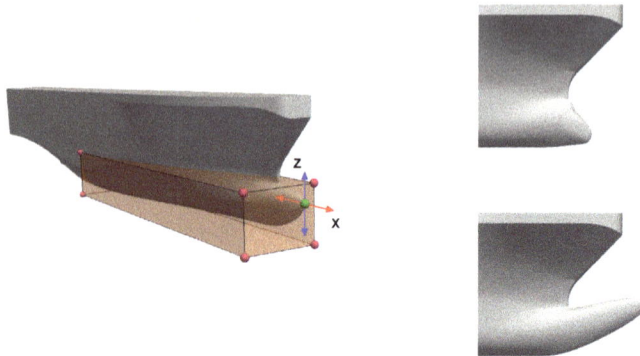

Figure 2. Left: undeformed geometry. The green sphere represents the control point used to deform the geometry; this control point can move in the plane XZ. The surface region within the selection box is free to deform. Right: two deformed geometries.

After generating the input geometry, we can move to the meshing phase. To mesh the computational domain, we used the meshing tools distributed with the open-source OpenFOAM library [26,43], namely, blockMesh and snappyHexMesh. These tools generate high-quality hexahedral-dominant meshes and can be easily parameterized. The whole meshing process is fully automatic and, to some extent, fault tolerant. During the meshing stage, we always monitored the quality of the far-field mesh, the quality of the inflation layers close to the body, and the transition between regions with different cell sizes. In the case of mesh quality problems (such as high non-orthogonality, high skewness, too aggressive expansion ratios between cells, or discontinuous inflation layers in critical areas), the mesh was automatically recomputed using more robust predefined parameters, which generates better meshes but at the cost of a higher cell count. The meshing parameters were determined using the original geometry, and two additional geometries representing the worst case scenarios (highly deformed bulbs). The possibility of using the meshing tools in parallel and concurrently allowed us to obtain fast outcomes.

All the numerical simulations were conducted using the open-source OpenFOAM library [26,43], which is a multi-physics solver based on the cell-centered finite volume method. This library has been extensively validated, counts with a wide community, is highly scalable in parallel architectures, and the simulation setup can be easily parameterized. As soon as the optimization loop was started, the simulations did not require continuous user supervision. Several quantities of interest (QoI) were sampled during each computation, and in the case of anomalies (such as high oscillations in the forces, unbounded quantities, or mass imbalance), the simulation parameters were automatically changed to a more robust and stable discretization scheme. The default and modified solver parameters were determined after conducting an extensive validation of the solver (as explained in Section 4). All the simulations were run in parallel, and as the optimization method used did not require the computation of gradients, we were able to run many simulations concurrently. The whole framework was deployed in an out-of-the-box workstation, with 16 cores and 128 GB of RAM. As all the tools used require only publicly available compilers and libraries, the same framework can be easily deployed in distributed memory HPC architectures.

At this point, we need to define the optimization strategy. One major obstacle to the use of optimization in CFD is the long CPU times of the simulations and, depending on the method used, the need for computing gradients or the need for running a large number of simulations to obtain the design space sensitivities. Therefore, due to the long computational times and the lack of analytic gradients, optimization in CFD is a slow iterative-converging task. With this in mind, we need to choose an optimization method that is reliable, computationally affordable, and will meet our deadlines.

Gradient-based methods have good converging rates, but they require the computation of the gradients (and depending o the method used, they might require the computation of the Hessians), and in the presence of a large design space these quantities can be very expensive to compute. Gradient-based methods also require some basic knowledge of the problem, as well as some input parameters (e.g., starting point in the parameter space, bounds of the design variables, iterative and gradients tolerances, how to compute the derivatives, and so on), and they do not give much information regarding the global behavior of the design space. Additionally, they do not guarantee the convergence to a global optimal value. On the other hand, global derivative-free methods have slow convergence rates, but they are more likely to converge to the global optimal value than local methods. Also, since they do not require the computation of the gradients, they behave quite well in the presence of numerical noise. However, to reach the optimal value, they need to perform a large number of function evaluations, which can impose serious time limitations, especially if we are running unsteady CFD simulations.

Another approach to conduct optimization is the so-called surrogate-based optimization (SBO), which is the method used in this study. The SBO method consists of constructing a mathematical model (also known as a surrogate, response surface, meta-model, emulator) from a limited number of observations (CFD simulations, in our case). After building the surrogate, the optimization can be performed at this level. Furthermore, during the design space exploration process, SBO provides a better global understanding of the problem, allowing the use of exploratory data analysis techniques and machine learning methods to get more insight, discover hidden patterns, and detect anomalies in the design space. In Figure 3, we illustrate the basic work-flow of the SBO method. The work-flow is rather simple; we start by conducting a limited number of simulations (design space exploration). After gathering the information at the design points, we can use those observations to construct the mathematical model or surrogate. At this point, we can validate the surrogate by comparing the values obtained from the high-fidelity simulations against the predicted values using the surrogate, or we can refine the surrogate by adding more training points. After building, validating, and improving the meta-model, we can use any optimization method to find the optimal value. Working at the surrogate level is orders of magnitude faster than working at the high-fidelity level [44].

It is important to mention that more sophisticated SBO techniques exist, such as efficient global optimization (EGO), gradient-enhanced meta-models, and adaptive sampling response surfaces [38, 44–47]. We did not use any of those state-of-the-art SBO approaches, because we were more interested in showing that complicated code coupling and complex optimization studies using traditional meta-models methods could be achieved using the Dakota library. However, the library gives the possibility to use EGO and gradient-enhanced Kriging emulators; if these methods are not enough, the user can add an external library to the optimization loop.

As already stated, surrogate models are inexpensive approximate models that are intended to capture the salient features of expensive high-fidelity experiments or observations. To construct the surrogate or meta-model, many methods are available, just to name a few: polynomial regression, Kriging interpolation, radial basis functions, neural networks, adaptive splines, and so on. In this work, we used Kriging interpolation, which is an interpolation method for which the interpolated values are modeled by a Gaussian process [44,48]. The implementation details of the Kriging interpolation method used in this work (universal Kriging), can be found in References [38,44,45,48–52]. In the Kriging interpolation method, the meta-model is forced to pass by all the observations, and in the presence of noise in the response, the surrogate can be smoothed. Kriging interpolation is well fitted for engineering design optimization problems which usually are nonlinear, multivariate, multi-modal, and noisy.

Sampling plan
Design space exploration

Observations
High-fidelity simulations

Construct surrogate

Validate surrogate
Improve surrogate

Optimize at the surrogate level
using any optimization method

Design sensitivities collection

Add new training points

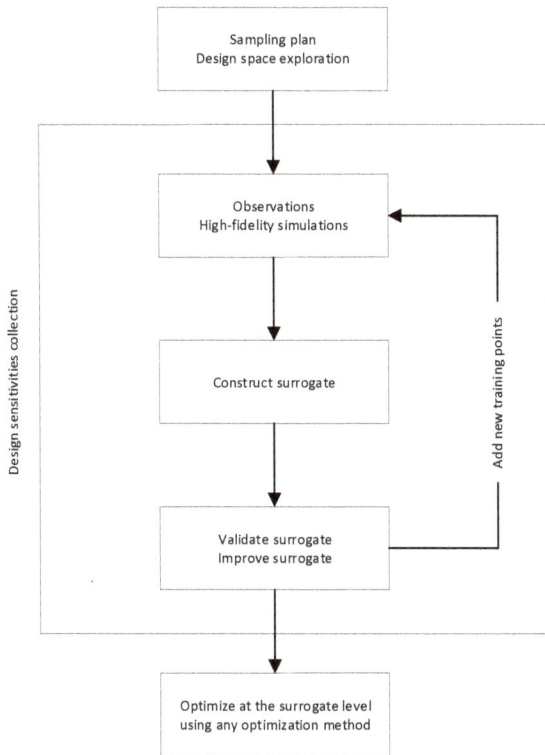

Figure 3. Surrogate-based optimization work-flow.

To get a better idea of how the SBO method works, let us use a toy function to go through every step of the work-flow. We will use the Branin function, which is defined as follows:

$$f(x,y) = \left(y - \frac{5.1}{4\pi^2}x^2 + \frac{5}{\pi}x - 6\right)^2 + 10\left(1 - \frac{1}{8\pi}\right)cos(x) + 10$$

$$s.t. \quad -5 \le x \le 10$$

$$s.t. \quad 0 \le y \le 15$$

(1)

this equation with the given bounds has three minima, located at

$$(x,y) = (-\pi, 12.275), (\pi, 2.275), (9.42478, 2.475)$$

(2)

where the minimum value in the three locations is equal to

$$f(x,y) = 0.397887$$

(3)

The Branin function (which is highly nonlinear and multi-modal) can easily represent the output of a CFD study. In SBO, the first step consists of exploring the design space in a precise and economical way. In Figure 4, we present the output of two sampling methods, namely, a full-factorial experiment and a space-filling experiment using Latin hypercube sampling (LHS) [44]. In each point illustrated in Figure 4 (bottom images), high-fidelity function evaluations were computed.

The next step consists of generating the surrogate model using the information gathered from the sampling plan. In Figure 4, we show the surrogates for both sampling experiments and the analytical solution. As it can be seen, there is no discernible difference between both models and the analytical solution, and this is an indication that the models are accurate. However, to construct a well-fitted surrogate, the full-factorial experiment needs more observation points (especially in high multi-dimensional cases, where the number of experiments increases to the power of the number of design variables), and this translates into higher computing times. It is clear that the quality of the surrogate depends on the number of observations; therefore, if at this point we observe differences between the meta-model and the analytical solution, we can add new points and compute a better surrogate. In SBO, the model can be trained as we gather the data, and we can even use multi-fidelity models and a mixture of physical experiments and computational experiments to construct the surrogate. Finally, after constructing the meta-model, we can conduct the optimization study at the surrogate level using any optimization method.

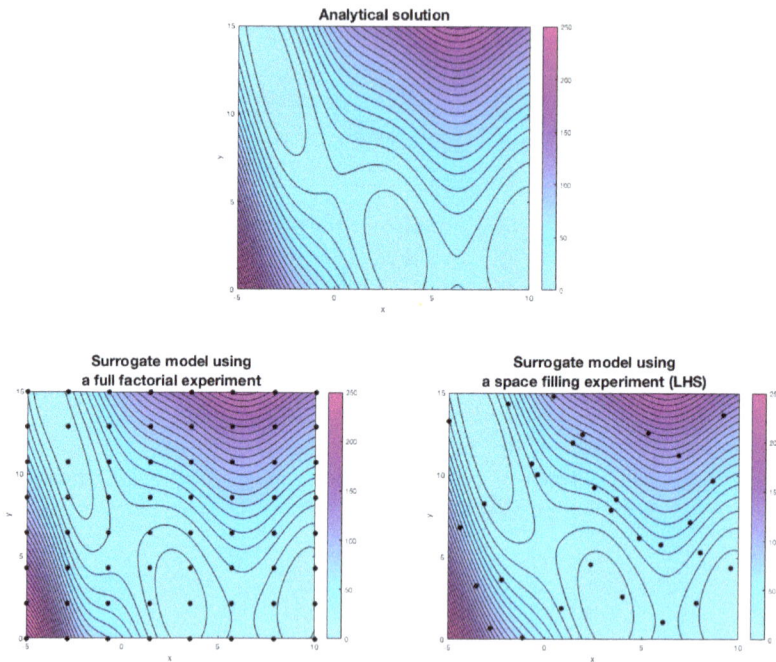

Figure 4. Top: analytical solution. Bottom-left: surrogate model using a full-factorial experiment with 64 training points. Bottom-right: surrogate model using a space-filling experiment with 30 training points (Latin hypercube sampling, LHS).

To demonstrate the advantages of using SBO, we compared its performance against the performance of a gradient-based method (method of feasible directions [53]), and a derivative-free method (single-objective genetic algorithm or SOGA [54]). All methods used were able to find the three minima; the main difference was the number of iterations used and the information provided by the user to start the optimization algorithm. The gradient-based method used 217 high-fidelity function evaluations, where we used a multi-start technique in which multiple start locations were given (this was needed in order to find the multiple minima), and we computed the gradients using forward differences. The derivative-free method used 1200 high-fidelity iterations and did not require any input information from the user (we used the recommended default parameters). Finally, the SBO method used 30 high-fidelity observations (space-filling experiment). These observations were used to construct the meta-model, then, at the surrogate level, derivative-free and gradient-based methods were used to find the minimum. The SBO method has the added value of showing information about the global design space.

As a final example, let us use the high-dimensional Rosenbrock function to demonstrate the ability of the SBO method in dealing with multivariate problems. The Rosenbrock function can be written in general form as

$$f(\mathbf{x}) = \sum_{i=1}^{d-1} \left[100 \left(x_{i+1} - x_i^2 \right)^2 + (x_i - 1)^2 \right]$$

$$s.t. \quad -2 \le \mathbf{x} \le 2$$

(4)

this equation with the given bounds has one global minimum, defined as follows:

$$f(\mathbf{x}) = 0, \text{ at } \mathbf{x} = (1, ..., 1)$$

(5)

To conduct the SBO study, we proceed in the same way as for the Branin function. At the high-fidelity level, we used two gradient-based methods, namely, the method of feasible directions [53] and the quasi-Newton BFGS update method [55]. Both methods converged to the global minimum, but the BFGS method showed a better convergence rate. At the surrogate level we used three methods: two gradient-based methods (same methods as in the high-fidelity study) and one derivative-free method, namely, the DIRECT method (DIvide a hyperRECTangle [56]). The surrogate was constructed using 500 observations, which roughly correspond to the same number of function evaluations in the best case of the high-fidelity optimization studies (refer to Table 1). It is noteworthy that as we have a mathematical representation of the design space, the gradients were computed analytically during the SBO study.

The outcome of this SBO study is presented in Table 1. As it can be seen in this table, at the surrogate level, we get values close to the global minimum; but, due to a noisy and under-fitted surrogate, the optimal value is under-predicted. We can also observe that the values of the design variables 4–6 are a little bit over-predicted, but in general they follow the trend of the high-fidelity values. One way to improve the results could be to construct a better surrogate by using a larger number of observations. However, as the number of observations will be much higher than the number of function evaluations in the best case of the high-fidelity optimization studies, the use of SBO is not very attractive, especially if the computation of the QoI is expensive. Another way to improve the surrogate is by using infilling, which consists of adding new points to the training set and then reconstructing the surrogate. For example, we can add a new point in the location of the predicted optimal value and a few new points in areas where high nonlinearities are observed or are close to the optimal value, recompute the surrogate, and find the new optimal value. It is worth mentioning that we also used 200 experiments to conduct the SBO study, but due to the coarseness of the space-filling sampling, the surrogate was under-fitted; therefore, the optimization studies failed to converge or give similar trends to the ones shown in Table 1.

Table 1. Outcome of the optimization study of the high-dimensional Rosenbrock function. In the table, observations refer to the number of experiments used to construct the surrogate. Note that the same starting point ($\mathbf{x} = 0$) was used for all the design variables in the gradient-based optimization studies. In the table, DV stands for design variable, HF stands for high-fidelity simulations, SBO stands for surrogate-based optimization, MFD stands for method of feasible directions, QN stands for quasi-Newton BFGS method, and DR stands for division of rectangles derivative-free method.

-	HF-MFD	HF-QN	SBO-MFD	SBO-QN	SBO-DR
DV-1	0.999	0.999	0.958	0.958	0.993
DV-2	0.999	0.999	0.992	0.994	1.007
DV-3	0.998	0.999	1.004	0.999	0.998
DV-4	0.997	0.998	1.115	1.091	1.037
DV-5	0.995	0.996	1.208	1.189	1.058
DV-6	0.989	0.992	1.317	1.251	1.108
QoI	0.00003	0.00002	−48	−48	−44
Function Evaluations	1238	420	-	-	-
Observations	-	-	500	500	500

Another advantage of SBO is that we can use exploratory data analysis and machine learning techniques to interrogate the data obtained during the design space exploration stage. These techniques can be used for knowledge extraction and anomaly detection. In Figure 5, we show one of the many plots that can be used to visualize the data [57,58]. This plot is called a scatter matrix, and in one single illustration, it shows the correlation information, the data distribution (using histograms and scatterplots), and regression models of the responses of the QoI. The reader should be aware that, in the context of design space exploration, exploratory data analysis and machine learning methods should not be used with biased data, e.g., data coming from a gradient-based or derivative-free optimization study.

By conducting a quick inspection of the scatterplot matrix displayed in Figure 5, we can see evidence that the data is distributed uniformly in the design space (meaning that the sampling plan is unbiased), and this is evidenced by the diagonal of the plot. By looking at the scatterplot of the experiments (lower triangular part of the matrix), we see the distribution of the data in the design space. If at this point we observe regions in the design space that remain unexplored, we can simply add new training points to cover those areas. In the case of outliers (anomalies), we can remove them from the dataset with no major inconvenience. However, we should be aware that outliers are telling us something, so it is a good idea to investigate the cause and effect of the outliers. In the upper triangular part of the plot, the correlation information is shown (Spearman correlation, in this case). This information tells us how correlated or uncorrelated the data are. For example, by looking at the last row of the plot that shows the response of the QoI, if we see a strong correlation between two variables, it is clear that this variable cannot be excluded from the study. The opposite is also true, that is, uncorrelated variables can be excluded from the study; therefore, the complexity of the problem is reduced. One should be aware that for data exclusion using correlation information, we should use the response of the QoI and not the design space distribution of the sampling plan. For a well-designed experiment, the design space distribution is uncorrelated. Additionally, the last row of the scatter matrix plot also shows the regression model (a quadratic model, in this case). As can be seen, this simple plot can be used to gather a deep understanding of the problem.

To close the optimization framework discussion, we would like to stress the fact that the optimization loop implemented is fault tolerant, so in the event of hardware or software failure, the optimization task can be restarted from the last saved state. Moreover, during the optimization loop, all the data monitored are made available immediately to the user, even when running multiple simulations at the same time (real-time data analytics). Therefore, anomalies and trends can be detected in real time, and corrections/decisions can be made.

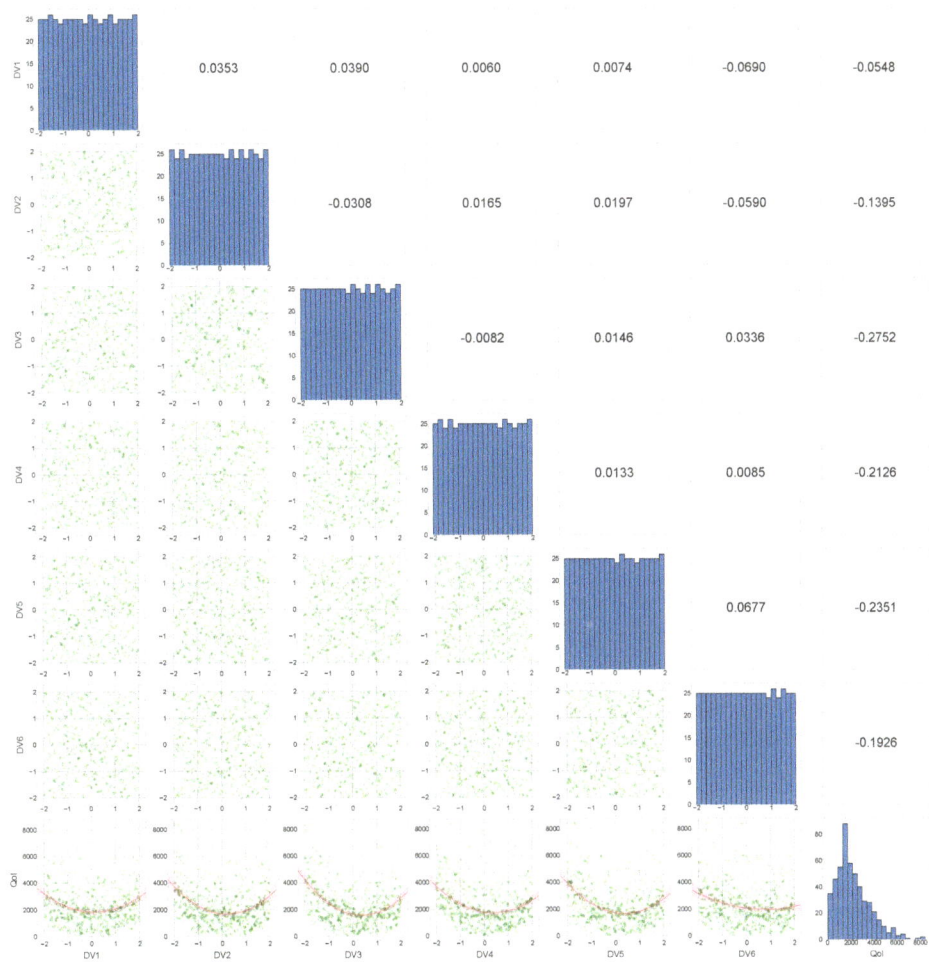

Figure 5. Scatterplot matrix of the high-dimensional Rosenbrock function space exploration study ($d = 6$). The Spearman correlation is shown in the upper triangular part of the matrix. In the diagonal of the matrix, the histograms showing the data distribution are displayed. In the lower triangular part of the matrix, the data distribution is shown using scatterplots. In the last row of the matrix plot, the response of the quantity of interest (QoI) in function of the design variables is illustrated, together with a quadratic regression model.

3. Description of the CFD Model

All the numerical simulations described in this manuscript were conducted using the open-source OpenFOAM library [26,43] (version 5.0). This toolbox is based on the cell-centered finite volume method, and consists of a series of numerical discretization schemes, linear systems solvers, velocity–pressure coupling methods, and physical models that can be used to solve multi-physics problems. To find the approximated numerical solution of the governing equations, and to deal with the physics of segregated multi-phase flows, we used the solver interFoam, which is distributed with OpenFOAM. This solver uses the volume of fluid (VoF) phase-fraction method to resolve the interface between phases and has extensive turbulence modeling capabilities. In this work, we used the $k - \omega$ SST turbulence model, as described in Reference [59].

3.1. Reference Geometry and Modeling Assumptions

The baseline geometry used in this study was provided by the ship manufacturer Fincantieri and is illustrated in Figure 6. The model is scaled 1/20 in reference to the full-scale prototype, and experimental data from towing tank tests were also provided. Due to the intellectual property of Fincantieri on the prototype and the experimental data, this information cannot be disclosed. Therefore, all the reference values are presented using non-dimensional numbers.

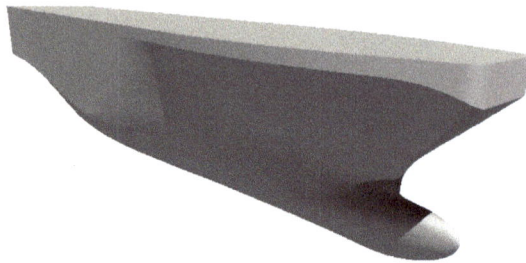

Figure 6. The baseline geometry.

During the experimental campaign, the following Froude number values were used: 0.294, 0.312, 0.331, 0.349, 0.367, and 0.386. The Froude number is defined as follows:

$$Fr = \frac{V}{\sqrt{gL}} \tag{6}$$

where V is the velocity, g is the gravity, and L is a reference length. Hereafter, we used the length between perpendiculars or L_{PP} (in naval design, L_{PP} is defined as the distance measured along the load waterline between the after and fore perpendiculars) as the reference length.

The setup of the towing tank tests allowed for the model to dynamically adjust its trim angle (or pitch angle). Conversely, the simulations reported in this manuscript were performed with a fixed trim condition. Even if this simplification can affect the accuracy of the force prediction, it can be considered minor in the context of this optimization study. This is chiefly due to the fact that the bulb shape modifications can be assumed to be uninfluential to the final dynamic trim. Furthermore, low trim angle values are recorded in the experimental data. These values are less than two degrees for the most critical case, that is, high Froude number (or off-design conditions). In addition, the use of a rigid-body motion solver would have considerably increased the computational requirements and CPU time.

Hereafter, we summarize the main assumptions used during this study:

1. All simulations were performed at model scale (1/20).
2. No propeller nor appendages were modeled (bare hull).
3. Only half of the hull was simulated (we used symmetry).
4. A fixed trim condition of zero degrees was imposed.
5. All the simulations were conducted in calm water conditions and no incoming waves.
6. The thermophysical properties of the working fluids are constant (refer to Table 2).

7. The simulations were conducted at the same Froude number as in the towing tank experiments.

Table 2. Phases transport properties used in this study.

Phase	Density ρ $[\frac{kg}{m^3}]$	Kinematic Viscosity ν $[\frac{m^2}{s}]$
Water	998.3	1.02×10^{-6}
Air	1.2	1.48×10^{-5}

As we are conducting a bulbous bow optimization study, we must define the design variables. In this study, we used the parameters proposed by Kracht [20]. Kracht defined six parameters to characterize a bulbous bow, namely, length, depth, and breadth (all linear); the cross-section; and lateral and volumetric parameters (which are nonlinear). Hereafter, the bulb geometry was parameterized using the length parameter C_{LPR} (or protrusion) and depth parameter C_{ZB} (or immersion). These parameters are defined as follows (as illustrated in Figure 7):

$$C_{LPR} = \frac{L_{PR}}{L_{PP}}, \qquad C_{ZB} = \frac{Z_B}{T_{FT}}, \tag{7}$$

These two parameters were enough to deform the geometry, and as we only used two design variables, the design space was greatly simplified. The choice of these two parameters (in reference to Figure 7), was driven by the interest of only moving point B in the plane *x-z*, as recommended by Fincantieri. This together with the suggested bounds ($0.031 \leq C_{LPR} \leq 0.131$ and $0.283 \leq C_{ZB} \leq 1.038$) will ensure that the bulbous bow volume variations remain within the acceptable values, without the need for introducing additional design variables and nonlinear constraints.

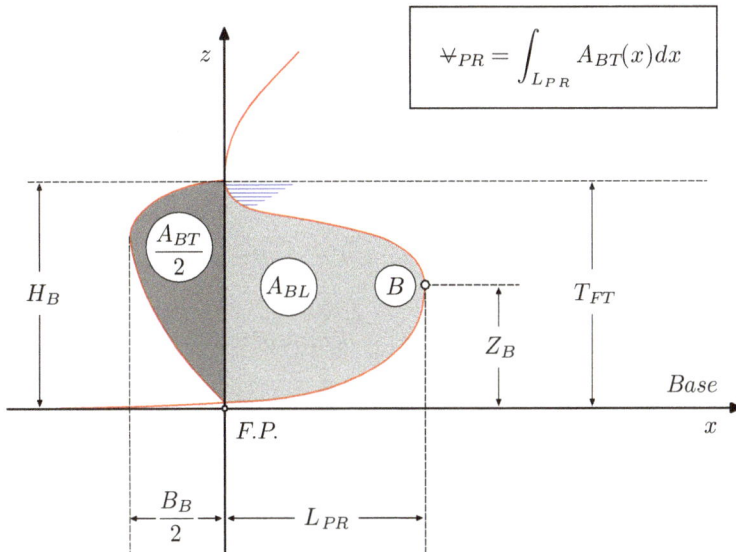

Figure 7. Sketch showing the characteristic lengths used to define the geometry parameters. Adapted from Reference [20].

3.2. Computational Domain and Boundary Conditions

The computational domain is aligned with the reference system and is made up of a rectangular box in which half of the hull geometry was placed (as depicted in Figure 8). In the domain, the *x-max* face represents the inlet (where air and water enter the domain at a given velocity), the *x-min* face is the outlet (where air and water go out of the domain and the water level is maintained at the same level of the inlet level), the *y-max* face is the symmetry plane, the *y-min* face is a lateral slip wall, the *z-min* face is the bottom of the domain modeled as a slip wall, and the *z-max* face is the top of the domain, which is defined as an opening where only the air phase is allowed to enter or exit the domain. The inlet and outlet patches are located $2.5 \times L_{PP}$ and $6 \times L_{PP}$ away from the bow and stern. The top and bottom faces are placed at $1 \times L_{PP}$ and $3 \times L_{PP}$ away from the sea level. The lateral face is placed at $5 \times L_{PP}$ away from the symmetry plane to avoid lateral reflection of the Kelvin waves. In general, the boundaries were placed far enough of the hull surface so there are no significant gradients normal to the boundaries or wave reflection. When setting the domain, we followed the practical guidelines given in Reference [60].

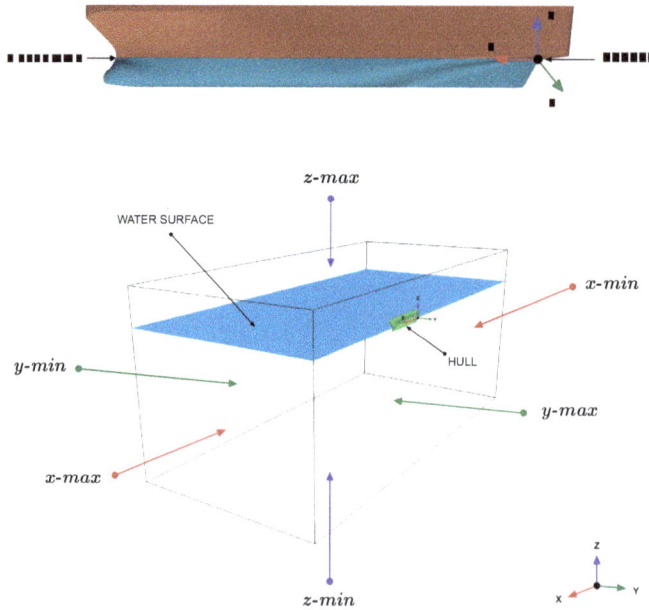

Figure 8. Top: coordinate system. Bottom: computational domain and boundary faces.

The hull was modeled as a no-slip wall, where we imposed wall function boundary conditions for the turbulence variables. The forces on the hull were computed by integrating the viscous and pressure forces over the hull surface. Finally, all the simulations were started using uniform initial conditions for both phases. In the domain, the water phase is located below the origin of the coordinates system, whereas the air phase is located above the coordinate system origin. All the turbulence variables were initialized following the guidelines given by Wilcox [59].

3.3. Numerical Schemes

Hereafter, we describe the numerical setup used with OpenFOAM. During this study, a robust, stable, and high-resolution numerical scheme was used. The cell-centered values of the variables are interpolated at the face locations using a second-order centered differences scheme for the diffusive terms. The convective terms are discretized by means of a second-order linear-upwind scheme, and to prevent spurious oscillations, a multi-dimensional slope limiter was used. To resolve the interface between the two phases, the second-order accurate bounded total variation diminishing (TVD) scheme of van Leer was used. For computing the gradients at the cell centers, the least squares cell-based reconstruction method was employed. The pressure–velocity coupling was achieved by means of the pressure-based PISO method. For turbulence modeling, the $\kappa - \omega$ SST turbulence model with wall functions was used [59].

As stated earlier, one of the biggest restrictions when conducting optimization studies in CFD is the long simulation times, especially if unsteady simulations are involved. To overcome this hurdle, we aimed at accelerating the solution of the governing equations by looking for steady-state solutions. In this study, we used the local time-stepping (LTS) method. In this method, the governing equations are integrated in time using the largest possible time step for each cell. As a result, the convergence to the steady state is considerably accelerated; however, the transient solution is no longer time accurate. The stability and accuracy of the method are driven by the local CFL number of each cell, which was fixed to 0.1 during the initial iterations and then gradually increased until reaching a value of 0.9. To avoid further instabilities due to large conservation errors caused by sudden changes in the time-step of each cell, the local time-step was smoothed and damped across the domain.

4. Calibration and Validation of the Solver

The purpose of this section is twofold. First, we validate the solver against the experimental data available, and secondly, we calibrate the parameters of the solver and the mesher. In other words, we look for the best parameters to be used during the optimization study, and in the case of meshing or convergence problems, these parameters are automatically adjusted without user intervention.

In Table 3, we list the cell count and average y^+ of the coarse and fine mesh used in this study. We used at least five inflation layers to resolve the boundary layer, and we targeted for a y^+ value between 40 and 200. The results obtained from transient simulations and steady simulations using the LTS method are displayed in Figure 9. The results illustrated correspond to the base geometry and a Froude number equal to 0.331. Due to the intellectual property of the experimental data, from this point on, all the results are normalized using as a reference the experimental value corresponding to this operating condition.

Table 3. Mesh cell count and average y^+.

Mesh	Number of cells	Average y^+
Coarse	≈ 800000	≈ 80
Fine	≈ 2800000	≈ 7

In Figure 9, we can observe that the outcome of the steady simulations using both meshes are in good agreement with the experimental values. Both solutions reached a steady state approximately after 4000 iterations. Nevertheless, we let the simulations run for longer times to ascertain that we reached good iterative convergence. Regarding the unsteady simulations, we can see that the simulation using the fine mesh reached periodic iterative convergence after approximately 40,000 iterations. On the other hand, the unsteady simulation using the coarse mesh reached a periodic iterative convergence after about 20,000 iterations. However, the results are slightly over-predicted due to numerical diffusion and turbulence model uncertainties. For the unsteady simulations, the time-step was chosen in such a way that the maximum CFL number is equal to 2, whereas, for the steady simulations (where we used the LTS method), the CFL number was gradually increased until reaching the value of 0.9.

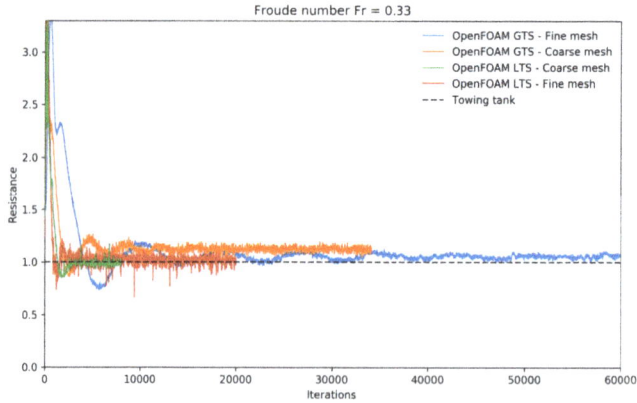

Figure 9. Comparison of the normalized resistance as a function of the iteration number for the coarse and fine meshes. In the legend, LTS (local time-stepping) corresponds to the steady solution, and GTS (global time-stepping, i.e., a time-step identical for all control volumes) corresponds to the transient solutions.

In Table 4, we show the percentage error (computed with respect to the experimental data) and the CPU time measured at 8000 iterations for each simulation plotted in Figure 9. From these results, it becomes clear why it is important to find steady solutions when conducting optimization studies in CFD. If we had conducted this optimization study using an unsteady solver, the computing time of each simulation would have been about 2 times the computing time of the corresponding steady simulation (for the same number of iterations); therefore, the total computing time of the optimization loop would have been at least 2 times higher.

Table 4. Percentage error (with respect to the experimental data) and CPU time of the benchmark cases. All the simulations were run in parallel with four processors. In all cases, the reported CPU time was measured at 8000 iterations.

Benchmark Case	Percentage Error	CPU Time (seconds)
Steady - Coarse mesh	1.8%	≈ 12000
Steady - Fine mesh	1.1%	≈ 20000
Unsteady - Coarse mesh	6.8%	≈ 27000
Unsteady - Fine mesh	4.3%	≈ 35000

In Figure 10, we present the normalized resistance as a function of the Froude number. The results reported were obtained using the coarse mesh and the LTS method. In the figure, a good agreement with the experimental results is observed up to a value of $Fr = 0.349$, then, the percentage error starts to increase, and this is mainly due to the fixed trim condition assumption used in this study. Moreover, the prototype is not meant to operate at such high Froude numbers, which corresponds to off-design conditions. In the light of this optimization study, where we know there are many uncertainties involved, errors of this order of magnitude are acceptable given the assumptions that were taken. Also, these discrepancies can be neglected because it is reasonable to consider that this error is constant for each bulb shape to be simulated and is, therefore, irrelevant for the optimization loop.

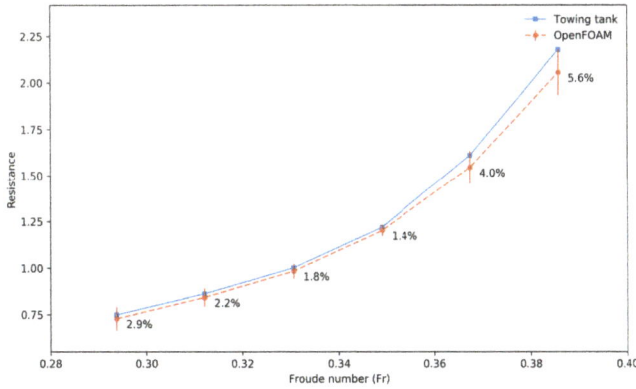

Figure 10. Comparison of the normalized resistance as a function of the Froude number. The numbers next to the markers indicate the percentage error between the numerical results and the experimental data.

Based on these results, we decided to use the coarse mesh as the base mesh, and in the case of mesh quality problems, the domain mesh was automatically recomputed using improved meshing parameters. As for the iterative method concerns, all the simulations were executed using the LTS method, and if proper iterative convergence was reached, the simulations were automatically stopped at 8000 iterations or until statistically stable results were obtained. These choices allowed us to obtain fast outcomes with good accuracy.

5. Results and Discussion

Hereafter, we discuss the results obtained from the surrogate-based optimization study. All the simulations presented in this section were conducted at a Froude number of 0.294. We would like to stress that to get the optimization loop started, we only need one input geometry. Then, this geometry is deformed by the shape morpher to generate new candidates. Also, the whole optimization framework was deployed using asynchronous simulations: that is, we executed many parallel simulations at the same time. Everything was controlled using the application coupling interface provided by Dakota. The average time of each simulation was between 4 and 5 hours.

As discussed previously, the first step when conducting SBO studies is to generate a sampling plan of the design space. The sampling plan should cover the design space in a uniform and well-distributed way, without leaving large areas unexplored; also, it should be cheap to compute. It is clear that the more information we gather during the initial sampling, the better the surrogate will be, but at the cost of longer computing times. In this study, we used a full-factorial experiment, where we conducted 25 experiments uniformly spaced in the design space. In Figure 11, we illustrate the sampling plan, where the uniform spaced points represent the locations where the high-fidelity computations were computed. The sampling plan illustrated in Figure 11 was chosen because it is inexpensive to compute (as we reduced the number of design variables to two), and it explores design points without missing too much information between observations.

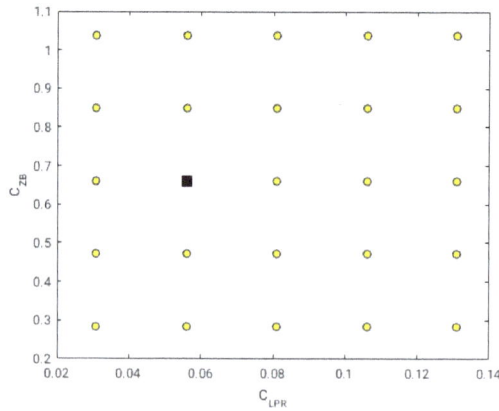

Figure 11. Sampling plan used in this study. The square represents the original geometry, whereas the circles represent the different geometry variations. Every point in the sampling plan represents a location where a high-fidelity computation was computed. All the simulations were conducted at a Froude number of 0.294.

The next step in our work-flow is to construct the surrogate (or meta-model) after conducting the high-fidelity simulations at the sampling locations. In this study, we used Kriging interpolation [38,44,45,48–52], which has been demonstrated to be a very reliable and accurate interpolation method when conducting SBO studies. The only drawback of this method (if it can be considered a drawback) is that it forces the meta-model to pass by all the points of the sampling plan; therefore, in the presence of numerical noise, we can have surrogates with many valleys and peaks that can make the optimization task difficult when using gradient-based methods. Nevertheless, as we are working at the surrogate level, we can use derivative-free methods which behave better in the presence of numerical noise.

In Figure 12, we illustrate two surrogates obtained during this study. In the figure, the surrogate in the left image was constructed using Kriging interpolation, and the one in the right image was build using a second-order polynomial function. The color scale represents the resistance variation in reference to the original geometry (where negative values indicate resistance reduction and positive values indicate resistance increment). As it can be seen, the Kriging model managed to capture the nonlinearities of the design space; we can even distinguish a valley region where the optimal solution (or solutions) might be located. On the other hand, the polynomial model is very smooth; however, it is not very accurate (it does not capture the nonlinearities). Nevertheless, it gives us a rough idea of how the design space behaves.

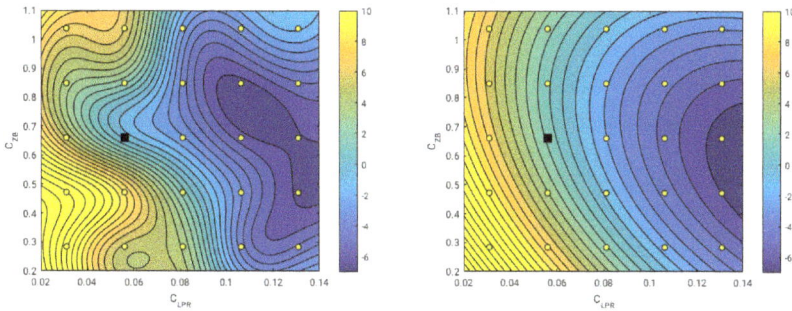

Figure 12. Left: surrogate model constructed using Kriging interpolation. Right: surrogate model constructed using quadratic polynomial interpolation. In the image, the uniform spaced points represent the locations where the high-fidelity computations were computed. The square represents the starting geometry and the circles represent the new geometries. In both images, the color scale represents the resistance variation in reference to the original geometry (where negative values indicate resistance reduction, and positive values indicate resistance increment).

At this stage, we already have a surrogate of the design space, or in other words, a mathematical representation of the design space. Our next tasks consist of validating the surrogate and conducting the optimization study at the surrogate level. The validation merely consists of computing the predicted value at the surrogate level in a location that has not been included in the sampling plan. Then, this value can be compared with the value obtained using high-fidelity simulations in the same location, and the percentage error can be computed. The optimization consists of doing the actual optimization at the surrogate level. Remember, as we are working using a mathematical representation of the design space, we can use any optimization method, disregarding how expensive it can be.

In this study, we found a global minimum and a local minimum; this situation is illustrated in Figure 13. These results suggest that there is a bifurcation in the valley region, where the global minimum is obtained when increasing the value of the design variable C_{ZB}, and the local minimum is obtained when decreasing the value of the design variable C_{ZB}. Notice also that as we increase the value of the design variable C_{ZB}, smaller increments of the design variable C_{LPR} are required in order to obtain larger resistance reduction. In Figure 13 (right image), we show the iterative path followed by the optimization algorithm when starting from two different initial positions, where each optimization task used about 40 iterations (function evaluations and gradient evaluations). As we used a gradient-based method to optimize the surrogate (method of feasible directions [53]), it is clear that we need to start the optimization algorithm from different initial locations if we want to find the global and local minima. If we were to perform this study by directly resorting to high-fidelity simulations, the number of simulations would be much higher than the number of simulations used during the sampling plan.

As previously mentioned, the SBO method gives the possibility of conducting an initial screening (or visualization) of the design space. This initial screening can be extremely helpful, especially when dealing with high multi-dimensional optimization problems. By screening the surrogate, designers can explore and exploit the surrogate in a more efficient way. Different optimal candidates can be immediately identified through the subjective opinion of the designer or on the basis of external constraints (such as manufacturing process, increased weight due to the added surface, the volume of the new bulb, structural loads, clearances, and so on), without the need of resorting to high-fidelity simulations (with the exception of the validation of the surrogate).

To demonstrate the validity of the surrogate, in Table 5, we show the numerical values of the resistance reduction and the percentage error between the surrogate value and the high-fidelity simulations outcome. The values presented in this table were measured in five different points; namely, the global minimum, the local minimum, and three points in the proximity of the global and local minima. From the table, we observe that all values listed effectively reduce the resistance. We also observe that the percentage error between the surrogate level and the high-fidelity level is about 2% for the global and local minima cases. On the other hand, for the values measured in the vicinity of the global and local minima, the percentage error is much larger, and this is presumably because in this region the nonlinearities are more significant.

Table 5. Resistance reduction and percentage error (with respect to the high-fidelity simulations) of five cases not included in the original sampling plan.

C_{LPR}	C_{ZB}	Resistance Reduction	Percentage Error	Note
0.119	0.660	$\approx 5\%$	$\approx 7\%$	-
0.131	0.566	$\approx 5\%$	$\approx 7\%$	-
0.131	0.755	$\approx 5\%$	$\approx 6\%$	-
0.109947	0.762845	$\approx 7\%$	$\approx 2\%$	Global minima
0.14	0.515651	$\approx 6\%$	$\approx 2\%$	Local minima

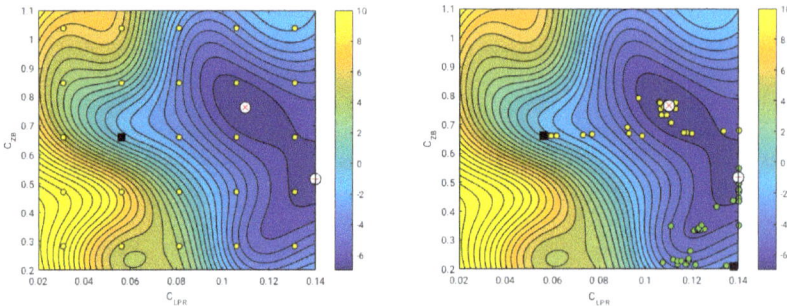

Figure 13. Left: the $\times$ symbol represents the global minimum, and the $+$ symbol represents the local minimum. The square represents the starting geometry and the circles represent the new geometries. Right: the two squares represent two different starting points of the optimization algorithm. The yellow circles represent the path followed by the gradient-based algorithm when starting from the topmost position, whereas the green circles represent the path followed by the gradient-based algorithm when starting from the bottom-most position. In both images, the color scale represents the resistance variation in reference to the original geometry (where negative values indicate resistance reduction, and positive values indicate resistance increment).

We can also improve the surrogate by adding new points to it (this is known as infilling). For example, we can add the points listed in Table 5 (C_{LPR} and C_{ZB}) and recompute a new surrogate (and optimal values). This situation is illustrated in Figure 14, where the triangle symbols represent the new training points. In the surrogate pictured in the figure, we are able to capture with more details the behavior of the design space close to the region where the global minimum is located. Also, we still can see the bifurcation, where there is a clear optimal solution when we increase the value of C_{ZB}; however, when decreasing the value C_{ZB}, the location of the new minimum is not very clear. It may be necessary to add a new training point in this region in order to get a better surrogate. Nevertheless, the resistance reduction values in this region are smaller; therefore, we can focus our attention on the cases where we increase the value of C_{ZB}.

For completeness, in Figure 15, we show the shape of the optimal bulbs at the design conditions corresponding to the global minimum and the local minimum. From all the results presented in this section, it is clear that surrogate-based optimization is an effective design tool for engineering design. However, the user should be cautious when constructing the surrogate, because the whole method depends on the quality of the meta-model. It is always recommended to validate the surrogate, do infilling using the optimal solutions, and add a couple of extra points in areas where high nonlinearities are observed or are close to the optimal values.

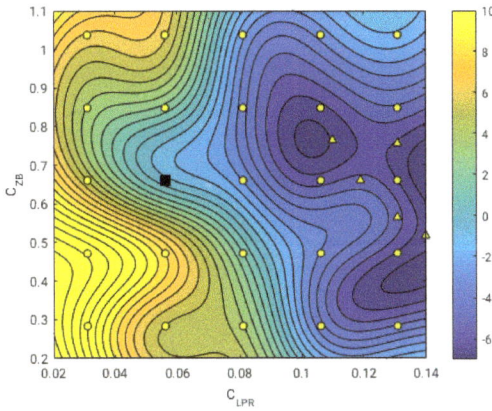

Figure 14. Improved surrogate using infilling. The square represents the starting geometry and the circles represent the new geometries. The triangles represent the infill points used. The color scale represents the resistance variation in reference to the original geometry (where negative values indicate resistance reduction, and positive values indicate resistance increment).

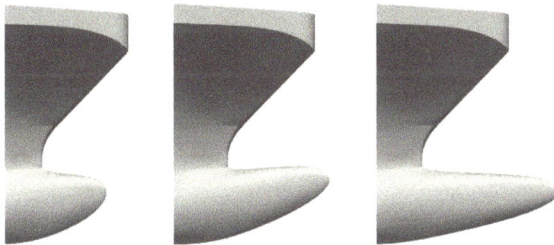

Figure 15. Bulbous bow shapes. Left: original shape. Middle: shape obtained at global minimum. Right: shape obtained at local minimum.

6. Conclusions and Future Perspectives

In this manuscript, we present an open-source optimization framework to perform industrial optimization studies. The optimization loop implemented allows for asynchronous simulations (i.e., many simulations can be run at the same time), and each simulation can be run in parallel; this allows to considerably reduce the output time of the optimization iterations. The optimization loop is fault tolerant and software agnostic, and it can be interfaced with any application able to interact using input/output files via a command line interface. The code coupling capabilities were provided by the library Dakota [28], and all the tools used in this work are open-source and free.

As for the optimization strategy concerns, we used surrogate-based optimization; this technique is very well suited to engineering design, as it allows designers and engineers to efficiently optimize and explore the design space from a limited number of training observations, without resorting to too many expensive high-fidelity simulations. To demonstrate the usability and flexibility of the proposed framework, we tested it in a practical case related to the naval industry, where we aimed at optimizing the shape of a bulbous bow to minimize the hydrodynamic resistance. We must highlight that the framework discussed in this work is general enough so that it may be used in any engineering design application.

During the optimization at the surrogate level, we found a global minimum. The shape of the optimal solution at the global minimum corresponds to a longer bulb and slightly shifted upwards. The resistance reduction at the global minimum design conditions is on the order of 7% at the surrogate level. The difference in resistance reduction between the global minimum at the surrogate level and the corresponding high-fidelity value is on the order of 2% (which is deemed more than acceptable). However, as optimization can be very abstract (e.g., balance between aesthetics and usability, operational requirements related to clearances, constraints related to weather conditions, ease of manufacturing new shapes, and so on), we further explored the meta-model and we found a local minimum corresponding to a configuration where we decreased the value of the design variable C_{ZB} in reference to the original geometry. At this design condition, the resistance reduction is on the order of 6%, with a difference between the local minimum at the surrogate level, and the corresponding high-fidelity value on the order of 2%. At this point, the optimal solution depends on the designer criteria and external constraints. These results demonstrate that design candidates showing considerable resistance reduction can be found using surrogate-based optimization, with very affordable computational times for the overall optimization task.

Finally, the results presented are limited to fixed trim, two design variables of the bulbous bow, and a single Froude number for the optimization study. We envisage the extension of the current work to rigid body simulations, and the exploration of the design space at different Froude numbers to determine if there is a dependence between protrusion and immersion of the bulbous bow and Froude number. We also look upon using a more complex parameterization of the bulbous bow geometry and conducting multi-objective optimization.

Author Contributions: Conceptualization, A.C., D.V., J.G. and J.P.; Methodology, A.C. and J.G.; Software, A.C. and J.G.; Validation, A.C. and J.G.; Formal Analysis, A.C.; Investigation, A.C., D.V., J.G. and J.P.; Resources, J.G.; Data Curation, A.C. and J.G.; Writing—Original Draft Preparation, A.C. and J.G.; Writing—Review & Editing, D.V., J.G. and J.P.; Visualization, A.C. and J.G.; Supervision, J.P.

Acknowledgments: We thank Fincantieri for making the solid model and experimental data available and for the support provided. We also thanks the DTLM (http://www.dltm.it), for letting us use their HPC facilities and the support provided.

Conflicts of Interest: The authors declare no conflict of interest.

References

1. Gaggero, S.; Tani, G.; Villa, D.; Viviani, M.; Ausonio, P.; Travi, P.; Bizzarri, G.; Serra, F. Efficient and multi-objective cavitating propeller optimization: An application to a high-speed craft. *Appl. Ocean Res.* **2017**, *64*, 31–57. [CrossRef]

2. Vesting, F.; Bensow, R.E. On surrogate methods in propeller optimization. *Ocean Eng.* **2014**, *88*, 214–227. [CrossRef]

3. Paik, B.G.; Kim, G.D.; Kim, K.Y.; Seol, H.S.; Hyun, B.S.; Lee, S.G.; Jung, Y.R. Investigation on the performance characteristics of the flexible propellers. *Ocean Eng.* **2013**, *73*, 139–148. [CrossRef]

4. Gaggero, S.; Gonzalez-Adalid, J.; Sobrino, M. Design and analysis of a new generation of CLT propellers. *Appl. Ocean Res.* **2016**, *59*, 424–450. [CrossRef]

5. Vernengo, G.; Bonfiglio, L.; Gaggero, S.; Brizzolara, S. Physics-Based Design by Optimization of Unconventional Supercavitating Hydrofoils. *J. Ship Res.* **2016**, *60*, 187–202. [CrossRef]
6. Harries, S. *Parametric Design and Hydrodynamic Optimization of Ship Hull Forms*; Mensch-und-Buch-Verlag: Berlin, Germany, 1998.
7. Campana, E.; Peri, D.; Tahara, Y.; Stern, F. Shape optimization in ship hydrodynamics using computational fluid dynamics. *Comput. Methods Appl. Mech. Eng.* **2006**, *196*, 634–651. [CrossRef]
8. Tahara, Y.; Peri, D.; Campana, E.; Stern, F. Computational fluid dynamics-based multi-objective optimization of a surface combatant using a global optimization method. *J. Mar. Sci. Technol.* **2008**, *13*, 95–116. [CrossRef]
9. Biliotti, I.; Brizzolara, S.; Viviani, M.; Vernengo, G.; Ruscelli, D.; Galliussi, M.; Guadalupi, D.; Manfredini, A. Automatic parametric hull form optimization of fast naval vessels. In Proceedings of the 11th International Conference on Fast Sea Transportation, Honolulu, HI, USA, 26–29 September 2011.
10. Vernengo, G.; Brizzolara, S.; Bruzzone, D. Resistance and sea-keeping optimization of a fast multi-hull passenger ferry. *Int. J. Offshore Polar Eng.* **2015**, *25*, 26–34.
11. Tahara, Y.; Peri, D.; Campana, E.; Stern, F. Single- and multi-objective design optimization of a fast multi-hull ship: Numerical and experimental results. *J. Mar. Sci. Technol.* **2011**, *16*, 412–433. [CrossRef]
12. Vernengo, G.; Brizzolara, S. Numerical investigation on the hydrodynamic performance of fast SWATHs with optimum canted struts arrangements. *Appl. Ocean Res.* **2017**, *63*, 76–89. [CrossRef]
13. Serani, A.; Leotardi, C.; Iemma, U.; Campana, E.; Fasano, G.; Diez, M. Parameter selection in synchronous and asynchronous deterministic particle swarm optimization for ship hydrodynamics problems. *Appl. Soft Comput.* **2016**, *49*, 313–334. [CrossRef]
14. Chen, X.; Diez, M.; Kandasamy, M.; Zhang, Z.; Campana, E.; Stern, F. High-fidelity global optimization of shape design by dimensionality reduction, meta-models and deterministic particle swarm. *Eng. Optim.* **2015**, *47*, 473–494. [CrossRef]
15. Peri, D.; Tinti, F. A multi-start gradient-based algorithm with surrogate model for global optimization. *Commun. Appl. Ind. Math.* **2012**, *3*, 1–22.
16. Demo, N.; Tezzele, M.; Gustin, G.; Lavini, G.; Rozza, G. Shape optimization by means of proper orthogonal decomposition and dynamic mode decomposition. In Proceedings of the NAV 2018, Trieste, Italy, 20-22 June 2018; pp. 212–219.
17. Umetani, N.; Bickel, B. Learning Three-dimensional Flow for Interactive Aerodynamic Design. *ACM Trans. Graph.* **2018**, *37*. [CrossRef]
18. Li, J.; Bouhlel, M.A.; Martins, J.R.R.A. A data-based approach for fast airfoil analysis and optimization. In Proceedings of the 2018 AIAA/ASCE/AHS/ASC Structures, Structural Dynamics, and Materials Conference, Kissimmee, FL, USA, 8–12 January 2018.
19. Matejka, J.; Glueck, M.; Bradner, E.; Hashemi, A.; Grossman, T.; Fitzmaurice, G. Dream Lens: Exploration and Visualization of Large-Scale Generative Design Datasets. In Proceedings of the ACM CHI Conference on Human Factors in Computing Systems, Montreal, QC, Canada, 21–26 April 2018.
20. Kracht, A.M. Design of bulbous bow. *SNAME Trans.* **1978**, *86*, 197–217.
21. Huang, F.; Yang, C. Hull form optimization of a cargo ship for reduced drag. *J. Hydrodyn. Ser. B* **2016**, *28*, 173–183. [CrossRef]
22. Lu, Y.; Chang, X.; Hu, A. A hydrodynamic optimization design methodology for a ship bulbous bow under multiple operating conditions. *Eng. Appl. Comput. Fluid Mech.* **2016**, *10*, 330–345. [CrossRef]
23. Mahmood, S.; Huang, D. Computational fluid dynamics based bulbous bow optimization using a genetic algorithm. *J. Mar. Sci. Appl.* **2012**, *11*, 286–294. [CrossRef]
24. Luo, W.; Lan, L. Design Optimization of the Lines of the Bulbous Bow of a Hull Based on Parametric Modeling and Computational Fluid Dynamics Calculation. *Math. Comput. Appl.* **2017**, *22*, 4. [CrossRef]
25. Bolbot, V.; Papanikolaou, A. Parametric, multi-objective optimization of ship's bow for the added resistance in waves. *Ship Technol. Res.* **2016**, *63*, 171–180. [CrossRef]
26. OpenFOAM Web Page. Available online: http://openfoam.com (accessed on 12 October 2018).
27. Mimmo Web Page. Available online: http://optimad.github.io/mimmo (accessed on 12 October 2018).
28. Dakota Web Page. Available online: https://dakota.sandia.gov/ (accessed on 12 October 2018).
29. Spisso, I. Parametric and Optimization Study: OpenFOAM and Dakota. In Proceedings of the Workshop on HPC Enabling of OpenFOAM for CFD Applications, Bologna, Italy, 26–28 November 2012.

30. Geremia, P. Industrial design optimization using open source tools. In Proceedings of the 6th OpenFOAM®
     Workshop, University Park, PA, USA, 13–16 June 2011.
31. Geremia, P.; de Villiers, E. Open source software tools for powertrain optimization. In Proceedings of the
     6th OpenFOAM® Workshop, University Park, PA, USA, 13–16 June 2011.
32. Skuric, V.; Uroic, T.; Rusche, H. Multi-objective optimization of a tube bundle heat exchanger. In
     Proceedings of the 9th OpenFOAM® Workshop, Zagreb, Croatia, 23–26 June 2014.
33. Byrne, J.; Cardiff, P.; Brabazon, A.; O'Neill, M. Evolving parametric aircraft models for design exploration.
     J. Neurocomput. **2014**, *142*, 39–47. [CrossRef]
34. Ohm, A.; Tetursson, H. *Automated CFD Optimization of a Small Hydro Turbine for Water Distribution Networks*;
     Chalmers University of Technology: Göteborg, Sweden, 2017.
35. Oh, J.T.; Chien, N.B. Optimization design by coupling computational fluid dynamics and genetic algorithm.
     In *Computational Fluid Dynamics*; IntechOpen: London, UK, 2018.
36. Lee, W.H.; Oh, S.H.; Lee, S.D. Deciding optimal parameter for internal wavemaker using coupling of
     Dakota and OpenFOAM. In Proceedings of the 13th OpenFOAM® Workshop, Shanghai, China, 24–29
     June 2018.
37. Slotnick, J.; Khodadoust, A.; Alonso, J.; Darmofal, D.; Gropp, W.; Lurie, E.; Mavriplis, D. *CFD Vision 2030
     Study: A Path To Revolutionary Computational Aerosciences*; NASA: Hampton, VA, USA, 2014.
38. Adams, B.M.; Ebeida, M.S.; Eldred, M.S.; Geraci, G.; Jakeman, J.D.; Maupin, K.A.; Monschke, J.A.; Swiler,
     L.P.; Stephens, J.A.; Vigil, D.M.; et al. *Dakota, A Multilevel Parallel Object-Oriented Framework for Design
     Optimization, Parameter Estimation, Uncertainty Quantification, and Sensitivity Analysis: Version 6.5 User
     Manual*; Sandia National Laboratories: Albuquerque, NM, USA, 2014.
39. Scardigli, A.; Arpa, R.; Chiarini, A.; Telib, H. Enabling of Large Scale Aerodynamic Shape Optimization
     through POD-Based Reduced-Order Modeling and Free Form Deformation. In Proceedings of the
     EUROGEN 2015, Glasgow, UK, 14–16 September 2015.
40. Crane, K.; Weischedel, C.; Wardetzky, M. Geodesics in Heat: A New Approach to Computing Distance
     Based on Heat Flow. *ACM Trans. Graph.* **2013**, *32*, 1–11. [CrossRef]
41. Buhmann, M.D. *Radial Basis Function: Theory and Implementations*; Cambrige University Press: Cambrige,
     UK, 2003.
42. Salmoiraghi, F.; Scardigli, A.; Telib, H.; Rozza, G. Free-form deformation, mesh morphing and
     reduced-order methods: enablers for efficient aerodynamic shape optimisation. *Int. J. Comput. Fluid Dyn.*
     2018. [CrossRef]
43. Weller, H.G.; Tabor, G.; Jasak, H.; Fureby, C. A tensorial approach to computational continuum mechanics
     using object-oriented techniques. *Comput. Phys.* **1998**, *12*, 620–631. [CrossRef]
44. Forrester, A.; Sobester, A.; Keane, A. *Engineering Design via Surrogate Modeling. A Practical Guide*; Wiley:
     New York, NY, USA, 2008.
45. Dalbey, K.R. *Efficient and Robust Gradient Enhanced Kriging Emulators*; Sandia National Laboratories:
     Albuquerque, NM, USA, 2013.
46. Jones, D.R.; Schonlau, M.; Welch, W.J. Efficient global optimization of expensive black-box functions. *J.
     Glob. Optim.* **1998**, *13*, 455–492. [CrossRef]
47. Iuliano, E.; Perez, E.A. *Application of Surrogate-Based Global Optimization to Aerodynamic Design*; Springer:
     Basel, Switzerland, 2016.
48. Oliver, M.; Webster, R. Kriging: A Method of Interpolation for Geographical Information Systems. *Int. J.
     Geogr. Inf. Syst.* **1990**, *4*, 313–332. [CrossRef]
49. Adams, B.M.; Ebeida, M.S.; Eldred, M.S.; Geraci, G.; Jakeman, J.D.; Maupin, K.A.; Monschke, J.A.; Swiler,
     L.P.; Stephens, J.A.; Vigil, D.M.; et al. *Dakota, A Multilevel Parallel Object-Oriented Framework for Design
     Optimization, Parameter Estimation, Uncertainty Quantification, and Sensitivity Analysis: Version 6.5 Theory
     Manual*; Sandia National Laboratories: Albuquerque, NM, USA, 2014.
50. Dalbey, K.R.; Giunta, A.A.; Richards, M.D.; Cyr, E.C.; Swiler, L.P.; Brown, S.L.; Eldred, M.S.; Adams, B.M.
     Surfpack User's Manual Version 1.1; Sandia National Laboratories: Albuquerque, NM, USA, 2013.
51. Giunta, A.A.; Watson, L. A comparison of approximation modeling techniques: polynomial versus
     interpolating models. In Proceedings of the 7th AIAA/USAF/NASA/ISSMO Symposium on
     Multidisciplinary Analysis and Optimization, St. Louis, MO, USA, 2–4 September 1998.

52. Romero, V.J.; Sqiler, L.P.; Giunta, A.A. Construction of response surfaces based on progressive lattice-sampling experimental designs. *Struct. Saf.* **2004**, *26*, 201–219. [CrossRef]
53. Zoutendijk, G. *Methods of Feasible Directions: A Study in Linear and Non-Linear Programming*; Elsevier: Amsterdam, The Netherlands, 1960.
54. Gen, M.; Cheng, R. *Genetic Algorithms and Engineering Optimization*; Wiley-Interscience: New York, NY, USA, 2000.
55. Vassiliadis, V.S.; Conejeros, R. Broyden family methods and the BFGS update. In *Encyclopedia of Optimization*; Floudas, C.A., Pardalos, P.M., Eds.; Springer: Boston, MA, USA, 2008.
56. Jones, D.R.; Perttunen, C.D.; Stuckman, B.E. Lipschitzian optimization without the Lipschitz constant. *J. Optim. Theory Appl.* **1993**, *79*, 157–181. [CrossRef]
57. Guerrero, J.; Bailardi, G.; Kifle, H. Visual storytelling and data visualization in numerical simulations. In Proceedings of the 11th OpenFOAM® Workshop, Guimarães, Portugal, 26–30 June 2016.
58. Guerrero, J. Opportunities and Challenges in CFD Optimization: Open Source Technology and the Cloud. In Proceedings of the Sixth Symposium on OpenFOAM® in Wind Energy, Gotland, Sweden, 13–14 June 2018.
59. Wilcox, D.C. *Turbulence Modeling for CFD*; DCW Industries: La Cañada Flintridge, CA, USA, 2010.
60. Practical Guidelines for Ship Resistance CFD. Available online: https://ittc.info/media/4198/75-03-02-04.pdf (accessed on 12 October 2018).

© 2018 by the authors. Licensee MDPI, Basel, Switzerland. This article is an open access article distributed under the terms and conditions of the Creative Commons Attribution (CC BY) license (http://creativecommons.org/licenses/by/4.0/).

Mathematical and Computational Applications

MDPI

Review

A Survey of Recent Trends in Multiobjective Optimal Control—Surrogate Models, Feedback Control and Objective Reduction

Sebastian Peitz * and Michael Dellnitz

Chair of Applied Mathematics, Faculty for Computer Science, Electrical Engineering and Mathematics, Paderborn University, Warburger Str. 100, 33098 Paderborn, Germany; dellnitz@uni-paderborn.de
* Correspondence: speitz@math.upb.de

Received: 15 May 2018; Accepted: 31 May 2018; Published: 1 June 2018

Abstract: Multiobjective optimization plays an increasingly important role in modern applications, where several criteria are often of equal importance. The task in multiobjective optimization and multiobjective optimal control is therefore to compute the set of optimal compromises (the Pareto set) between the conflicting objectives. The advances in algorithms and the increasing interest in Pareto-optimal solutions have led to a wide range of new applications related to optimal and feedback control, which results in new challenges such as expensive models or real-time applicability. Since the Pareto set generally consists of an infinite number of solutions, the computational effort can quickly become challenging, which is particularly problematic when the objectives are costly to evaluate or when a solution has to be presented very quickly. This article gives an overview of recent developments in accelerating multiobjective optimal control for complex problems where either PDE constraints are present or where a feedback behavior has to be achieved. In the first case, surrogate models yield significant speed-ups. Besides classical meta-modeling techniques for multiobjective optimization, a promising alternative for control problems is to introduce a surrogate model for the system dynamics. In the case of real-time requirements, various promising model predictive control approaches have been proposed, using either fast online solvers or offline-online decomposition. We also briefly comment on dimension reduction in many-objective optimization problems as another technique for reducing the numerical effort.

Keywords: multiobjective optimization; optimal control; model order reduction; model predictive control

MSC: 90C29; 49M37

---

1. Introduction

There is hardly ever a situation where only one goal is of interest at the same time. When performing a purchase for example, we want to pay a low price while getting a high quality product. In the same manner, multiple goals are present in most technical applications, maximizing quality versus minimizing the cost being only one of many examples. This dilemma leads to the field of multiobjective optimization, where we want to optimize all relevant objectives simultaneously. However, this is obviously impossible as the above example illustrates. Generically, the different objectives contradict each other such that we are forced to choose a compromise. While we are usually satisfied with one optimal solution in the scalar-valued setting, there exists in general an infinite number of optimal compromises in the situation where multiple objectives are present. The set of these compromise solutions is called the Pareto set, the corresponding points in the objective space form the Pareto front.

Since the solution to a Multiobjective Optimization Problem (MOP) is a set, it is significantly more expensive to compute than the optimum of a single objective problem, and many researchers devote their work to the development of algorithms for the efficient numerical approximation of Pareto sets. These advances have opened up new challenging application areas for multiobjective optimization. In optimal control, the optimization variable is not finite-dimensional, but rather a function, typically depending on time. The goal is to steer a dynamical system in such a way that one (or multiple) objective is minimized. Two particularly challenging control problems are feedback control and control problems constrained by Partial Differential Equations (PDEs). In the first case, the time for computing the Pareto set is strictly limited, often to a small fraction of a second. In the latter case, even the solution of single objective problems is often extremely time consuming so that the development of new algorithmic ideas is necessary to make these problems computationally feasible.

In fact, in situations like these, surrogate models or dimension reduction techniques are a promising approach for significantly reducing the computational effort and thereby enabling real-time applicability. This article gives an overview of recent advances in surrogate modeling for multiobjective optimal control problems, where the approach is to replace the underlying system dynamics by a reduced order model, which can be solved much faster. On the other hand, it introduces an approximation error, which has to be taken into account when analyzing convergence properties. The article is structured as follows. In Section 2, we are going to review some basics about multiobjective optimization, including the most popular solution methods. The two main challenges are addressed individually in the next sections, starting with expensive models in Section 3. There are also surrogate modeling approaches for MOPs that directly provide a mapping from the control variable to the objective function values. Since there already exist extensive surveys for this case (cf. [1–3], for instance), these are summarized only very briefly. In Section 4, real-time feedback control is discussed. Finally, we briefly discuss the question of dimension reduction in the number of objectives in Section 5 before concluding with a summary of further research directions in Section 6.

2. Multiobjective Optimization

In this section, the concepts of multiobjective optimization and Pareto optimality will be introduced and some widely-used solution methods will be summarized. More detailed introductions to multiobjective optimization can be found in, e.g., [4,5].

2.1. Theory

In multiobjective optimization, we want to minimize multiple objectives at the same time. Consequently, the fundamental difference from scalar optimization is that the objective function $J : \mathcal{U} \to \mathbb{R}^k$ is vector-valued. Hence, the general problem is of the form:

$$\min_{u \in \mathcal{U}} J(u) = \min_{u \in \mathcal{U}} \begin{pmatrix} J_1(u) \\ \vdots \\ J_k(u) \end{pmatrix} \tag{MOP}$$

$$\text{s.t.} \quad g_i(u) \leq 0, \quad i = 1, \dots, l,$$
$$h_j(u) = 0, \quad j = 1, \dots, m,$$

where $u \in \mathcal{U}$ is the control variable and $g : \mathcal{U} \to \mathbb{R}^l$, $g(u) = (g_1(u), \dots, g_l(u))^\top$ and $h : \mathcal{U} \to \mathbb{R}^m$, $h(u) = (h_1(u), \dots, h_m(u))^\top$, are inequality and equality constraints, respectively. The space of the control variables $\mathcal{U}$ is also called the decision space (according to the term decision variable for u in classical multiobjective optimization), and the objective function maps u to the objective space. Depending on the problem setup, $\mathcal{U}$ can either be finite-dimensional, i.e., $\mathcal{U} = \mathbb{R}^n$, or some appropriate function space.

Remark 1. *It is common in finite-dimensional optimization to use the notation x for the control or optimization variable and F for the objective function. In contrast to that, u and J are more common for control problems. In order to unify the notation, the latter will be used throughout this article for all optimization and optimal control problems.*

In contrast to classical optimization, in optimal control, we have to compute an input in such a way that a dynamical system behaves optimally with respect to some specified cost functional. Hence, we have the system dynamics as an additional constraint, very often in the form of ordinary (ODEs) or partial differential equations (PDEs):

$$\dot{y}(x,t) = G(y(x,t), u(t)), \quad (x,t) \in \Omega \times (t_0, t_e],$$

$$a(x,t)\frac{\partial y}{\partial n}(x,t) + b(x,t)y(x,t) = c(x,t), \quad (x,t) \in \Gamma \times (t_0, t_e], \quad \text{(PDE)}$$

$$y(x,t_0) = y_0(x), \quad x \in \Omega,$$

where the domain of interest $\Omega \subset \mathbb{R}^{n_x}$ is a connected open set with spatial dimension n_x and the boundary is denoted by $\Gamma = \partial\Omega$ with outward normal vector n. The coefficients $a(x,t)$, $b(x,t)$ and $c(x,t)$ in the boundary condition are given by the problem definition. The operator G is a partial differential operator describing the evolution of the system. The cost functional $\hat{J} : \mathcal{U} \times \mathcal{Y} \to \mathbb{R}^k$ of an optimal control problem consequently depends on the control u, as well as the system state y, which results in a multiobjective optimal control problem:

$$\min_{u \in \mathcal{U}, y \in \mathcal{Y}} \hat{J}(u,y) = \min_{u \in \mathcal{U}, y \in \mathcal{Y}} \begin{pmatrix} \int_{t_0}^{t_e} C_1(y(x,t), u(t)) \, dt + \Phi_1(y(x,t_e)) \\ \vdots \\ \int_{t_0}^{t_e} C_k(y(x,t), u(t)) \, dt + \Phi_k(y(x,t_e)) \end{pmatrix}$$

$$\text{s.t.} \quad \text{(PDE)}$$
$$g_i(y,u) \le 0, \quad i = 1,\dots,l,$$
$$h_j(y,u) = 0, \quad j = 1,\dots,m. \qquad \text{(MOCP)}$$

There are articles on multiobjective optimal control that specifically address the implications of multiple objectives for optimal control [6,7]; see also [8] for a short survey of methods. However, for many problems, there exists a unique solution y for every u such that (MOCP) can be simplified by introducing a so-called control-to-state operator $\mathcal{S} : \mathcal{U} \to \mathcal{Y}$; see [9] for details. By setting $\hat{J}(u,y) = \hat{J}(u, \mathcal{S}u) = J(u)$, the problem is transformed into (MOP). For this reason, we will from now on only consider (MOP).

In the situation where $\mathcal{U}$ is a function space (i.e., in the case of optimal control), the problem can be numerically transformed into a high-, yet finite-dimensional problem in a direct solution method via discretization, cf. [10,11]. This results in a large number of control variables, which can be very challenging on its own in multiobjective optimization. If the system dynamics are governed by a PDE, then the spatial discretization of the state y results in an even higher number of unknowns, which can easily reach several millions or more [12].

In contrast to single objective optimization problems, there exists no total order of the objective function values in $\mathbb{R}^k$ with $k \ge 2$ (unless they are not conflicting). Therefore, the comparison of values is defined in the following way [4]:

Definition 1. *Let $v, w \in \mathbb{R}^k$. The vector v is less than w (denoted by $v < w$), if $v_i < w_i$ for all $i \in \{1,\dots,k\}$. The relation $\le$ is defined in an analogous way.*

A consequence of the lack of a total order is that we cannot expect to find isolated optimal points. Instead, the solution to (MOP) is the set of optimal compromises (also called the Pareto set or set of non-dominated points):

Definition 2. *Consider the multiobjective optimization problem* (MOP). *Then:*

1. *a point u^* dominates a point u, if $J(u^*) \leq J(u)$ and $J(u^*) \neq J(u)$.*
2. *a feasible point u^* is called globally Pareto optimal if there exists no feasible point $u \in \mathcal{U}$ dominating u^*. The image $J(u^*)$ of a globally Pareto optimal point u^* is called a globally Pareto optimal value. If this property holds in a neighborhood $U(u^*) \subset \mathcal{U}$, then u^* is called locally Pareto optimal.*
3. *the set of non-dominated feasible points is called the Pareto set $\mathcal{P}_S$ and its image the Pareto front $\mathcal{P}_F$.*

A consequence of Definition 2 is that for each point that is contained in the Pareto set (the red line in Figure 1a), one can only improve one objective by accepting a trade-off in at least one other objective. Figuratively speaking, in a two-dimensional problem, we are interested in finding the "lower left" boundary of the feasible set in objective space (cf. Figure 1b).

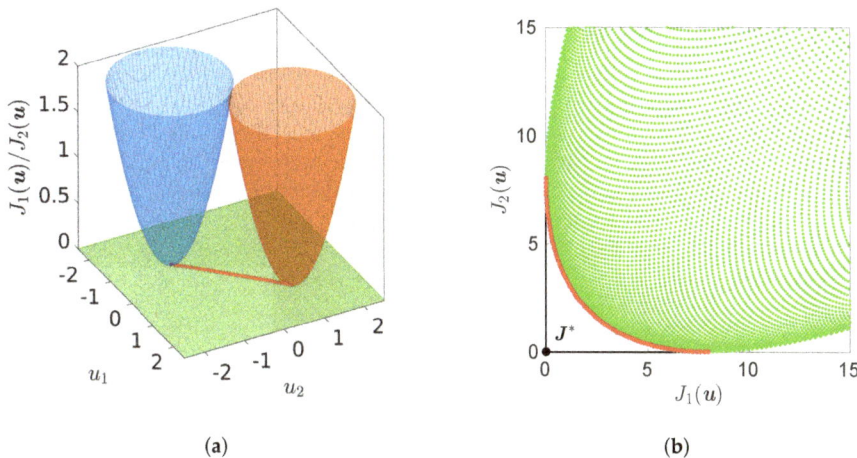

(a)   (b)

Figure 1. The red lines are the Pareto set (**a**) and Pareto front (**b**) of an exemplary multiobjective optimization problem (two paraboloids) of the form $\min_{u \in \mathbb{R}} J(u)$, $J : \mathbb{R}^2 \to \mathbb{R}^2$. The point $J^* = (0,0)^\top$ is called the utopian point.

Similar to single objective optimization, a necessary condition for optimality is based on the gradients of the objective functions. The first order conditions were independently discovered by Karush in 1939 [13] and by Kuhn and Tucker in 1951 [14]. Due to this, they are widely known as the Karush–Kuhn–Tucker (KKT) conditions:

Theorem 1 ([14]). *Let u^* be a Pareto-optimal point of Problem* (MOP), *and assume that $\nabla h_j(u^*)$ for $j = 1, \ldots, m$ and $\nabla g_s(u^*)$ for $s = 1, \ldots, l$ are linearly independent. Then, there exist non-negative scalars $\alpha_1, \ldots, \alpha_k \geq 0$ with $\sum_{i=1}^{k} \alpha_i = 1$, $\gamma \in \mathbb{R}^m$ and $\mu \in \mathbb{R}^l$ such that:*

$$\sum_{i=1}^{k} \alpha_i \nabla J_i(u^*) + \sum_{j=1}^{m} \gamma_j \nabla h_j(u^*) + \sum_{s=1}^{l} \mu_s \nabla g_s(u^*) = 0,$$

$$h_j(u^*) = 0, \quad j = 1, \ldots, m,$$
$$g_s(u^*) \leq 0, \quad s = 1, \ldots, l,$$ (KKT)
$$\mu_s g_s(u^*) = 0, \quad s = 1, \ldots, l,$$
$$\mu_s \geq 0, \quad s = 1, \ldots, l.$$

The set of points satisfying these conditions is called the set of substationary points $\mathcal{P}_{S,\text{sub}}$ [4]. Obviously, $\mathcal{P}_{S,\text{sub}}$ is a superset of the Pareto set $\mathcal{P}_S$. Many algorithms for MOPs compute the set of substationary points, in particular gradient-based methods, as we will see in the next section. This set can be reduced to the Pareto set in a consecutive step by performing a (comparatively inexpensive) non-dominance test.

2.2. Solution Methods

Many researchers in multiobjective optimization focus their attention on developing efficient algorithms for the computation of Pareto sets. Algorithms for solving MOPs can be compiled into several fundamentally different categories of approaches. The first category is based on scalarization techniques, where ideas from single objective optimization theory are extended to the multiobjective situation. All scalarization techniques have in common that the Pareto set is approximated by a finite set of Pareto-optimal points, which are computed by solving scalar subproblems. Consequently, the resulting solution methods involve solving multiple optimization problems consecutively. Scalarization can be achieved by various approaches such as the weighted-sum method, the ϵ-constraint method, normal boundary intersection or reference point methods [4,5,15].

Continuation methods make use of the fact that under certain conditions, the Pareto set is a smooth manifold of dimension $k-1$ [16]. This means that one can compute the tangent space in each point of the set, and a predictor step is performed in this space. The resulting point then has to be corrected to a Pareto-optimal solution using a descent method [17].

Another prominent approach is based on evolutionary algorithms [18,19], where the underlying idea is to evolve an entire population of solutions during the optimization process. Significant advances have been made concerning Multiobjective Evolutionary Algorithms (MOEAs) in recent years [20,21] (see also [22] for a survey) such that they are nowadays the most popular choice for solving MOPs due to the applicability to very complex problems and being easy to use in a black box fashion. Since convergence rates can be relatively slow for MOEAs, they can be coupled with locally fast methods close to the Pareto set. These approaches are known as memetic algorithms; see, e.g., [23–26].

Set-oriented methods provide an alternative deterministic approach to the solution of MOPs. Utilizing subdivision techniques, the desired Pareto set is approximated by a nested sequence of increasingly refined box coverings [27–29]. This way, a superset is computed, which converges to the desired solution, even in situations where the Pareto set is disconnected. However, their complexity depends on both the dimension of the Pareto set, as well as the decision space dimension. Due to this, one has to take additional steps to apply these algorithms for multiobjective optimal control problems.

Depending on the method of choice, gradient information can be used to accelerate convergence. While this is widely accepted in scalar-valued optimization, this is less the case when multiple objectives are present [30]. Nonetheless, many approaches exist where gradients are exploited, for example in order to create sequences converging to single points [31–33], to compute the entire set of valid decent directions [30,34], to obtain superlinear or quadratic convergence [35,36] or in combination with evolutionary approaches (memetic algorithms) [37–41]. In many of the gradient-based methods, the descent direction for all objectives is a convex combination of the individual gradients:

$$q(u) = \sum_{i=1}^{k} \hat{\alpha}_i \nabla J_i(u). \tag{1}$$

Here, $\hat{\alpha}$ is a fixed weight vector, which is determined in such a way that:

$$\langle q(u), \nabla J_i(u) \rangle < 0 \quad \forall \, i = 1, \ldots, k, \tag{2}$$

see, e.g., [31,32]. In the unconstrained case, there exists no u satisfying (2) only if $\|q(u)\| = 0$, which implies that u is substationary, cf. (KKT).

3. Surrogate Models

The ever-increasing computational capabilities allow us to analyze more and more complicated systems with a very large number of degrees of freedom, also in the context of optimal control, where practical problems range from process control [42,43] and energy management [15] over space mission design [44] to mobility and autonomous driving [45–48].

The above-mentioned examples can all be described by ordinary differential equations with a finite-dimensional state space $\mathcal{Y}$. In contrast to these problems, many phenomena in physics such as mechanical strain, heat flow, electromagnetism, fluid flow or even multi-physics simulations are governed by partial differential equations. Using numerical discretization schemes for the approximation of the spatial domain (such as finite elements or finite volumes) results in a very large number of degrees of freedom and a heavy computational burden. For more complex systems (such as turbulent flows [49]), simulating the dynamics is already very costly. Consequently, optimal control of these systems is all the more challenging, and considering multiple objectives further increases the cost. Due to this reason, only a few problems have been addressed directly; see, e.g., [50–53]. A method exploiting the special structure in the system dynamics has been proposed in [54], and a special case of Pareto-optimal solutions, namely Nash equilibria, has been computed in [55,56] (When not using a priori selection methods such as scalarization or Nash equilibria, a decision maker selects the appropriate solution. This is called Multi-Criteria Decision Making (MCDM) [57] and is an entire area of research on its own. Thus, we will not go into further details about the decision making process here.).

A very popular approach to circumvent the problem of prohibitively large computational cost is the use of surrogate models. Here, the exact objective function $J(u)$ is replaced by a surrogate $J^r(u)$, where the superscript r stands for reduced. In many situations, this surrogate function can be evaluated faster by several orders of magnitude. The challenge here is to find a good trade-off between acceleration and model accuracy, and many approaches have been proposed over the past two decades.

In optimal control, there are two fundamentally different ways for model reduction. The first case, which is equivalently applicable to multiobjective optimization problems, is to directly derive a surrogate for the objective function, i.e., $J^r : \mathcal{U} \to \mathbb{R}^k$ is constructed by polynomials, radial basis functions or other means. In optimal control, an alternative way of reducing the computational effort is by introducing a reduced model for the system dynamics:

$$J^r(u) = \hat{J}^r(u, y) = \hat{J}(u, \mathcal{S}^r u),$$

where the reduced control-to-state operator $\mathcal{S}^r$ indicates that the model reduction is due to a surrogate model for the system dynamics.

In both situations, we cannot expect that $J^r(u) = J(u)$ holds for all $u \in \mathcal{U}$. Instead, we introduce an error, which has to be taken into account. This is closely related to questions concerning uncertainty and noise. In this context, many researchers have addressed inaccuracies. In [58], the notion of ϵ-efficiency (cf. Definition 3) was first introduced in order to handle uncertain objective values. Uncertainty has also been considered in the context of multiobjective evolutionary computation (see, e.g., [59–61]. Alternative). Methods such as probabilistic [62,63], deterministic [64] or set-oriented approaches [65,66] have also been proposed. The special case of many-objective optimization is covered in [67], and applications are addressed in [68,69]. A different approach to uncertainties is via robust algorithms. Several examples from multiobjective optimization, as well as optimal control can be found in [43,70–72].

In the following, we will first introduce some results concerning inaccuracies in multiobjective optimization in Section 3.1 and then give an overview of existing methods for both of the above-mentioned approaches, i.e., surrogate models for the objective function (Section 3.2) or for the system dynamics (Section 3.3). Since the first approach has already been covered extensively in several surveys [1–3], we only give a brief overview of the existing methods and the corresponding references.

3.1. Inaccuracies and ϵ-Dominance

When using surrogate models in order to accelerate the solution process, we have to accept an error both in the objective function, as well as the respective gradients. Furthermore, inaccuracies may occur due to stochastic processes or due to unknown model parameters. In these situations, the objective function and the corresponding gradients are only known approximately, which has to be taken into account.

Suppose now that we only have approximations $J^r(u), \nabla J^r(u)$ of the objectives $J_i(u)$ and the gradients $\nabla J_i(u), i = 1, \ldots, k$, respectively. Furthermore, let us assume that upper bounds $\epsilon, \kappa \in \mathbb{R}^k$ for these errors are known:

$$\|J_i^r(u) - J_i(u)\|_2 \leq \epsilon_i \quad \forall u \in \mathcal{U}, \tag{3}$$

$$\|\nabla J_i^r(u) - \nabla J_i(u)\|_2 \leq \kappa_i \quad \forall u \in \mathcal{U}. \tag{4}$$

In this situation, we need to replace the dominance property 2 by an inexact version, also known as ϵ-dominance (see also [58,69]):

Definition 3 ([66]). *Consider the multiobjective optimization problem* (MOP) *where the objective function* $J(u)$ *is only known approximately according to* (3). *Then:*

1. *a point* u^* *confidently dominates a point* u, *if* $J^r(u^*) + \epsilon \leq J^r(u) - \epsilon$ *and* $J_i^r(u^*) + \epsilon_i < J_i^r(u) - \epsilon_i$ *for at least one* $i \in 1, \ldots, k$.
2. *The set of almost non-dominated points, which is a superset of the Pareto set* $\mathcal{P}_S$, *is defined as:*

$$\mathcal{P}_{S,\epsilon} = \left\{ u^* \in \mathcal{U} \,\middle|\, \nexists u \in \mathcal{U} \text{ which confidently dominates } u^* \right\}. \tag{5}$$

The concept of ϵ-dominance is visualized in Figure 2. Theoretically, the true point could be anywhere inside the box defined by ϵ such that in the cases (a)–(c), the lower left point does not confidently dominate the other point. The necessary condition is violated for one component in (a) and (b), respectively, and for both components in (c). The gray points in Figure 2a show a possible realization of the true points in which no point is dominated by the other. In (d), the orange point confidently dominates the black one, and in (e), we see the implications for the computation of Pareto fronts. Due to the inexactness, the number of points that are not confidently dominated is larger than in the exact case. This is also evident in Figure 3, where the exact and the inexact solution of an example problem from production [32] have been computed with an extension of the subdivision technique presented in [27], cf. [66] for details. Here, inexactness is introduced due to uncertainties in pricing. ϵ-dominance can be used for the development of algorithms for MOPs with uncertainties [59,61–63,65,67,68], for accelerating expensive MOPs [60,66,73], as well as for increasing the number of compromise solutions for the decision maker [63–65,69].

When considering gradient-based methods, inaccuracies in the gradients (Equation (4)) have to be taken into account. Since only approximations of the true gradient are known, these result in an inexact descent direction $q^r(u)$. The inaccuracy in the gradients introduces an upper bound for the angle between the individual gradients $\nabla J_i^r(u)$ and the descent direction $q^r(u)$. This is equivalent to a lower bound $\hat{\alpha}_{min}$ for the weight vector, i.e., $\hat{\alpha}_i \in [0,1]$ in Equation (1) has to be replaced by $\hat{\alpha}_i^r \in [\hat{\alpha}_{min,i}, 1]$ for $i = 1, \ldots, k$:

$$q^r(u) = \sum_{i=1}^{k} \hat{\alpha}_i^r \nabla J_i^r(u); \qquad (6)$$

see [66] for a detailed discussion. The additional constraint ensures that if $q^r(u)$ is a descent direction for the inexact problem, it is also a descent direction for the true problem. Furthermore, we obtain a criterion for the accuracy up to which we can compute $\mathcal{P}_{S,sub}$ based on inexact gradient information.

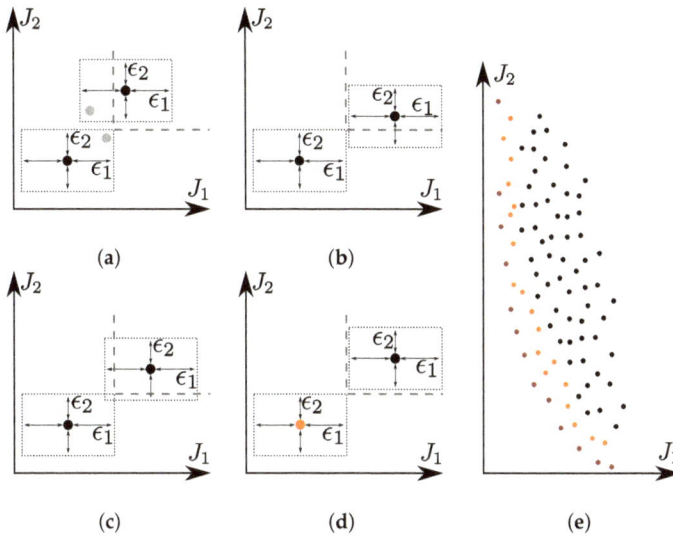

Figure 2. Example for the ϵ-dominance property. A point-wise comparison is illustrated in (a)–(d). The uncertainties are marked by the dashed boxes. Only in case (d), the lower left point confidently dominates the other point. In (e), the entirety of Pareto fronts for the exact problem ($\mathcal{P}_F$) and the inexact problem ($\mathcal{P}_{F,\epsilon}$) are shown in red and orange, respectively.

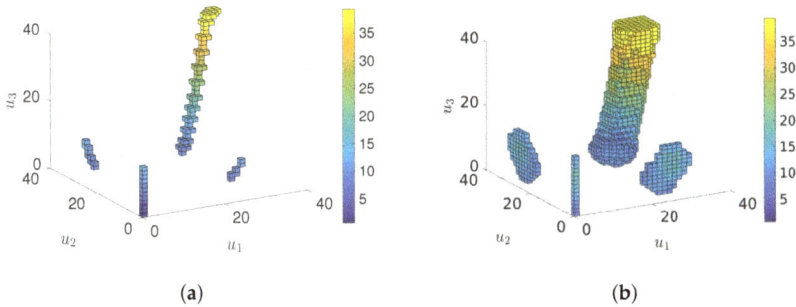

Figure 3. Example for an Multiobjective Optimization Problem (MOP) from production [32] where inexactness is introduced due to uncertainties in pricing. The Pareto set $\mathcal{P}_S$ for the exact problem is shown in (a), and the inexact set $\mathcal{P}_{S,\epsilon}$ is shown in (b).

Theorem 2 ([66]). *Consider the multiobjective optimization problem* (MOP) *without constraints. Only approximate gradients according to* (4) *are available, and consequently, the descent direction is also only known approximately according to Equation* (6). *Assume that* $\|q^r(u)\|_2 \neq 0$ *and* $\|\nabla J_i^r(u)\|_2 \neq 0$ *for* $i = 1, \ldots, k$. *Let:*

$$\hat{\alpha}_{\min,i} = \frac{1}{\|\nabla J_i^r(u)\|_2^2}\left(\|q^r(u)\|_2 \kappa_i - \sum_{\substack{j=1 \\ j \neq i}}^{k}\hat{\alpha}_j\left(\nabla J_i^r(u) \cdot \nabla J_i^r(u)\right)\right), \quad i = 1, \ldots, k.$$

Then, the following statements are valid:

(a) *If* $\sum_{i=1}^{k}\hat{\alpha}_{\min,i} > 1$ *then there exists no direction* $q(u)$ *with:*

$$\langle q(u), \nabla J_i(u)\rangle < 0 \quad \forall\, i = 1, \ldots, k,$$

i.e., no descent direction for the exact problem.

(b) *All points u with* $\sum_{i=1}^{k}\hat{\alpha}_{\min,i} = 1$ *are contained in the set:*

$$\mathcal{P}_{S,\kappa} = \left\{u \in \mathbb{R}^n \;\middle|\; \left\|\sum_{i=1}^{k}\hat{\alpha}_i\nabla J_i(u)\right\|_2 \leq 2\|\kappa\|_\infty\right\}. \tag{7}$$

A combination of Theorem 2 with the subdivision algorithm from [27] is shown in Figure 4. The algorithm constructs a nested sequence of increasingly refined box coverings, which converges to the set of substationary points where in the unconstrained case, $\|q(u^*)\| = 0$ holds for all $u^* \in \mathcal{P}_{S,\text{sub}}$. The set $\mathcal{P}_{S,\text{sub}}$ is shown in red in Figure 4a. Due to the inexactness, we can no longer guarantee $\|q(u^*)\| = 0$. Instead, we obtain the set $\mathcal{P}_{S,\kappa}$, which is shown in green. The background is colored according to the optimality condition $\|q(u)\|$ of the exact problem, and the dashed white line indicates the error bound (7) from the above theorem.

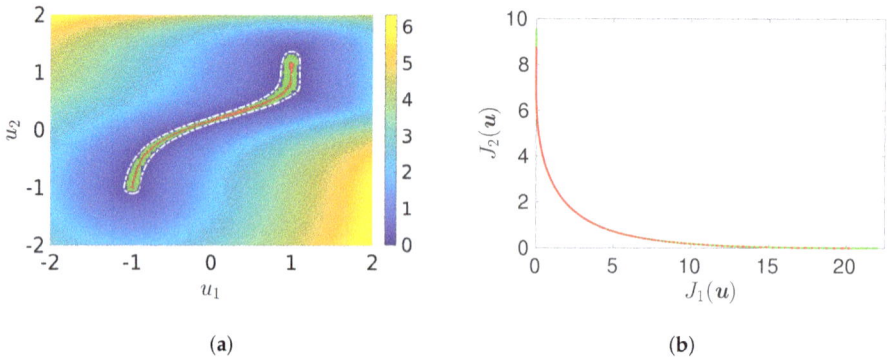

(a)

(b)

Figure 4. Exact and inexact solutions ($\mathcal{P}_S$ and $\mathcal{P}_{S,\kappa}$) for a simple example with $J : \mathbb{R}^2 \to \mathbb{R}^2$, cf. [66] for details. (**a**) The sets $\mathcal{P}_S$ and $\mathcal{P}_{S,\kappa}$ (for a random error with $\kappa = (0.01, 0.01)^\top$) are shown in red and green, respectively. The background is colored according to the optimality condition $\|q(u)\|$, which has to be zero for all substationary points. The dashed white line shows the error bound as derived in Theorem 2. (**b**) The corresponding Pareto fronts.

3.2. Surrogate Models for the Objective Function

The most straight-forward approach for introducing a surrogate model is to directly construct a map from the control to the decision space. This means that only the essential input-output behavior is covered, whereas internal states, as well as the system dynamics are neglected. Using such a meta model, one can very quickly obtain the objective function value for every u. This approach is equally applicable in finite-dimensional multiobjective optimization and has been used extensively in this context. To this end, we will only briefly cover the main questions that have to be addressed when using such models.

In special cases, one can exploit some structure in the problem formulation that yields a simplified analytic expression. However, in most cases, even if the equations can be written down in closed form, this approach requires a deep understanding of the underlying system, which is often hard to obtain. Moreover, small changes in the problem setup may result in having to repeat this tedious process all over. Due to these reasons, data-based approaches are used much more frequently. They are often easy to apply and much more general. In these approaches, the original objective function (or even a real-world experiment) is evaluated for a small number of inputs that are contained in the set $\mathcal{U}_{ref} = \{u^1, \ldots, u^p\}$, where p is the number of function evaluations (or experiments). The data points $J(u^1)$–$J(u^p)$ are then used to fit the meta model such as, e.g., the coefficients of a polynomial basis. Obviously, the choice of suitable ansatz functions is essential for the success of a meta modeling strategy. Popular choices are:

- Response Surface Models (RSM);
- Radial Basis Functions (RBF);
- statistical models such as Kriging or Gauss regression;
- machine learning methods such as artificial neural networks or support vector machines;

See [1] for an extensive survey in the context of multiobjective optimization. Additional surveys can be found in [74], where different statistical methods are compared, in [3,75] in relation to MOEAs or in [76], where RSM and RBF are compared for crashworthiness problems.

Besides selecting the correct meta model, questions concerning the training dataset have to be answered:

1. How large does the set $\mathcal{U}_{ref}$ have to be?
2. How can we pick the correct elements for $\mathcal{U}_{ref}$?
3. Do we define $\mathcal{U}_{ref}$ in advance or online during the model building process?

In addition to the meta model, Point 1 significantly depends on the problem under consideration. The more non-linear a problem is, the more data points are generally required to accurately construct a meta model. Obviously, the number also depends on Point 2, the locations u^1–u^k of these evaluations. The question of choosing the correct location is closely related to the field of optimum experimental design or Design of Experiments (DoE); see, e.g., [77,78]. The relevant question there is how to optimally pick a set of experiments such that the overall approximation error becomes minimal. Such approaches have successfully been coupled with multiobjective optimization in [79,80]. Point 3 depends both on the meta modeling approach, as well as the problem under consideration. In some situations (e.g., for real-world experiments), it may not be possible to iteratively determine the experiments. Instead, a batch approach has to be used. In computer experiments, flexibility is often higher such that an interplay between model building and high-fidelity evaluations can help to further reduce the number of experiments. In the context of machine learning, this process is also known as active learning [81,82].

Many algorithms combining multiobjective optimization and meta modeling have been proposed, and there exists a vast literature concerning this topic. Table 1 on page 17 mentions both survey articles and a (non-exhaustive) selection of popular methods.

3.3. Surrogate Models for the Dynamical System

The above-mentioned meta modeling methods are widely studied and have been successfully applied in a large variety of multiobjective optimization problems. In the context of control, there exists an alternative option, which is to derive a surrogate model for the system dynamics, i.e., to replace the high-fidelity control-to-state operator $\mathcal{S}$ by a less expensive surrogate model $\mathcal{S}^r$. Since the largest part of the computational effort is due to solving the dynamical system, by this, the reduced cost function $J^r(u) = \hat{J}^r(u,y) = \hat{J}(u, S^r u)$ is also much less expensive to evaluate.

The use of Reduced Order Models (ROMs) is not limited to control, but has been used successfully in a large variety of multi-query problems [12], and several extensive surveys have been written on reduced order modeling and using ROMs in prediction, uncertainty quantification or optimization; see [83–85]. For nonlinear problems, the two most widely-used approaches are the reduced basis method and Proper Orthogonal Decomposition (POD) (also known as Principal Component Analysis (PCA) or Karhunen–Loève decomposition).

3.3.1. ROMS via Proper Orthogonal Decomposition or the Reduced Basis Method

Numerically solving a PDE is generally realized by discretizing the spatial domain with a numerical mesh using finite differences, finite elements or finite volumes. By this, the infinite-dimensional state space $\mathcal{Y}$ is transformed into a finite-dimensional space $\mathcal{Y}^N$ via Galerkin projection:

$$y(x,t) \approx y^N(x,t) = \sum_{i=1}^{N} z_i^N(t)\phi_i(x). \tag{8}$$

Here, N denotes the number of degrees of freedom, and $\phi_i(x)$ are basis functions with local support such as indicator functions or hat functions. This transforms the PDE into an N-dimensional ordinary differential equation for the coefficients z. For complex domains, as well as complex dynamics, the dimension can easily reach the order of millions such that solving the problem in $\mathcal{Y}^N$ can quickly become very expensive, which is particularly challenging in the multi-query case.

The general concept in projection-based model reduction is therefore to find an appropriate space $\mathcal{Y}^r$ with dimension $\ell \ll N$ in which the system dynamics can nevertheless be approximated with sufficient accuracy. The two most common approaches to do this are the Reduced Basis (RB) method and Proper Orthogonal Decomposition (POD). In both cases, we compute s so-called snapshots of the high-dimensional system and then use the dataset $\{y_1^N, \ldots, y_s^N\}$ to construct a reduced basis $\psi = \{\psi_1, \ldots, \psi_\ell\}$ such that:

$$y(x,t) \approx y^N(x,t) \approx y^r(x,t) = \sum_{i=1}^{\ell} z_i^\ell(t)\psi_i(x).$$

The following example describes this approach in more detail. For an extensive introduction, the reader is referred to [86,87].

Example 1 (Heat equation). *Suppose we want to solve the time-dependent heat equation on a domain Ω with homogeneous Neumann conditions on the boundary Σ:*

$$\begin{aligned} y_t(x,t) - \lambda \Delta y(x,t) &= 0, & (x,t) &\in \Omega \times (t_0, t_e], \\ y(x,0) &= y_0(x), & x &\in \Omega, \\ y_n(x,t) &= 0, & (x,t) &\in \Sigma \times (t_0, t_e]. \end{aligned} \tag{9}$$

Here, y is the temperature, and λ is the heat conductivity. The subscripts t and n indicate the derivatives with respect to time and the outward normal vector of the boundary, respectively. We now derive the weak form of (9) by multiplying with a test function φ and integrating over the domain Ω. Using Gauss's theorem, we obtain the following equation:

$$\int_\Omega y_t(\cdot,t)\varphi - \lambda \nabla y(\cdot,t) \cdot \nabla \varphi \, dx = 0, \qquad (x,t) \in \Omega \times (t_0, t_e]. \tag{10}$$

If we want to solve (10) *using the finite element method, we have to insert the Galerkin ansatz* (8) *into* (10) *and individually take each of the basis functions as a test function. By this, we obtain the following system of equations:*

$$\int_\Omega \sum_{i=1}^N z_{i,t}^N(t)\phi_i\phi_j - \lambda z_i^N(t)\nabla\phi_i \cdot \nabla\phi_j \, dx = 0, \qquad j = 1,\ldots,N. \tag{11}$$

Introducing the mass matrix $M \in \mathbb{R}^{N \times N}$ and the stiffness matrix $K \in \mathbb{R}^{N \times N}$ with:

$$M_{i,j} = \int_\Omega \phi_i\phi_j \, dx, \qquad K_{i,j} = \int_\Omega \nabla\phi_i \cdot \nabla\phi_j \, dx,$$

this yields the following N-dimensional linear system:

$$M z_t^N(t) - \lambda K z^N(t) = 0.$$

If we now want to compute a reduced order model instead of a high-dimensional finite element approximation, we can apply the same procedure, except that now, we have to use the reduced basis in the Galerkin ansatz, as well as for test functions.

The most important difference between RB and POD is the area of application, although this is not a strict separation. RB is mostly applied to parameter-dependent, yet time-independent (i.e., elliptic) problems, whereas POD (introduced in [88]) is applied to time-dependent problems described by parabolic or hyperbolic PDEs. Consequently, in RB, the snapshots $\{y_1^N,\ldots,y_s^N\}$ are solutions corresponding to parameters $\{u_1,\ldots,u_s\}$, and in POD, they are snapshots in time, collected at the time instants $\{t_0,\ldots,t_{s-1}\}$. Using an equidistant time grid h, the snapshots are taken at $\{t_0, t_0 + h \ldots, t_0 + (s-1)h\}$. In RB, the snapshots $\{y_1^N,\ldots,y_s^N\}$ often directly serve as the basis ψ. For time-dependent problems, this can cause numerical difficulties since some snapshots might be very similar (e.g., for very slow systems or periodic dynamics) such that the snapshot matrix $S = (y_1^N,\ldots,y_s^N)$ is ill-conditioned. Due to this, a singular value decomposition is performed on S, and the ℓ leading left singular vectors are taken as the basis ψ. This results in an orthonormal basis, which can be shown to be optimal with respect to the L^2 projection error [87,88]. Furthermore, the truncation error is given by the sum over the neglected singular values:

$$\epsilon_\ell = \frac{\sum_{i=\ell+1}^S \sigma_i}{\sum_{j=1}^S \sigma_j}.$$

Whereas the error between the infinite-dimensional solution and the solution via a standard discretization approach can be neglected in many situations, the error of the ROM depends on several factors such as the reference data, the basis size and the parameter or control for which the ROM is evaluated. Consequently, this error can be significant such that proper care has to be taken. The most common approach is to derive bounds either for the error of the reduced state $\|y^r - y^N\|$ or, in the case of optimal control, of the optimal solution $\|u^r - u^N\|$ obtained using the ROM; see, e.g., [89–93] for POD or [94–97] for RB methods. In addition, there are other measures that can be taken such as deriving balanced input-output behavior [98,99] or introducing additional terms [100] or modifications [101] in the POD-based ROM. For more detailed introductions to RB and POD, the reader is referred to [97] and [87,88], respectively.

3.3.2. Optimal Control Using Surrogate Models

There is a rich literature on optimal control of PDEs using surrogate models. The approaches can be summarized into three main categories:

1. build a model once,
2. construction of regular updates in a trust region approach,
3. construction of regular updates using error estimators.

Whereas the first category is the most efficient one (see, e.g., [102] for optimal control of the Navier–Stokes equations), it is in general not possible to prove the convergence of the resulting algorithm.

In the second approach (which was developed by Fahl [103] for POD-based ROMs and one objective), one defines a trust region within which the current surrogate model is considered as trustworthy; see Figure 5 for an illustration. The ROM-based optimal control problem is then solved with the additional constraint that the solution has to remain within the trust region, i.e., $\|u^i - u^{\text{ref}}\| \leq \delta^i$, where i is the current step of the iterative optimization scheme and δ^i is the current trust region radius. After having obtained u^i, the high-dimensional system is evaluated, and the improvement of the full system is determined:

$$\rho = \frac{|J^N(u^i) - J^N(u^{i-1})|}{|J^r(u^i) - J^r(u^{i-1})|}.$$

If ρ is close to one, then the ROM is sufficiently accurate, and the iterate u^i is accepted. We then use the high-dimensional solution to construct the next ROM at $u^{\text{ref}} = u^i$. If, on the other hand, ρ is close to zero, then the ROM accuracy was bad, and the iterate u^i is rejected. Instead, the trust region radius δ^i is reduced, and the optimal control problem is solved again. Using the Trust Region (TR-POD) approach, one can ensure convergence to the optimal solution of the high-dimensional problem. In the case of the Navier–Stokes equations, this has been shown for different problem setups [103,104].

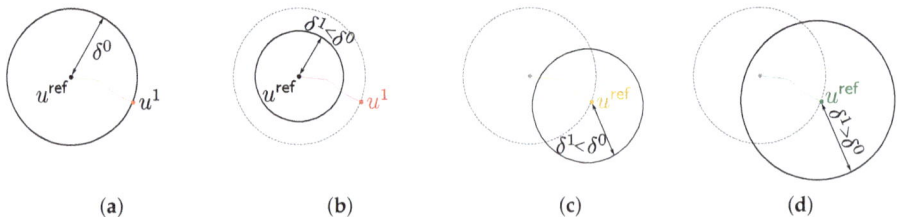

(a)	(b)	(c)	(d)

Figure 5. Trust region method. (a) The Reduced Order Model (ROM)-based optimal control problem is solved within the trust region δ^0. (b) If the improvement is poor for the full system (i.e., ρ is small), then the trust region radius is reduced, and we repeat the computation with the same problem. (c) If the improvement is acceptable (intermediate values of ρ), then we compute a new model and proceed with a smaller trust region $\delta^1 < \delta^0$. (d) If the improvement is good (i.e., $\rho \approx 1$), then the trust region radius is increased.

In the third approach, error estimators Δ^J for the current iterate u^i are required. By evaluating $\Delta^J(u^i)$, it is possible to efficiently estimate the error between the high- and the low-dimensional solution, since:

$$|J^r(u^i) - J^N(u^i)| < \Delta^J.$$

If this error estimate is larger than some prescribed upper bound ϵ, then the ROM has to be updated using data from a high-dimensional solution. Detailed information on error estimates can be found in, e.g., [91–97].

3.4. ROM-Based Multiobjective Optimal Control of PDEs

All three approaches for using ROMs in optimal control have recently been extended to multiobjective optimal control problems. Besides different ROM techniques, different algorithms for MOPs have been used, as well, such that a variety of methods has evolved, each of which is well suited for certain situations.

3.4.1. Scalarization

A natural and widely-used approach to MOPs is via scalarization. By this, the vector of objectives is synthesized into a scalar objective function, and the MOP is transformed into a sequence of scalar optimization problems for different scalarization parameters. In terms of ROM-based optimal control, this is advantageous because many techniques from scalar-valued optimal control can be extended. The main difference is now that the objective function may have a more complicated structure. In [105,106], the weighted sum method has been used in combination with RB in order to solve MOPs constrained by elliptic PDEs. In the weighted sum method, scalarization is achieved via convex combination of the individual objectives using the weight vector α:

$$\min_{u \in \mathcal{U}} \bar{J}(u) = \min_{u \in \mathcal{U}} \sum_{i=1}^{k} \alpha_i J_i(u). \tag{12}$$

The weighted sum method is probably the most straight-forward approach for including ROMs in MOPs. However, the method has strong limitations in the situation of non-convex problems, where it is impossible to compute the entire Pareto set [5].

A more advanced approach is the so-called reference point method [5] (cf. Figure 6 for an illustration), where the distance to an infeasible target point T with $T < J(u)$ has to be minimized:

$$\min_{u \in \mathcal{U}} \bar{J}(u) = \min_{u \in \mathcal{U}} \|T - J_i(u)\|. \tag{13}$$

By adjusting the target, we can move along the Pareto front and hence obtain an approximately equidistant covering of the front. The reference point method has been coupled with all three of the above-mentioned ROM approaches. In [107], it was used for multiobjective optimal control of the Navier–Stokes equations using one reduced model (cf. Figure 7a). Here, the objectives are to stabilize a periodic solution (the well-known von Kármán vortex street) and to minimize the control cost at the same time. In [73], the trust region framework by Fahl (TR-POD, [103]) was extended (cf. Figure 7b–c for a heat flow problem with a tracking and a cost minimization objective). The third ROM approach was used in [108,109]. The difficulty here is that the minimization of the distance to the target point results in a more complicated objective function, which has to be treated carefully.

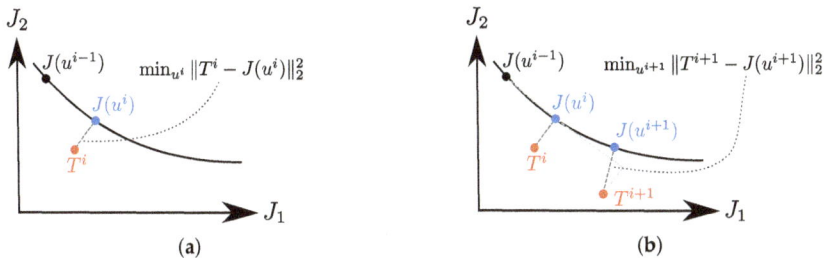

Figure 6. Reference point method. (a) Determination of a Pareto-optimal solution by solving (13). (b) Determination of the consecutive point on the Pareto front by adjusting the target and solving the next scalar problem.

Scalarization techniques generally have the same limitations on the decision space dimension as scalar-valued optimal control problems. This means that very efficient techniques (both direct and indirect) exist for high-dimensional controls. However, the number of objectives is limited because the parametrization in the scalarization step becomes extremely tedious, and it is almost impossible to obtain a good approximation of the entire Pareto set for more than three objectives.

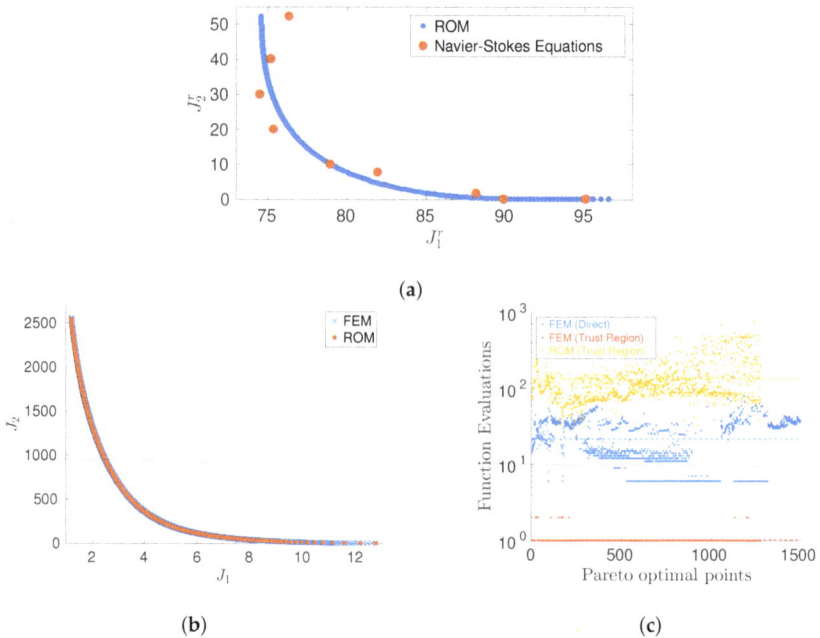

(a)

(b)

(c)

Figure 7. (**a**) Pareto front for an MOCP involving the Navier–Stokes equations (flow stabilization vs. cost), solved by coupling of an ROM (created once in advance) with the reference point method. Although we observe acceptable agreement, convergence cannot be guaranteed. (**b**) Pareto front for a heat flow MOCP (reference tracking vs. cost), solved by the TR-POD approach coupled with the reference point method. Convergence is achieved while reducing the number of expensive finite element (FEM) evaluations by a factor of ≈ 22, cf. (**c**).

3.4.2. Set-Oriented Approaches with ϵ-Dominance

In contrast to scalarization, in set-oriented techniques, the Pareto set is approximated by a box covering [27–29]. Here, the limitations are contrary, meaning that the dimension of the decision space is rather limited, while the number of objectives does not pose any problems for the algorithms. In practice, however, the computational cost increases exponentially with the number of objectives (i.e., the dimension of the Pareto set) such that we are still limited to a moderate number of objectives.

First results coupling the subdivision algorithm developed in [27] with error estimates for POD-based ROMs have recently appeared [110,111]. In the subdivision algorithm, the decision space is divided into boxes, which are alternately subdivided and selected. In the subdivision step, each existing box is subdivided into two smaller boxes. In the selection step, all boxes are eliminated that are dominated, i.e., they do not cover any part of the Pareto set. Numerically, this is realized by representing a box by a finite number of sample points and then marking a box as dominated if all sample points are dominated by samples from another box in the covering; see [27] for details.

The subdivision algorithm can be extended to inaccuracies by replacing the strict dominance test by an ϵ-dominance test as presented in Section 3.1 (see also Figure 3). After fixing an upper bound ϵ, we have to ensure that the surrogate models we use do not violate this bound anywhere in the control domain. Since this cannot be achieved with a single ROM, one has to use multiple, locally valid ROMs instead (cf. Figure 8c). The covering by local ROMs is managed in such a way that all points in the neighborhood around the reference u^{ref} at which the data for the ROM were collected satisfy the prescribed error bound ϵ. This way, the number of solutions of the high-dimensional system can be reduced significantly. A comparison between the exact and the ROM-based solution is shown in Figure 8 for a semilinear heat flow MOCP with two tracking type objectives and a cost minimization objective. Due to the ROM approach, the number of evaluations of the FEM model could be reduced by a factor of ≈ 1000.

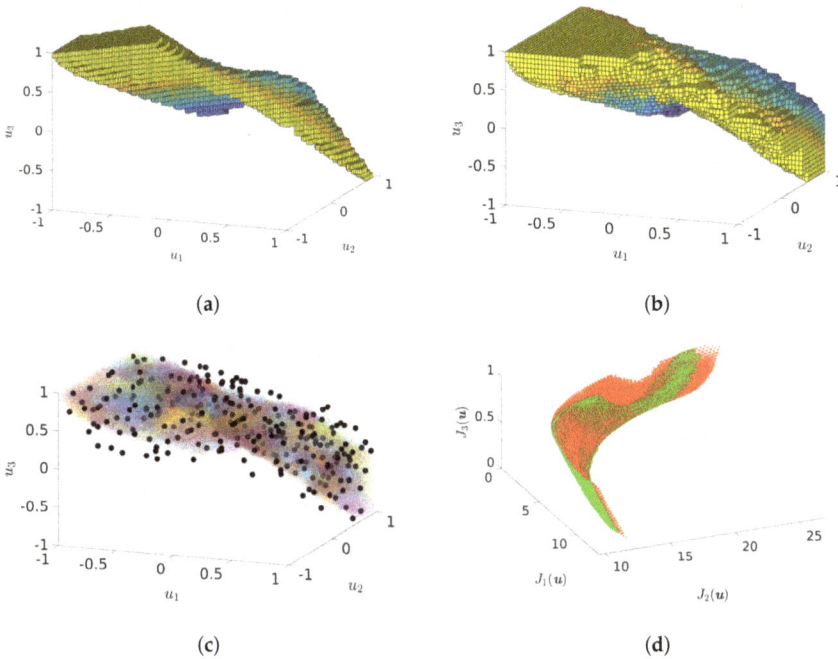

Figure 8. (**a**) Pareto set of a semilinear heat flow MOP with four controls (coloring according to u_4), solved directly including an FEM model in the subdivision algorithm. (**b**) Pareto set of the same problem, solved with localized ROMs. (**c**) The reference controls for which the local ROMs have been computed are shown in black, and the colored dots are sample points at which the objective function was evaluated. The colorings denote assignments to a specific ROM. (**d**) The corresponding Pareto fronts, where the FEM solution is shown in green and the ROM solution in red.

3.5. Summary

Before moving on to feedback control, a summary of the relevant publications where multiobjective optimization and meta modeling interact is given in Table 1. The references are categorized into surveys, algorithms using surrogate models for the objective function, the system dynamics and specific reduction approaches for MOPs. Furthermore, some applications are referenced.

Table 1. Overview of publications (in chronological order) where surrogate modeling and multiobjective optimization are combined. MOEA, Multiobjective Evolutionary Algorithm; RSM, Response Surface Model; POD, Proper Orthogonal Decomposition; TR, Trust Region.

Surveys	
Tabatabei et al. [2], Chugh et al. [3]	Extensive surveys on meta modeling for MOEAs
Voutchkov and Keane [74], Knowles and Nakayama [1], Jin [75]	Surveys on meta modeling approaches from statistics (RSM, RBF) and machine learning in combination with MOEAs
Benner et al. [83], Taira et al. [85], Peherstorfer et al. [84]	Surveys on reduced order modeling of dynamical systems
Algorithms Using Meta Models for the Objective Function	
Ong et al. [112], Ray et al. [113]	Combination of RBF and MOEA
Chung and Alonso [114], Keane [115]	Combination of kriging models and MOEA
Karakasis and Giannakoglou [116]	RBF as an inexpensive pre-processing step in a MOEA
Knowles [117]	Combination of DoE and an interactive method
Zhang et al. [118]	Combination of Gaussian process models and scalarization
Telen et al. [79]	Combination of DoE and scalarization and MOEA
Chugh et al. [119]	Kriging model in combination with reference vector approach for MOPs
Meta Models Specifically Tailored to Multiobjective Optimization	
Shimoyama et al. [120]	Kriging surrogate for hypervolume approximation (MOEA)
Pan et al. [121]	Surrogate model for dominance relations with uncertainties
Algorithms Using Surrogate Models for the System Dynamics	
Iapichino et al. [105]	Combination of POD and weighted sum
Banholzer et al. [108,109]	Combination of POD and reference point method
Iapichino et al. [106]	Combination of RB and weighted sum
Peitz [73]	Combination of TR-POD and reference point method
Beermann et al. [110,111]	Combination of POD and set-oriented method
Applications	
Albunni et al. [53]	POD and MOEA applied to the Maxwell equation
Ma and Qu [80]	MO of a switched reluctance motor by coupling RSM and MOEA (particle swarm optimization)
Peitz et al. [107]	POD-based multiobjective optimal control of the Navier–Stokes equations via scalarization and set-oriented methods
Wang et al. [122]	MOEA with multi-fidelity surrogate-management and offline-online decomposition applied to a trauma system

4. Feedback Control

Even when the objective function is not very expensive to evaluate, MOCPs often have a large computational cost; see, e.g., [123] for various examples. This becomes a limiting factor in situations where the solution time is critical as is the case in real-time applications. Due to the increasing computational power, as well as the advances in algorithms, Model Predictive Control (MPC) (see [124,125] for extensive introductions) has become a very powerful and widely-used method for realizing model-based feedback control of complex systems.

In MPC, an optimal control problem is solved in a short time horizon (the prediction horizon) while the real system (the plant) is running. Then, the first entry of the optimal control is applied to the plant, and the process is repeated with the time frame moving forward by one sample time h;

see Figure 9 for an illustration. This way, a closed-loop behavior is achieved. On the downside, we have to solve the optimal control problem within the sample time h. This can be in the order of seconds or minutes (in the case of chemical processes) down to a few microseconds, for example in power electronics applications.

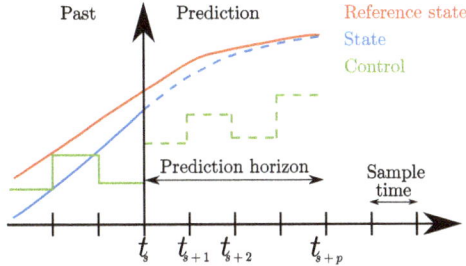

Figure 9. Sketch of the MPC method. Due to the real-time constraints, the optimization problem has to be solved faster than the sample time h.

In many MPC problems, stabilization of the system with respect to some reference state is the most important aspect. Nevertheless, there exist variations where stability is not an issue such that other (more economic) objectives can be pursued. These methods are known as economic MPC [126,127]. Another variation is the so-called explicit MPC [128], where the optimal control is computed in advance for a large number of different states and stored in a library. This way, the computational effort is shifted to an offline phase, and during operation, we only have to select the optimal control from the library.

As in multiobjective optimal control, there are numerous applications where multiple objectives are interesting in feedback control. This means that we would have to solve the problem (MOCP) with $t_0 = t_s$ and $t_e = t_{s+p}$ repeatedly within the sample time h. It is immediately clear that even the simplest MOCPs cannot be solved fast enough to allow for real-time applicability. Consequently, efficient algorithms have to be developed, which can be divided into approaches where one Pareto optimal solution is computed online and approaches with an offline phase during which the MOCP is solved (in this article, algorithms where Artificial Neural Networks (ANN) have to be trained beforehand are nonetheless assigned to the first category if the optimization is performed entirely online).

When implementing a multiobjective MPC (MOMPC) algorithm, one should keep in mind that, regardless of the algorithm used, the resulting trajectory need not be Pareto optimal, even if each single step is, cf. [129]. A remedy to this issue is presented in [130], where the selection of compromise solutions is restricted to a part of the Pareto front that is determined in the first MPC step. Due to this, upper bounds for the objective function values can be guaranteed.

4.1. Online Multiobjective Optimization

In the classical MPC framework, the optimal control problem is solved online within the sample time h. Since it is in general impossible to approximate the entire Pareto set sufficiently accurately within this time frame, there are three alternatives:

1. compute a single Pareto-optimal solution according to some predefined preference,
2. compute only a rough approximation of the Pareto set,
3. compute an arbitrary Pareto-optimal control that satisfies additional constraints (e.g., stability).

In the first approach, the objective function is scalarized using, for instance, the weighted sum method (12) or the reference point method (13). In this situation, well-established approaches from scalar-valued MPC exist on which one can build. First results using the weighted sum method have appeared in [131]. In [132,133], the authors use the same scalarization for an MPC problem with convex objective functions. In this situation, it is guaranteed that any Pareto-optimal solution can be computed using weighted sums, and the weights can be adapted online according to a decision maker's preference. Due to the convexity, stability can be proven for the resulting MPC algorithm. In [134], this approach is extended by providing gradient information of the objectives with respect to the weight vector α. This way, the weights can be adapted in such a way that a desired change in the objective space is realized. For non-convex problems, the weighted sum method is incapable of computing the entire Pareto set. Therefore, in [135], a variation of the reference point method is applied, where the target T is the utopian point J^*, i.e., the vector of the individual minima. This way, also non-convex problems can be treated. In fact, due to the reference point method, the objective function is always convex [109], which can be exploited during the optimization. Alternative scalarization methods are the ϵ-constraint method [136] or lexicographic ordering [137].

A disadvantage of a priori scalarization is that it is often difficult to select the scalarization parameter in such a way that a desired trade-off solution is obtained, and the remedy proposed in [134] is only applicable to a specific class of problems. Therefore, an alternative approach is to quickly compute a rough approximation of the entire Pareto set and then select the desired control online. Such methods have been proposed by many authors. The general approach is to us an MOEA and stop the computation after a few iterations. In the next step, one of these suboptimal solutions is selected. This selection is realized by specifying a weight vector for the objectives in [138–141] and by the satisficing trade-off method in [142].

As a third option, we can compute a single Pareto-optimal point without specifying which one we are specifically interested in as long as it satisfies additional constraints such as the stability of the system. Approaches of this type have been developed in [136,143], where a game theoretic approach is used.

4.2. Offline-Online Decomposition

A well-known trick to avoid heavy online computations is to introduce an offline-online decomposition (very similar to meta modeling approaches where surrogate models are constructed before solving the MOCP). This means that the Pareto set is computed beforehand, and in the online phase, an optimal compromise is selected according to a decision maker's preference or some heuristic based on the system state or the environment.

Many of the approaches that fall into this category use a standard feedback controller instead of MPC; see [144] for a short review concerning methods using scalarization and offline PID controller optimization. In the offline phase, a Pareto set is computed for the controller parameters. Possible objectives are, among many others, overshooting behavior, energy efficiency or robustness. Algorithms of this type have been proposed in [145–147] using MOEAs and in [148–150] using set-oriented methods.

An alternative approach is motivated by explicit MPC, i.e., the idea of solving many MOPs offline such that the correct solution can be extracted from a library in the online phase. Such a method has been proposed in [48]. In the offline phase, one has to identify all possible scenarios that can occur in the online phase. Such a scenario consists of both system states, as well as constraints. This results in a large number of MOPs that have to be solved. In order to reduce this number, symmetries in the problem are exploited. To this end, a concept known as motion primitives [151,152] is extended. In short, this means that if:

$$\arg\min_{u \in \mathcal{U}} \mathrm{MOP}_1 = \arg\min_{u \in \mathcal{U}} \mathrm{MOP}_2,$$

where MOP_1 and MOP_2 are two problem instances from the offline library, then we only have to solve one of the problems in order to have a Pareto-optimal solution for both. Moreover, if two problem instances vary only slightly, one can use a previously-computed solution as a good initial guess for the next MOCP to further decrease the computational effort [73]. In the online phase, the correct Pareto set is selected from the library (according to the system state and the constraints), and an optimal compromise is selected according so a decision maker's preference α. In contrast to the affine linear solutions, which can be computed in explicit MPC for linear-quadratic problems, one has to rely on interpolation between solutions in the nonlinear setting.

4.2.1. Example: Autonomous Driving

We here want to demonstrate the superiority of multiobjective approaches over scalar-valued MPC using the example of autonomously-driving electric vehicles [47,48,153]. The problem there is to find the set of optimal engine torque profiles such that the velocity is maximized while the energy consumption is minimized. Additional constraints have to be taken into account such as speed limits or stop signs. The system dynamics are described by a four-dimensional, highly nonlinear ODE for the vehicle velocity, the battery state of charge and two battery voltage drops, cf. [153] for details. Numerical investigations reveal several symmetries in the system such that the only relevant state for a scenario is the current velocity, whereas all other states only have a minor influence on the solution of the MOP. Consequently, the velocity, as well as the constraints form the above-mentioned scenarios; see Figure 10a for an illustration. For example, a scenario could be that the current velocity is 60 km/h and that the speed limit is currently increasing from 50–100 km/h, cf. Scenario (II) in Figure 10a. We then solve the MOP for this scenario and store the Pareto set in a library. By discretizing the velocity into steps of 0.1 (i.e., $v(t_0) = \ldots, 59.9, 60.0, 60.1, \ldots$), we have to solve 1727 MOPs in total in the offline phase.

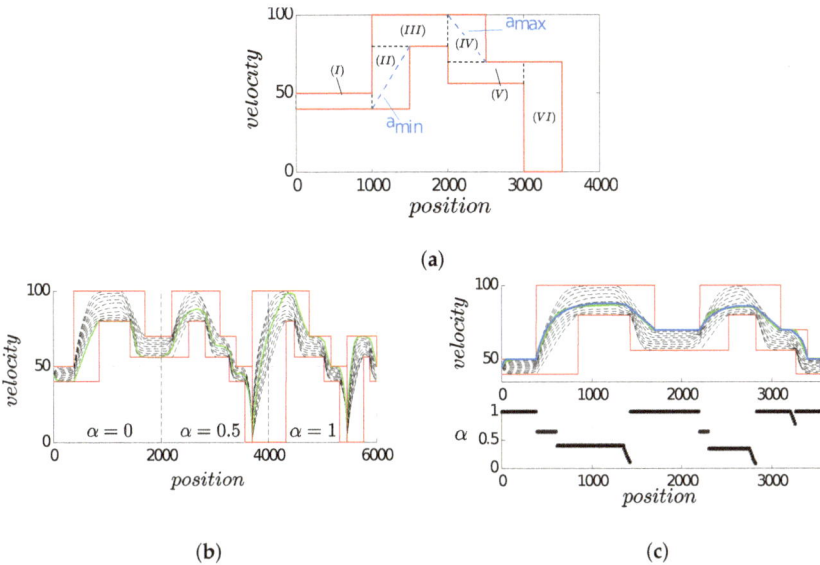

(a)

(b)

(c)

Figure 10. Results for the offline-online MPC approach from [48]. (**a**) Different constraint scenarios ((I)–(VI)), i.e., constant velocity, acceleration, deceleration and stopping. (**b**) Example track driven with the MPC algorithm. The red lines define the velocity bounds; the black dashed lines are trajectories corresponding to a constant weight α; and the green line is a trajectory where the weight is changed from 0 (energy efficient) over 0.5 (average) to 1 (fast). (**c**) Comparison between the MPC algorithm (coupled with a simple heuristic for the weighting) and the global optimum obtained via dynamic programming.

Table 2. Overview of publications (in chronological order) for multiobjective feedback control.

Algorithms without Offline Phase: Computation of Single Points	
Kerrigan et al. [131], Wojsznis et al. [154]	Scalarization via Weighted Sum (WS)
Kerrigan and Maciejowski [155], He et al. [137]	Scalarization via lexicographic ordering
Bemporad and Muñoz de la Peña [132,133]	Scalarization via WS for convex objectives, guaranteed stability for large gain vs. noise robust stabilizing objectives
Geisler and Trächtler [134]	WS, online adaptation of weights using gradient information
Maestre et al. [143]	Scalarization via game-theoretic approach
Zavala and Flores-Tlacuahuac [135]	Scalarization via reference point approach
Hackl et al. [129]	Scalarization via WS for Linear Time-Invariant (LTI) systems
Zavala [136]	Scalarization via ϵ-constraint: economic objective, stability as constraint
Grüne and Stieler [130]	Economic objectives, performance bounds via selection criterion
Algorithms without Offline Phase: Approximation of the Entire Pareto Set	
Laabidi et al. [138,140], Garcìa et al. [141]	ANN for state prediction, optimization via MOEA, selection of Pareto point via WS
Bouani et al. [139]	ANN for state prediction, comparison of two MOEAs and WS for MOP
Nakayama et al. [142]	Few MOEA iterations online, selection via satisficing trade-off method
Algorithms with Offline Phase	
Fonseca [145], Herreros et al. [146]	Offline computation of Pareto optimal controller parameters using MOEA
Scherer et al. [156]	Robust control using a common Lyapunov function for multiple stability criteria
Ben Aicha et al. [147]	Offline computation of Pareto optimal controllers parameters via EA and WS, online selection according to geometric mean of objectives
Krüger et al. [148]	Offline computation of Pareto optimal controllers parameters via Set oriented methods, parametric model reduction for increased efficiency
Hernández et al. [149], Xiong et al. [150]	Offline computation of Pareto-optimal controllers parameters via simple cell mapping
Peitz et al. [48]	Offline-online decomposition similar to explicit MPC
Applications	
Zambrano and Camacho [157]	MOMPC of a solar refrigeration plant via scalarization
Porfírio et al. [158]	MOMPC of an industrial splitter using a min-max reformulation
Pedersen and Yang [159]	MO PID controller design for magnetic levitation systems via MOEA
Li et al. [160]	Multiobjective adaptive cruise control for vehicles
Hu et al. [161]	MOMPC of high-power converters via WS
Núñez et al. [162]	MOMPC of dynamic pickup and delivery problems using MOEA
Peitz et al. [163]	MOMPC of an industrial laundry, scalarization of a traveling salesman problem via WS

In the online phase, we now select the relevant Pareto set from the library and, according to a decision maker's preference, apply one of the Pareto-optimal controls to the electric vehicle. This is done repeatedly such that a feedback loop is realized. The result is illustrated in Figure 10b for an example track, where the black lines correspond to constant weighting of the two criteria and the green line corresponds to a varying weight. This way, a flexible cruise control is established where the driver can quickly adapt, for instance, to changing energy requirements.

As has been mentioned before, an alternative to interactively choosing a weight is to implement some heuristic that automatically chooses a weight based on the current situation. Such an approach is visualized in Figure 10c, where the weighting depends on the vehicle velocity, as well as on current and future velocity constraints. For a simpler track, it is possible to compute a globally optimal solution for a scalarized objective using dynamic programming. We see that with the heuristic, the MOMPC approach yields trajectories close to the global optimum while only having finite horizon information.

4.3. Summary

We again conclude the section by giving a summary of publications where multiobjective optimization is applied in a real-time context, cf. Table 2. The publications are divided into four categories. The first two contain algorithms where the MOP is solved online. The categories differ in whether a single point is computed or the entire set is approximated. Consequently an offline phase is not required except in the case where surrogate models are trained in order to accelerate the online computations. The third category then contains the methods with an offline optimization phase, and some applications are mentioned in the fourth category.

5. Reduction Techniques for Many-Objective Optimization Problems

Another important restricting factor in multiobjective optimization is the number of objectives [164]. For MOPs with four or more objectives, the term Many-Objective Optimization (MaOP) has been coined, and over the past few years, many researchers have dedicated their work to address MaOPs and the issues arising from the curse of dimensionality, cf. [165] for an overview, [166,167] for new concepts for identifying non-dominated solutions and [168–172] for evolutionary approaches.

A popular approach for MaOPs is comprised of interactive methods [173–178]. These methods do not compute the entire set of optimal compromises, but instead interactively explore the Pareto set. Starting at the current Pareto-optimal solution, a decision maker can choose in which direction to proceed, i.e., which objective to improve at the expense of some other, currently less important objective. The approach in [178], for example, allows for Pareto-optimal movements both in the decision and objective space. One of the main advantages of interactive methods is the reduced computational effort, especially in the presence of many criteria, since it is not affected significantly by the dimension of the Pareto set. Moreover, this way, decision making from a vast number of Pareto-optimal solutions is avoided, which can be overwhelming for a decision maker. Consequently, interpretability and usability are increased.

Besides interactive methods, several reduction techniques have been proposed in the context of many-objective optimization, and although it is not the main theme of this review article, we want to give a brief overview of these reduction approaches since they also aim at increasing the efficiency of solving MOPs. These reduction techniques can be divided into two main categories. The first one is objective reduction, where the aim is to reduce the number of objectives while (approximately) preserving the Pareto set. The observation behind this is that not all objectives are of equal importance to the structure of the Pareto set, which is measured by the degree of conflict [179]. Consequently, when it is possible to identify the main contributors to the Pareto set, then one can solve a reduced problem taking into account only these most important objectives. Different approaches have been proposed for this identification step, all of which use a set of sample points. In [179], both exact and inexact algorithms are proposed for selecting a subset of objectives such that only those points in the Pareto set are lost that are worse in all remaining objectives by a constant δ or more. This approach is also exploited in [180]. In [181,182], POD (cf. Section 3.3) is used to identify such a subset, and a related concept is implemented in [183] using hyperplanes. In [184], an entropy-based approach is presented, and in [185], the relevant subset is selected multiple times within an evolutionary procedure.

A slightly different approach is pursued in [186], where the authors split large decision spaces into several smaller ones according to the relevance of decision variables for specific objectives.

The method proposed in [187] possesses characteristics of the first category, namely objective reduction, as well as of the second category, which is the exploitation of the structure of Pareto sets and front. Therein, first the corners of the Pareto front are identified, and this information is used to select the relevant objectives. Algorithms of the second type all exploit this hierarchical structure. This means that under certain assumptions, the Pareto front is bounded by the Pareto fronts of subproblems where one or more objectives have been neglected, cf. [188,189]. This way, the solution can be computed by a hierarchical approach where, starting with scalar problems, the boundary is computed before, finally, the interior is obtained. Very recently, results about the hierarchical structure of the Pareto set, i.e., in decision space, have appeared; see [73,190] for details. This approach is illustrated in Figure 11 where the solution to an MOP with four objectives is shown, as well as the Pareto sets of the subproblems with three and two objectives, respectively.

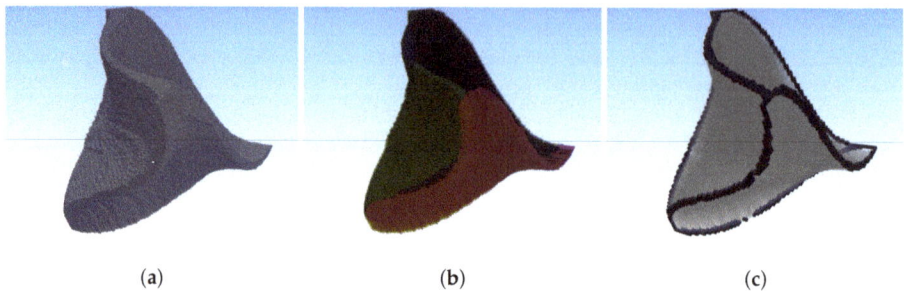

(a) (b) (c)

Figure 11. Visualization of the hierarchical structure of Pareto sets. (a) Pareto set of an example problem with $J : \mathbb{R}^3 \rightarrow \mathbb{R}^4$. (b) The four Pareto sets taking only three objectives into account form the boundary of the original Pareto set. (c) The Pareto sets in (b) are again bounded by the respective bi-objective subproblems.

6. Future Directions

This survey has given an overview of recent advances in the context of accelerating multiobjective optimization. These are surrogate models, feedback control and objective reduction techniques. Similar to almost every other field of science, it can be expected that the immense developments in data-based methods will also have a major impact on research in multiobjective optimization, in particular in the context of surrogate modeling. A very large number of researchers from the dynamical systems community is working on data-based methods using the Koopman operator, which is an infinite-dimensional, but linear operator describing the dynamics of observables [191,192]. Significant effort has been put into the development of numerical methods for approximating this operator from data; see, e.g., [193,194]. This way, the dynamics of observations can be reconstructed entirely from data and without any knowledge of the underlying system dynamics. In a way, this allows us to merge the two surrogate modeling categories from Sections 3.2 and 3.3 since we can approximate the dynamics not only of the state, but directly of the objectives. Several methods have recently been proposed to use the Koopman operator for data-based controller design, both in simulations [195–198], as well as experiments [199,200]. The results are very promising, such that it is just a matter of time until these methods are utilized for multiobjective optimal control.

In the same manner, machine learning techniques [201] will very likely gain more and more attention, both in multiobjective optimization, as well as optimal control. There are already many papers on this topic or related ones, and the number is growing quickly.

Math. Comput. Appl. **2018**, *23*, 30

Author Contributions: S.P. is responsible for the literature research and writing large parts of the paper. M.D. directed and supervised the research presented in the various examples and contributed to writing the final manuscript.

Acknowledgments: This work is supported by the Priority Programme SPP 1962 "Non-smooth and Complementarity-based Distributed Parameter Systems" of the German Research Foundation (DFG).

Conflicts of Interest: The authors declare no conflict of interest.

References

1. Knowles, J.; Nakayama, H. Meta-Modeling in Multiobjective Optimization. In *Multiobjective Optimization: Interactive and Evolutionary Approaches*; Branke, J., Deb, K., Miettinen, K., Slowinski, R., Eds.; Springer: Berlin/Heidelberg, Germany, 2008; pp. 245–284.
2. Tabatabaei, M.; Hakanen, J.; Hartikainen, M.; Miettinen, K.; Sindhya, K. A survey on handling computationally expensive multiobjective optimization problems using surrogates: Non-nature inspired methods. *Struct. Multidiscip. Optim.* **2015**, *52*, 1–25. [CrossRef]
3. Chugh, T.; Sindhya, K.; Hakanen, J.; Miettinen, K. A survey on handling computationally expensive multiobjective optimization problems with evolutionary algorithms. *Soft Comput.* **2017**, 1–30. [CrossRef]
4. Miettinen, K. *Nonlinear Multiobjective Optimization*; Springer Science and Business Media: Berlin, Germany, 2012.
5. Ehrgott, M. *Multicriteria Optimization*, 2nd ed.; Springer: Berlin/Heidelberg, Germany; New York, NY, USA, 2005.
6. Hindi, H.A.; Hassibi, B.; Boyd, S.B. Multiobjective $\mathcal{H}_2/\mathcal{H}_\infty$-Optimal Control via Finite Dimensional Q-Parametrization and Linear Matrix Inequalities. In Proceedings of the American Control Conference, Philadelphia, PA, USA, 24–26 June 1998; pp. 3244–3249.
7. Zhu, Q.J. Hamiltonian Necessary Conditions for a Multiobjective Optimal Control Problem with Endpoint Constraints. *SIAM J. Control Optim.* **2000**, *39*, 97–112. [CrossRef]
8. Gambier, A.; Badreddin, E. Multi-objective optimal control: An overview. In Proceedings of the 16th IEEE International Conference on Control Applciations, Singapore, 1–3 October 2007; pp. 170–175.
9. Tröltzsch, F. Optimal Control Of Partial Differential Equations. In *Graduate Studies in Mathematics*; American Mathematical Society: Providence, RI, USA, 2010; Volume 112.
10. Hinze, M.; Pinnau, R.; Ulbrich, M.; Ulbrich, S. *Optimization with PDE Constraints*; Springer Science+Business Media: Berlin, Germany, 2009.
11. Ober-Blöbaum, S. Discrete Mechanics and Optimal Control. Ph.D. Thesis, University of Paderborn, Paderborn, Germany, 2008.
12. Schilders, W.H.A.; van der Vorst, H.A.; Rommes, J. *Model Order Reduction*; Springer: Berlin/Heidelberg, Germany, 2008.
13. Karush, W. Minima of Functions of Several Variables with Inequalities as Side Constraints. Master's Thesis, University of Chicago, Chicago, IL, USA, 1939.
14. Kuhn, H.W.; Tucker, A.W. Nonlinear programming. In Proceedings of the 2nd Berkeley Symposium on Mathematical and Statsitical Probability, Oakland, CA, USA, 31 July–12 August 1950; University of California Press: Berkeley, CA, USA, 1951; pp. 481–492.
15. Romaus, C.; Böcker, J.; Witting, K.; Seifried, A.; Znamenshchykov, O. Optimal energy management for a hybrid energy storage system combining batteries and double layer capacitors. In Proceedings of the IEEE Energy Conversion Congress and Exposition, San Jose, CA, USA, 20–24 September 2009; pp. 1640–1647.
16. Hillermeier, C. *Nonlinear Multiobjective Optimization: A Generalized Homotopy Approach*; Birkhäuser: Basel, Switzerland, 2001.
17. Schütze, O.; Dell'Aere, A.; Dellnitz, M. On Continuation Methods for the Numerical Treatment of Multi-Objective Optimization Problems. Available online: http://drops.dagstuhl.de/opus/volltexte/2005/349/pdf/04461.SchuetzeOliver.Paper.349.pdf (accessed on 31 May 2018).
18. Deb, K. *Multi-Objective Optimization Using Evolutionary Algorithms*; John Wiley and Sons: Hoboken, NJ, USA, 2001; Volume 16.
19. Coello Coello, C.A.; Lamont, G.B.; Van Veldhuizen, D.A. *Evolutionary Algorithms for Solving Multi-Objective Problems*; Springer Science and Business Media: Berlin, Germany, 2007; Volume 2.

20. Zitzler, E.; Laumanns, M.; Thiele, L. SPEA2: Improving the Strength Pareto Evolutionary Algorithm. Available online: https://www.research-collection.ethz.ch/bitstream/handle/20.500.11850/145755/eth-24689-01.pdf (accessed on 31 May 2018).

21. Deb, K.; Pratap, A.; Agarwal, S.; Meyarivan, T. A Fast and Elitist Multiobjective Genetic Algorithm: NSGA-II. *IEEE Trans. Evol. Comput.* **2002**, *6*, 182–197. [CrossRef]

22. Zhou, A.; Qu, B.Y.; Li, H.; Zhao, S.Z.; Suganthan, P.N.; Zhang, Q. Multiobjective evolutionary algorithms: A survey of the state of the art. *Swarm Evol. Comput.* **2011**, *1*, 32–49. [CrossRef]

23. Ong, Y.S.; Lim, M.H.; Chen, X. Memetic Computation—Past, Present and Future. *IEEE Comput. Intell. Mag.* **2010**, *5*, 24–31. [CrossRef]

24. Neri, F.; Cotta, C.; Moscato, P. *Handbook of Memetic Algorithms*; Springer: Berlin, Germany, 2012; Volume 379.

25. Schütze, O.; Alvarado, S.; Segura, C.; Landa, R. Gradient subspace approximation: A direct search method for memetic computing. *Soft Comput.* **2017**, 21, 6331–6350. [CrossRef]

26. Schütze, O.; Martín, A.; Lara, A.; Alvarado, S.; Salinas, E.; Coello Coello, C.A. The directed search method for multi-objective memetic algorithms. *Comput. Optim. Appl.* **2016**, *63*, 305–332. [CrossRef]

27. Dellnitz, M.; Schütze, O.; Hestermeyer, T. Covering Pareto sets by Multilevel Subdivision Techniques. *J. Optim. Theory Appl.* **2005**, *124*, 113–136. [CrossRef]

28. Jahn, J. Multiobjective Search Algorithm with Subdivision Technique. *Comput. Optim. Appl.* **2006**, *35*, 161–175. [CrossRef]

29. Schütze, O.; Witting, K.; Ober-Blöbaum, S.; Dellnitz, M. Set Oriented Methods for the Numerical Treatment of Multiobjective Optimization Problems. In *EVOLVE—A Bridge between Probability, Set Oriented Numerics and Evolutionary Computation*; Tantar, E., Tantar, A.A., Bouvry, P., Del Moral, P., Legrand, P., Coello Coello, C.A., Schütze, O., Eds.; Studies in Computational Intelligence; Springer: Berlin/Heidelberg, Germany, 2013; Volume 447, pp. 187–219.

30. Bosman, P.A.N. On Gradients and Hybrid Evolutionary Algorithms. *IEEE Trans. Evol. Comput.* **2012**, *16*, 51–69. [CrossRef]

31. Fliege, J.; Svaiter, B.F. Steepest descent methods for multicriteria optimization. *Math. Methods Oper. Res.* **2000**, *51*, 479–494. [CrossRef]

32. Schäffler, S.; Schultz, R.; Weinzierl, K. Stochastic Method for the Solution of Unconstrained Vector Optimization Problems. *J. Optim. Theory Appl.* **2002**, *114*, 209–222. [CrossRef]

33. Gebken, B.; Peitz, S.; Dellnitz, M. A Descent Method for Equality and Inequality Constrained Multiobjective Optimization Problems. *arXiv* **2017**, arXiv:1712.03005.

34. Bosman, P.A.N.; de Jong, E.D. Exploiting gradient information in numerical multi-objective evolutionary optimization. In Proceedings of the 2005 Conference on Genetic and Evolutionary Computation, Washington, DC, USA, 25–29 June 2005; pp. 755–762.

35. Fliege, J.; Grana Drummond, L.M.; Svaiter, B.F. Newton's method for multiobjective optimization. *SIAM J. Optim.* **2009**, *20*, 602–626. [CrossRef]

36. Fliege, J.; Vaz, A.I.F. A SQP Type Method for Constrained Multiobjective Optimization. Available online: http://www.optimization-online.org/DB_FILE/2015/05/4929.pdf (accessed 31 May 2018).

37. Brown, M.; Smith, R.E. Directed multi-objective optimization. *Int. J. Comput. Syst. Signals* **2005**, *6*, 3–17.

38. Harada, K.; Sakuma, J.; Kobayashi, S. Local Search for Multiobjective Function Optimization: Pareto Descent Method. In Proceedings of the 8th Annual Conference on Genetic and Evolutionary Computation (GECCO 06), Seattle, WA, USA, 8–12 July 2006; pp. 659–666.

39. Harada, K.; Sakuma, J.; Ono, I.; Kobayashi, S. Constraint-Handling Method for Multi-objective Function Optimization: Pareto Descent Repair Operator. In *Evolutionary Multi-Criterion Optimization*; Obayashi, S., Deb, K., Poloni, C., Hiroyasu, T., Murata, T., Eds.; Springer: Berlin/Heidelberg, Germany, 2007; pp. 156–170.

40. Custódio, A.L.; Madeira, J.F.A.; Vaz, A.I.F.; Vicente, L.N. Direct multisearch for multiobjective optimization. *SIAM J. Optim.* **2011**, *21*, 1109–1140. [CrossRef]

41. Désidéri, J.A. Mutiple-Gradient Descent Algorithm for Multiobjective Optimization. In Proceedings of the European Congress on Computational Methods in Applied Sciences and Engineering, Vienna, Austria, 10–14 September 2012; pp. 3974–3993.

42. Garduno-Ramirez, R.; Lee, K.Y. Multiobjective optimal power plant operation through coordinate control with pressure set point scheduling. *IEEE Trans. Energy Convers.* **2001**, *16*, 115–122. [CrossRef]

43. Logist, F.; Houska, B.; Diehl, M.; van Impe, J. Robust multi-objective optimal control of uncertain (bio)chemical processes. *Chem. Eng. Sci.* **2011**, *66*, 4670–4682. [CrossRef]

44. Ober-Blöbaum, S.; Ringkamp, M.; zum Felde, G. Solving Multiobjective Optimal Control Problems in Space Mission Design using Discrete Mechanics and Reference Point Techniques. In Proceedings of the 51st IEEE International Conference on Decision and Control, Maui, HI, USA, 10–13 December 2012; pp. 5711–5716.

45. Lu, J.; DePoyster, M. Multiobjective optimal suspension control to achieve integrated ride and handling performance. *IEEE Trans. Control Syst. Technol.* **2002**, *10*, 807–821.

46. Geisler, J.; Witting, K.; Trächtler, A.; Dellnitz, M. Multiobjective optimization of control trajectories for the guidance of a rail-bound vehicle. *IFAC Proc.* **2008**, *17*, 4380–4386. [CrossRef]

47. Dellnitz, M.; Eckstein, J.; Flaßkamp, K.; Friedel, P.; Horenkamp, C.; Köhler, U.; Ober-Blöbaum, S.; Peitz, S.; Tiemeyer, S. Multiobjective Optimal Control Methods for the Development of an Intelligent Cruise Control. In Progress in Industrial Mathematics at ECMI 2014; Russo, G., Capasso, V., Nicosia, G., Romano, V., Eds.; Springer: Berlin, Germany, 2017; pp. 633–641.

48. Peitz, S.; Schäfer, K.; Ober-Blöbaum, S.; Eckstein, J.; Köhler, U.; Dellnitz, M. A Multiobjective MPC Approach for Autonomously Driven Electric Vehicles. *IFAC PapersOnLine* **2017**, *50*, 8674–8679. [CrossRef]

49. Brunton, S.L.; Noack, B.R. Closed-Loop Turbulence Control: Progress and Challenges. *Appl. Mech. Rev.* **2015**, *67*, 1–48. [CrossRef]

50. Vemuri, V.R.; Cedeno, W. A New Genetic Algorithm for Multi-objective Optimization in Water Resource Management. In Proceedings of the IEEE International Conference on Evolutionary Computation, Perth, Australia, 29 November–1 December 1995; Volume 1, pp. 1–11.

51. Rosehart, W.D.; Cañizares, C.A.; Quintana, V.H. Multiobjective optimal power flows to evaluate voltage security costs in power networks. *IEEE Trans. Power Syst.* **2003**, *18*, 578–587. [CrossRef]

52. Lotov, A.V.; Kamenev, G.K.; Berezkin, V.E.; Miettinen, K. Optimal control of cooling process in continuous casting of steel using a visualization-based multi-criteria approach. *Appl. Math. Model.* **2005**, *29*, 653–672. [CrossRef]

53. Albunni, M.N.; Rischmuller, V.; Fritzsche, T.; Lohmann, B. Multiobjective Optimization of the Design of Nonlinear Electromagnetic Systems Using Parametric Reduced Order Models. *IEEE Trans. Magn.* **2009**, *45*, 1474–1477. [CrossRef]

54. Ober-Blöbaum, S.; Padberg-Gehle, K. Multiobjective optimal control of fluid mixing. *Proc. Appl. Math. Mech.* **2015**, *15*, 639–640. [CrossRef]

55. Ramos, A.M.; Glowinski, R.; Periaux, J. Nash Equilibria for the Multiobjective Control of Linear Partial Differential Equations 1. *J. Optim. Theory Appl.* **2002**, *112*, 457–498. [CrossRef]

56. Borzì, A.; Kanzow, C. Formulation and Numerical Solution of Nash Equilibrium Multiobjective Elliptic Control Problems. *SIAM J. Control Optim.* **2013**, *51*, 718–744. [CrossRef]

57. Triantaphyllou, E. Multi-criteria decision making methods. In *Multi-Criteria Decision Making Methods: A Comparative Study*; Springer: Berlin, Germany, 2000; pp. 5–21.

58. White, D.J. Epsilon efficiency. *J. Optim. Theory Appl.* **1986**, *49*, 319–337. [CrossRef]

59. Hughes, E.J. Evolutionary Multi-objective Ranking with Uncertainty and Noise. In *Evolutionary Multi-Criterion Optimization: First International Conference, Zurich, Switzerland, March 7–9*; Zitzler, E., Thiele, L., Deb, K., Coello Coello, C.A., Corne, D., Eds.; Springer: Berlin/Heidelberg, Germany, 2001; pp. 329–343.

60. Deb, K.; Mohan, M.; Mishra, S. Evaluating the epsilon-Domination Based Multi-Objective Evolutionary Algorithm for a Quick Computation of Pareto-Optimal Solutions. *Evol. Comput.* **2005**, *13*, 501–525. [CrossRef] [PubMed]

61. Basseur, M.; Zitzler, E. A Preliminary Study on Handling Uncertainty in Indicator-Based Multiobjective Optimization. In *Workshops on Applications of Evolutionary Computation*; Rothlauf, F., Branke, J., Cagnoni, S., Costa, E., Cotta, C., Drechsler, R., Lutton, E., Machado, P., Moore, J.H., Romero, J., et al., Eds.; Springer: Berlin/Heidelberg, Germany, 2006; pp. 727–739.

62. Teich, J. Pareto-Front Exploration with Uncertain Objectives. In *Evolutionary Multi-Criterion Optimization*; Zitzler, E., Thiele, L., Deb, K., Coello Coello, C.A., Corne, D., Eds.; Springer: Berlin/Heidelberg, Germany, 2001; pp. 314–328.

63. Schütze, O.; Coello Coello, C.A.; Tantar, E.; Talbi, E.G. Computing the Set of Approximate Solutions of an MOP with Stochastic Search Algorithms. In Proceedings of the 10th Annual Conference on Genetic and Evolutionary Computation, Atlanta, GA, USA, 12–16 July 2008; pp. 713–720.

64. Engau, A.; Wiecek, M.M. Generating ϵ-efficient solutions in multiobjective programming. *Eur. J. Oper. Res.* **2007**, *177*, 1566–1579. [CrossRef]

65. Hernández, C.; Sun, J.Q.; Schütze, O. Computing the Set of Approximate Solutions of a Multi-objective Optimization Problem by Means of Cell Mapping Techniques. In *EVOLVE—A Bridge between Probability, Set Oriented Numerics, and Evolutionary Computation IV: International Conference held at Leiden University, July 10–13, 2013*; Emmerich, M., Deutz, A., Schütze, O., Bäck, T., Tantar, E., Tantar, A.A., Moral, P.D., Legrand, P., Bouvry, P., Coello, C.A., Eds.; Springer International Publishing: Basel, Switzerland, 2013; pp. 171–188.

66. Peitz, S.; Dellnitz, M. Gradient-based multiobjective optimization with uncertainties. In *NEO 2016*; Maldonado, Y., Trujillo, L., Schütze, O., Riccardi, A., Vasile, M., Eds.; Springer: Berlin, Germany, 2018; Volume 731, pp. 159–182.

67. Farina, M.; Amato, P. A fuzzy definition of "optimality" for many-criteria optimization problems. *IEEE Trans. Syst. Man Cybern. Part A Syst. Hum.* **2004**, *34*, 315–326. [CrossRef]

68. Singh, A.; Minsker, B.S. Uncertainty-based multiobjective optimization of groundwater remediation design. *Water Resour. Res.* **2008**, *44*. [CrossRef]

69. Schütze, O.; Vasile, M.; Coello Coello, C.A. Computing the set of epsilon-efficient solutions in multi-objective space mission design. *J. Aerosp. Comput. Inf. Commun.* **2009**, *8*, 53–70. [CrossRef]

70. Deb, K.; Gupta, H.V. Searching for Robust Pareto-optimal Solutions in Multi-objective Optimization. In *Evolutionary Multi-Criterion Optimization*; Coello Coello, C.A., Hernández Aguirre, A., Zitzler, E., Eds.; Springer: Berlin/Heidelberg, Germany, 2005; pp. 150–164.

71. Xue, Y.; Li, D.; Shan, W.; Wang, C. Multi-objective Robust Optimization Using Probabilistic Indices. In Proceedings of the Third International Conference on Natural Computation, Haikou, China, 24–27 August 2007; Volume 4, pp. 466–470.

72. Dellnitz, M.; Witting, K. Computation of robust Pareto points. *Int. J. Comput. Sci. Math.* **2009**, *2*, 243–266. [CrossRef]

73. Peitz, S. Exploiting Structure in Multiobjective Optimization and Optimal Control. Ph.D. Thesis, Paderborn University, Paderborn, Germany, 2017.

74. Voutchkov, I.; Keane, A.J. Multiobjective optimization using surrogates. In Proceedings of the Adaptive Computing in Design and Manufacture, Bristol, UK, 25–27 April 2006; pp. 167–175.

75. Jin, Y. Surrogate-assisted evolutionary computation: Recent advances and future challenges. *Swarm Evol. Comput.* **2011**, *1*, 61–70. [CrossRef]

76. Fang, H.; Rais-Rohani, M.; Liu, Z.; Horstemeyer, M.F. A comparative study of metamodeling methods for multiobjective crashworthiness optimization. *Comput. Struct.* **2005**, *83*, 2121–2136. [CrossRef]

77. Sacks, J.; Welch, W.J.; Mitchell, T.J.; Wynn, H.P. Design and Analysis of Computer Experiments. *Statist. Sci.* **1989**, *4*, 409–435. [CrossRef]

78. Atkinson, A.C.; Donev, A.N.; Tobias, R.D. *Optimum Experimental Designs, With SAS*; Oxford University Press: Oxford, UK, 2007.

79. Telen, D.; Logist, F.; Van Derlinden, E.; Tack, I.; Van Impe, J. Optimal experiment design for dynamic bioprocesses: A multi-objective approach. *Chem. Eng. Sci.* **2012**, *78*, 82–97. [CrossRef]

80. Ma, C.; Qu, L. Multiobjective Optimization of Switched Reluctance Motors Based on Design of Experiments and Particle Swarm Optimization. *IEEE Trans. Energy Convers.* **2015**, *30*, 1144–1153. [CrossRef]

81. Cohn, D.A.; Ghahramani, Z.; Jordan, M.I. Active learning with statistical models. *J. Artif. Intell. Res.* **1996**, *4*, 129–145.

82. Yu, K.; Bi, J.; Tresp, V. Active Learning via Transductive Experimental Design. In Proceedings of the 23rd International Conference on Machine Learning, Pittsburgh, PA, USA, 25–29 June 2006; pp. 1081–1088.

83. Benner, P.; Gugercin, S.; Willcox, K. A Survey of Projection-Based Model Reduction Methods for Parametric Dynamical Systems. *SIAM Rev.* **2015**, *57*, 483–531. [CrossRef]

84. Peherstorfer, B.; Willcox, K.; Gunzburger, M.D. Survey of Multifidelity Methods in Uncertainty Propagation, Inference, and Optimization. In *ACDL Technical Report TR16-1*; Massachusetts Institute of Technology: Cambridge, MA, USA, 2016; pp. 1–57.

85. Taira, K.; Brunton, S.L.; Dawson, S.T.M.; Rowley, C.W.; Colonius, T.; McKeon, B.J.; Schmidt, O.T.; Gordeyev, S.; Theofilis, V.; Ukeiley, L.S. Modal Analysis of Fluid Flows: An Overview. *AIAA J.* **2017**, *55*, 4013–4041. [CrossRef]

86. Brenner, S.L.; Scott, R. *The Mathematical Theory of Finite Element Methods*, 3rd ed.; Springer Science and Business Media: Berlin, Germany, 2003.

87. Volkwein, S. Model reduction using proper orthogonal decomposition. In *Lecture Notes*; University of Konstanz: Konstanz, Germany, 2011; pp. 1–43.

88. Sirovich, L. Turbulence and the dynamics of coherent structures part I: Coherent structures. *Q. Appl. Math.* **1987**, *XLV*, 561–571. [CrossRef]

89. Kunisch, K.; Volkwein, S. Control of the Burgers Equation by a Reduced-Order Approach Using Proper Orthogonal Decomposition. *J. Optim. Theory Appl.* **1999**, *102*, 345–371. [CrossRef]

90. Kunisch, K.; Volkwein, S. Galerkin proper orthogonal decomposition methods for a general equation in fluid dynamics. *SIAM J. Numer. Anal.* **2002**, *40*, 492–515. [CrossRef]

91. Hinze, M.; Volkwein, S. Proper Orthogonal Decomposition Surrogate Models for Nonlinear Dynamical Systems: Error Estimates and Suboptimal Control. In *Reduction of Large-Scale Systems*; Benner, P., Sorensen, D.C., Mehrmann, V., Eds.; Springer: Berlin/Heidelberg, Germany, 2005; Volume 45, pp. 261–306.

92. Tröltzsch, F.; Volkwein, S. POD a-posteriori error estimates for linear-quadratic optimal control problems. *Comput. Optim. Appl.* **2009**, *44*, 83–115. [CrossRef]

93. Lass, O.; Volkwein, S. Adaptive POD basis computation for parametrized nonlinear systems using optimal snapshot location. *Comput. Optim. Appl.* **2014**, *58*, 645–677. [CrossRef]

94. Grepl, M.A.; Patera, A.T. A posteriori error bounds for reduced-basis approximations of parametrized parabolic partial differential equations. *ESAIM Math. Model. Numer. Anal.* **2005**, *39*, 157–181. [CrossRef]

95. Veroy, K.; Patera, A.T. Certified real-time solution of the parametrized steady incompressible Navier-Stokes equations: Rigorous reduced-basis a posteriori error bounds. *Int. J. Numer. Methods Fluids* **2005**, *47*, 773–788. [CrossRef]

96. Haasdonk, B.; Ohlberger, M. Reduced Basis Method for Finite Volume Approximations of Parametrized Evolution Equations. *ESAIM Math. Model. Numer. Anal.* **2008**, *42*, 277–302. [CrossRef]

97. Rozza, G.; Huynh, D.B.P.; Patera, A.T. Reduced Basis Approximation and a Posteriori Error Estimation for Affinely Parametrized Elliptic Coercive Partial Differential Equations. *Arch. Comput. Methods Eng.* **2008**, *15*, 229–275. [CrossRef]

98. Willcox, K.; Peraire, J. Balanced Model Reduction via the Proper Orthogonal Decomposition. *AIAA J.* **2002**, *40*, 2323–2330. [CrossRef]

99. Rowley, C.W. Model Reduction for Fluids, Using Balanced Proper Orthogonal Decomposition. *Int. J. Bifurc. Chaos* **2005**, *15*, 997–1013. [CrossRef]

100. Noack, B.R.; Papas, P.; Monkewitz, P.A. The need for a pressure-term representation in empirical Galerkin models of incompressible shear flows. *J. Fluid Mech.* **2005**, *523*, 339–365. [CrossRef]

101. Cordier, L.; Abou El Majd, B.; Favier, J. Calibration of POD reduced-order models using Tikhonov regularization. *Int. J. Numer. Methods Fluids* **2009**, *63*, 269–296. [CrossRef]

102. Bergmann, M.; Cordier, L.; Brancher, J.P. Optimal rotary control of the cylinder wake using proper orthogonal decomposition reduced-order model. *Phys. Fluids* **2005**, *17*, 1–21. [CrossRef]

103. Fahl, M. Trust-region Methods for Flow Control based on Reduced Order Modelling. Ph.D. Thesis, University of Trier, Trier, Germany, 2000.

104. Bergmann, M.; Cordier, L.; Brancher, J.P. Drag Minimization of the Cylinder Wake by Trust-Region Proper Orthogonal Decomposition. In *Active Flow Control*; King, R., Ed.; Springer: Berlin, Germany, 2007; pp. 309–324.

105. Iapichino, L.; Trenz, S.; Volkwein, S. Multiobjective optimal control of semilinear parabolic problems using POD. In *Numerical Mathematics and Advanced Applications (ENUMATH 2015)*; Karasözen, B., Manguoglu, M., Tezer-Sezgin, M., Goktepe, S., Ugur, Ö., Eds.; Springer: Berlin, Germany, 2016; pp. 389–397.

106. Iapichino, L.; Ulbrich, S.; Volkwein, S. Multiobjective PDE-Constrained Optimization Using the Reduced-Basis Method. *Adv. Comput. Math.* **2017**, *43*, 945–972. [CrossRef]

107. Peitz, S.; Ober-Blöbaum, S.; Dellnitz, M. Multiobjective Optimal Control Methods for Fluid Flow Using Model Order Reduction. *arXiv* **2015**, arXiv:1510.05819.

108. Banholzer, S.; Beermann, D.; Volkwein, S. POD-Based Bicriterial Optimal Control by the Reference Point Method. *IFAC-PapersOnLine* **2016**, *49*, 210–215. [CrossRef]

109. Banholzer, S.; Beermann, D.; Volkwein, S. POD-Based Error Control for Reduced-Order Bicriterial PDE-Constrained Optimization. *Annu. Rev. Control* **2017**, *44*, 226–237. [CrossRef]

110. Beermann, D.; Dellnitz, M.; Peitz, S.; Volkwein, S. Set-Oriented Multiobjective Optimal Control of PDEs using Proper Orthogonal Decomposition. In *Reduced-Order Modeling (ROM) for Simulation and Optimization*; Springer: Berlin, Germany, 2018; pp. 47–72.

111. Beermann, D.; Dellnitz, M.; Peitz, S.; Volkwein, S. POD-based multiobjective optimal control of PDEs with non-smooth objectives. *Proc. Appl. Math. Mech.* **2017**, *17*, 51–54. [CrossRef]

112. Ong, Y.S.; Nair, P.B.; Keane, A.J. Evolutionary Optimization of Computationally Expensive Problems via Surrogate Modeling. *AIAA J.* **2003**, *41*, 687–696. [CrossRef]

113. Ray, T.; Isaacs, A.; Smith, W. Surrogate Assisted Evolutionary Algorithm for Multi-Objective Optimization. In *Multi-Objective Optimization*; World Scientific: Singapore, 2011; pp. 131–151.

114. Chung, H.S.; Alonso, J. Mutiobjective Optimization Using Approximation Model-Based Genetic Algorithms. In Proceedings of the 10th AIAA/ISSMO Multidisciplinary Analysis and Optimization Conference, Albany, NY, USA, 30 August–1 September 2004.

115. Keane, A.J. Statistical Improvement Criteria for Use in Multiobjective Design Optimization. *AIAA J.* **2006**, *44*, 879–891. [CrossRef]

116. Karakasis, M.K.; Giannakoglou, K.C. Metamodel-Assisted Multi-Objective Evolutionary Optimization. In Proceedings of the Sixth Conference on Evolutionary and Deterministic Methods for Design, Optimization and Control with Applications to Industrial and Societal Problems, Munich, Germany, 12–14 September 2005.

117. Knowles, J. ParEGO: A hybrid algorithm with on-line landscape approximation for expensive multiobjective optimization problems. *IEEE Trans. Evol. Comput.* **2006**, *10*, 50–66. [CrossRef]

118. Zhang, Q.; Liu, W.; Tsang, E.; Virginas, B. Expensive Multiobjective Optimization by MOEA/D with Gaussian Process Model. *IEEE Trans. Evol. Comput.* **2010**, *14*, 456–474. [CrossRef]

119. Chugh, T.; Jin, Y.; Miettinen, K.; Hakanen, J.; Sindhya, K. A Surrogate-Assisted Reference Vector Guided Evolutionary Algorithm for Computationally Expensive Many-Objective Optimization. *IEEE Trans. Evol. Comput.* **2018**, *22*, 129–142. [CrossRef]

120. Shimoyama, K.; Jeong, S.; Obayashi, S. Kriging-Surrogate-Based Optimization Considering Expected Hypervolume Improvement in Non-Constrained Many-Objective Test Problems. In Proceedings of the 2013 IEEE Congress on Evolutionary Computation, Cancun, Mexico, 20–23 June 2013; pp. 658–665.

121. Pan, L.; He, C.; Tian, Y.; Wang, H.; Zhang, X.; Jin, Y. A Classification Based Surrogate-Assisted Evolutionary Algorithm for Expensive Many-Objective Optimization. *IEEE Trans. Evol. Comput.* **2018**. [CrossRef]

122. Wang, H.; Jin, Y.; Jansen, J.O. Data-Driven Surrogate-Assisted Multiobjective Evolutionary Optimization of a Trauma System. *IEEE Trans. Evol. Comput.* **2016**, *20*, 939–952. [CrossRef]

123. Logist, F.; Houska, B.; Diehl, M.; van Impe, J. Fast Pareto set generation for nonlinear optimal control problems with multiple objectives. *Struct. Multidiscip. Optim.* **2010**, *42*, 591–603. [CrossRef]

124. Allgöwer, F.; Zheng, A. *Nonlinear Model Predictive Control*; Birkhäuser: Basel, Switzerland, 2012; Volume 26.

125. Grüne, L.; Pannek, J. *Nonlinear Model Predictive Control*, 2rd ed.; Springer International Publishing: Basel, Switzerland, 2017.

126. Rawlings, J.B.; Amrit, R. Optimizing process economic performance using model predictive control. In *Nonlinear Model Predictive Control*; Springer: Berlin/Heidelberg, Germany, 2009; pp. 119–138.

127. Grüne, L.; Müller, M.A.; Faulwasser, T. Economic Nonlinear Model Predictive Control. *Found. Trends Syst. Control* **2018**, *5*, 1–98.

128. Alessio, A.; Bemporad, A. A survey on explicit model predictive control. In *Nonlinear Model Predictive Control: Towards New Challenging Applications*; Magni, L., Raimondo, D.M., Allgöwer, F., Eds.; Springer: Berlin/Heidelberg, Germany, 2009; Volume 384, pp. 345–369.

129. Hackl, C.M.; Larcher, F.; Dötlinger, A.; Kennel, R.M. Is multiple-objective model-predictive control "optimal"? In Proceedings of the IEEE International Symposium on Sensorless Control for Electrical Drives and Predictive Control of Electrical Drives and Power Electronics (SLED/PRECEDE), München, Germany, 17–19 October 2013; pp. 1–8.

130. Grüne, L.; Stieler, M. Performance Guarantees for Multiobjective Model Predictive Control. Universität Bayreuth, 2017. Available online: https://epub.uni-bayreuth.de/3359/ (accessed on 31 May 2018).

131. Kerrigan, E.C.; Bemporad, A.; Mignone, D.; Morari, M.; Maciejowski, J.M. Multi-objective Prioritisation and Reconfiguration for the Control of Constrained Hybrid Systems. In Proceedings of the American Control Conference, Chicago, Illinois, IL, USA, 28–30 June 2000; pp. 1694–1698.

132. Bemporad, A.; Muñoz de la Peña, D. Multiobjective model predictive control. *Automatica* **2009**, *45*, 2823–2830. [CrossRef]

133. Bemporad, A.; Muñoz de la Peña, D. Multiobjective Model Predictive Control Based on Convex Piecewise Affine Costs. In Proceedings of the European Control Conference, Budapest, Hungary, 23–26 August 2009; pp. 2402–2407.

134. Geisler, J.; Trächtler, A. Control of the Pareto optimality of systems with unknown disturbances. In Proceedings of the IEEE International Conference on Control and Automation, Christchurch, New Zealand, 9–11 December 2009; pp. 695–700.

135. Zavala, V.M.; Flores-Tlacuahuac, A. Stability of multiobjective predictive control: A utopia-tracking approach. *Automatica* **2012**, *48*, 2627–2632. [CrossRef]

136. Zavala, V.M. A Multiobjective Optimization Perspective on the Stability of Economic MPC. *IFAC-PapersOnLine* **2015**, *48*, 974–980. [CrossRef]

137. He, D.; Wang, L.; Sun, J. On stability of multiobjective NMPC with objective prioritization. *Automatica* **2015**, *57*, 189–198. [CrossRef]

138. Laabidi, K.; Bouani, F. Genetic algorithms for multiobjective predictive control. In Proceedings of the First International Symposium on Control, Communications and Signal Processing, Hammamet, Tunisia, 21–24 March 2004; pp. 149–152.

139. Bouani, F.; Laabidi, K.; Ksouri, M. Constrained Nonlinear Multi-objective Predictive Control. In Proceedings of the IMACS Multiconference on Computational Engineering in Systems Applications, Beijing, China, 4–6 October 2006; pp. 1558–1565.

140. Laabidi, K.; Bouani, F.; Ksouri, M. Multi-criteria optimization in nonlinear predictive control. *Math. Comput. Simul.* **2008**, *76*, 363–374. [CrossRef]

141. García, J.J.V.; Garay, V.G.; Gordo, E.I.; Fano, F.A.; Sukia, M.L. Intelligent Multi-Objective Nonlinear Model Predictive Control (iMO-NMPC): Towards the 'on-line' optimization of highly complex control problems. *Expert Syst. Appl.* **2012**, *39*, 6527–6540. [CrossRef]

142. Nakayama, H.; Yun, Y.; Shirakawa, M. Multi-objective Model Predictive Control. In Multiple Criteria Decision Making for Sustainable Energy and Transportation Systems; Ehrgott, M., Naujoks, B., Stewart, T.J., Wallenius, J., Eds.; Springer: Berlin/Heidelberg, Germany, 2010; pp. 277–287.

143. Maester, J.M.; Muñoz de la Peña, D.; Camacho, E.F. Distributed model predictive control based on a cooperative game. *Optim. Control Appl. Methods* **2011**, *32*, 153–176. [CrossRef]

144. Gambier, A. MPC and PID control based on Multi-Objective Optimization. In Proceedings of the American Control Conference, Seattle, WA, USA, 11–13 June 2008; pp. 4727–4732.

145. Fonseca, C.M.M. Multiobjective Genetic Algorithms with Application to Control Engineering Problems. Ph.D. Thesis, University of Sheffield, Sheffield, UK, 1995.

146. Herreros, A.; Baeyens, E.; Perán, J.R. MRCD: A genetic algorithm for multiobjective robust control design. *Eng. Appl. Artif. Intell.* **2002**, *15*, 285–301. [CrossRef]

147. Ben Aicha, F.; Bouani, F.; Ksouri, M. Automatic Tuning of GPC synthesis parameters based on Multi-Objective Optimization. In Proceedings of the XIth International Workshop on Symbolic and Numerical Methods, Modeling and Applications to Circuit Design, Gammarth, Tunisia, 5–6 October 2010; pp. 1–5.

148. Krüger, M.; Witting, K.; Trächtler, A.; Dellnitz, M. Parametric Model-Order Reduction in Hierarchical Multiobjective Optimization of Mechatronic Systems. In Proceedings of the 18th IFAC World Congress 2011, Milan, Italy, 28 August–2 September 2011; Elsevier: Oxford, UK, 2011; Volume 18, pp. 12611–12619.

149. Hernández, C.; Naranjani, Y.; Sardahi, Y.; Liang, W.; Schütze, O.; Sun, J.Q. Simple cell mapping method for multi-objective optimal feedback control design. *Int. J. Dyn. Control* **2013**, *1*, 231–238. [CrossRef]

150. Xiong, F.R.; Qin, Z.C.; Xue, Y.; Schütze, O.; Ding, Q.; Sun, J.Q. Multi-objective optimal design of feedback controls for dynamical systems with hybrid simple cell mapping algorithm. *Commun. Nonlinear Sci. Numer. Simul.* **2014**, *19*, 1465–1473. [CrossRef]

151. Kobilarov, M. Discrete Geometric Motion Control of Autonomous Vehicles. Ph.D. Thesis, University of Southern California, Los Angeles, CA, USA, 2008.

152. Flaßkamp, K.; Ober-Blöbaum, S.; Kobilarov, M. Solving optimal control problems by using inherent dynamical properties. *Proc. Appl. Math. Mech.* **2010**, *10*, 577–578. [CrossRef]

153. Eckstein, J.; Peitz, S.; Schäfer, K.; Friedel, P.; Köhler, U.; Hessel-von Molo, M.; Ober-Blöbaum, S.; Dellnitz, M. A Comparison of two Predictive Approaches to Control the Longitudinal Dynamics of Electric Vehicles. *Procedia Technol.* **2016**, *26*, 465–472. [CrossRef]

154. Wojsznis, W.; Mehta, A.; Wojsznis, P.; Thiele, D.; Blevins, T. Multi-objective optimization for model predictive control. *ISA Trans.* **2007**, *46*, 351–361. [CrossRef] [PubMed]

155. Kerrigan, E.C.; Maciejowski, J.M. Designing model predictive controllers with prioritised constraints and objectives. In Proceedings of the IEEE International Symposium on Computer Aided Control System Design, Glasgow, UK, 20 September 2002, pp. 33–38.

156. Scherer, C.; Gahinet, P.; Chilali, M. Multiobjective Output-Feedback Control via LMI Optimization. *IEEE Trans. Autom. Control* **1997**, *42*, 896–911. [CrossRef]

157. Zambrano, D.; Camacho, E.F. Application of MPC with multiple objective for a solar refrigeration plant. In Proceedings of the IEEE International Conference on Control applications, Glasgow, UK, 18–20 September 2002; pp. 1230–1235.

158. Porfírio, C.R.; Almeida Neto, E.; Odloak, D. Multi-model predictive control of an industrial C3/C4 splitter. *Control Eng. Pract.* **2003**, *11*, 765–779. [CrossRef]

159. Pedersen, G.K.M.; Yang, Z. Multi-objective PID-controller tuning for a magnetic levitation system using NSGA-II. In Proceedings of the 8th Annual Conference on Genetic and Evolutionary Computation, Seattle, WA, USA, 8–12 July 2006; pp. 1737–1744.

160. Li, S.; Li, K.; Rajamani, R.; Wang, J. Model Predictive Multi-Objective Vehicular Adaptive Cruise Control. *IEEE Trans. Control Syst. Technol.* **2011**, *19*, 556–566. [CrossRef]

161. Hu, J.; Zhu, J.; Lei, G.; Platt, G.; Dorrell, D.G. Multi-objective model-predictive control for high-power converters. *IEEE Trans. Energy Convers.* **2013**, *28*, 652–663.

162. Núñez, A.; Cortés, C.E.; Sáez, D.; De Schutter, B.; Gendreau, M. Multiobjective model predictive control for dynamic pickup and delivery problems. *Control Eng. Pract.* **2014**, *32*, 73–86. [CrossRef]

163. Peitz, S.; Gräler, M.; Henke, C.; Hessel-von Molo, M.; Dellnitz, M.; Trächtler, A. Multiobjective Model Predictive Control of an Industrial Laundry. *Procedia Technol.* **2016**, *26*, 483–490. [CrossRef]

164. Schütze, O.; Lara, A.; Coello Coello, C.A. On the influence of the Number of Objectives on the Hardness of a Multiobjective Optimization Problem. *IEEE Trans. Evol. Comput.* **2011**, *15*, 444–455. [CrossRef]

165. Fleming, P.J.; Purshouse, R.C.; Lygoe, R.J. Many-Objective Optimization: An Engineering Design Perspective. In Proceedings of the International Conference on Evolutionary Multi-Criterion Optimization, Guanajuato, Mexico, 9–11 March 2005; pp. 14–32.

166. Kukkonen, S.; Lampinen, J. Ranking-Dominance and Many-Objective Optimization. In Proceedings of the IEEE Congress on Evolutionary Computation, Singapore, 25–28 September 2007; pp. 3983–3990.

167. Bader, J.; Zitzler, E. HypE : An Algorithm for Fast Hypervolume-Based Many-Objective Optimization. *Evol. Comput.* **2011**, *19*, 45–76. [CrossRef] [PubMed]

168. Purshouse, R.C.; Fleming, P.J. On the Evolutionary Optimization of Many Conflicting Objectives. *IEEE Trans. Evol. Comput.* **2007**, *11*, 770–784. [CrossRef]

169. Ishibuchi, H.; Tsukamoto, N.; Nojima, Y. Evolutionary Many-Objective Optimization: A short Review. In Proceedings of the 2008 IEEE Congress on Evolutionary Computation, Hong Kong, China, 1–6 June 2008; pp. 2419–2426.

170. Von Lücken, C.; Barán, B.; Brizuela, C. A survey on multi-objective evolutionary algorithms for many-objective problems. *Comput. Optim. Appl.* **2014**, *58*, 707–756. [CrossRef]

171. Yang, S.; Li, M.; Liu, X.; Zheng, J. A Grid-Based Evolutionary Algorithm for Many-Objective Optimization. *IEEE Trans. Evol. Comput.* **2013**, *17*, 721–736. [CrossRef]

172. Li, B.; Li, J.; Tang, K.; Yao, X. Many-Objective Evolutionary Algorithms: A Survey. *ACM Comput. Surv. (CSUR)* **2015**, *48*, 13. [CrossRef]

173. Alves, M.J.; Clímaco, J. A review of interactive methods for multiobjective integer and mixed-integer programming. *Eur. J. Oper. Res.* **2007**, *180*, 99–115. [CrossRef]

174. Monz, M.; Küfer, K.H.; Bortfeld, T.R.; Thieke, C. Pareto navigation—Algorithmic foundation of interactive multi-criteria IMRT planning. *Phys. Med. Biol.* **2008**, *53*, 985–998. [CrossRef] [PubMed]

175. Eskelinen, P.; Miettinen, K.; Klamroth, K.; Hakanen, J. Pareto navigator for interactive nonlinear multiobjective optimization. *OR Spectr.* **2010**, *32*, 211–227. [CrossRef]

176. Cuate, O.; Lara, A.; Schütze, O. A Local Exploration Tool for Linear Many Objective Optimization Problems. In Proceedings of the 13th International Conference on Electrical Engineering, Computing Science and Automatic Control, Mexico City, Mexico, 26–30 September 2016.

177. Cuate, O.; Derbel, B.; Liefooghe, A.; Talbi, E.G. An Approach for the Local Exploration of Discrete Many Objective Optimization Problems. In Proceedings of the 9th International Conference Evolutionary Multi-Criterion Optimization, Münster, Germany, 19–22 March 2017; Trautmann, H., Rudolph, G., Klamroth, K., Schütze, O., Wiecek, M., Jin, Y., Grimme, C., Eds.; Springer International Publishing: Basel, Switzerland, 2017; pp. 135–150.

178. Martin, A.; Schütze, O. Pareto Tracer: A predictor-corrector method for multi-objective optimization problems. *Eng. Optim.* **2018**, *50*, 516–536. [CrossRef]

179. Brockhoff, D.; Zitzler, E. Objective Reduction in Evolutionary Multiobjective Optimization: Theory and Applications. *Evol. Comput.* **2009**, *17*, 135–166. [CrossRef] [PubMed]

180. Gu, F.; Liu, H.L.; Cheung, Y.M. A Fast Objective Reduction Algorithm Based on Dominance Structure for Many Objective Optimization. In *Simulated Evolution and Learning*; Shi, Y., Tan, K.C., Zhang, M., Tang, K., Li, X., Zhang, Q., Tan, Y., Middendorf, M., Jin, Y., Eds.; Springer International Publishing: Basel, Switzerland, 2017; pp. 260–271.

181. Deb, K.; Saxena, D.K. On Finding Pareto-Optimal Solutions Through Dimensionality Reduction for Certain Large-Dimensional Multi-Objective Optimization Problems. *Kangal Report* **2005**. Available online: http://citeseerx.ist.psu.edu/viewdoc/summary?doi=10.1.1.461.3039 (accessed on 31 May 2018).

182. Saxena, D.K.; Duro, J.A.; Tiwari, A.; Deb, K.; Zhang, Q. Objective reduction in many-objective optimization: Linear and nonlinear algorithms. *IEEE Trans. Evol. Comput.* **2013**, *17*, 77–99. [CrossRef]

183. Li, Y.; Liu, H.; Gu, F. An objective reduction algorithm based on hyperplane approximation for many-objective optimization problems. In Proceedings of the 2016 IEEE Congress on Evolutionary Computation, Vancouver, BC, Canada, 24–29 July 2016; pp. 2470–2476.

184. Wang, H.; Yao, X. Objective reduction based on nonlinear correlation information entropy. *Soft Comput.* **2016**, *20*, 2393–2407. [CrossRef]

185. Bandyopadhyay, S.; Mukherjee, A. An Algorithm for Many-Objective Optimization With Reduced Objective Computations: A Study in Differential Evolution. *IEEE Trans. Evol. Comput.* **2015**, *19*, 400–413. [CrossRef]

186. Wang, H.; Jiao, L.; Shang, R.; He, S.; Liu, F. A Memetic Optimization Strategy Based on Dimension Reduction in Decision Space. *Evol. Comput.* **2010**, *23*, 69–100. [CrossRef] [PubMed]

187. Singh, H.K.; Isaacs, A.; Ray, T. A Pareto Corner Search Evolutionary Algorithm and Dimensionality Reduction in Many-Objective Optimization Problems. *IEEE Trans. Evol. Comput.* **2011**, *15*, 539–556. [CrossRef]

188. Mueller-Gritschneder, D.; Graeb, H.; Schlichtmann, U. A successive approach to compute the bounded pareto front of practical multiobjective optimization problems. *SIAM J. Optim.* **2009**, *20*, 915–934. [CrossRef]

189. Motta, R.D.S.; Afonso, S.M.B.; Lyra, P.R.M. A modified NBI and NC method for the solution of N-multiobjective optimization problems. *Struct. Multidiscip. Optim.* **2012**, *46*, 239–259. [CrossRef]

190. Gebken, B.; Peitz, S.; Dellnitz, M. On the hierarchical structure of Pareto critical sets. *arXiv* **2018**, arXiv:1803.06864.

191. Budišić, M.; Mohr, R.; Mezić, I. Applied Koopmanism. *Chaos* **2012**, *22*, 047510. [CrossRef] [PubMed]

192. Rowley, C.W.; Mezić, I.; Bagheri, S.; Schlatter, P.; Henningson, D.S. Spectral analysis of nonlinear flows. *J. Fluid Mech.* **2009**, *641*, 115–127. [CrossRef]

193. Schmid, P.J. Dynamic mode decomposition of numerical and experimental data. *J. Fluid Mech.* **2010**, *656*, 5–28. [CrossRef]

194. Tu, J.H.; Rowley, C.W.; Luchtenburg, D.M.; Brunton, S.L.; Kutz, J.N. On Dynamic Mode Decomposition: Theory and Applications. *J. Comput. Dyn.* **2014**, *1*, 391–421.

195. Proctor, J.L.; Brunton, S.L.; Kutz, J.N. Dynamic mode decomposition with control. *SIAM J. Appl. Dyn. Syst.* **2015**, *15*, 142–161. [CrossRef]

196. Korda, M.; Mezić, I. Linear predictors for nonlinear dynamical systems: Koopman operator meets model predictive control. *arXiv* **2016**, arXiv:1611.03537.

197. Peitz, S.; Klus, S. Koopman operator-based model reduction for switched-system control of PDEs. *arXiv* **2017**, arXiv:1710:06759.

198. Peitz, S. Controlling nonlinear PDEs using low-dimensional bilinear approximations obtained from data. *arXiv* **2018**, arXiv:1801.06419.

199. Abraham, I.; De La Torre, G.; Murphey, T.D. Model-Based Control Using Koopman Operators. *arXiv* **2017**, arXiv:1709.01568.

200. Hanke, S.; Peitz, S.; Wallscheid, O.; Klus, S.; Böcker, J.; Dellnitz, M. Koopman Operator Based Finite-Set Model Predictive Control for Electrical Drives. *arXiv* **2018**, arXiv:1804.00854.

201. Duriez, T.; Brunton, S.L.; Noack, B.R. *Machine Learning Control—Taming Nonlinear Dynamics and Turbulence*; Springer: Berlin/Heidelberg, Germany, 2017.

© 2018 by the authors. Licensee MDPI, Basel, Switzerland. This article is an open access article distributed under the terms and conditions of the Creative Commons Attribution (CC BY) license (http://creativecommons.org/licenses/by/4.0/).

Mathematical and Computational Applications

MDPI

Article

A (p,q)-Averaged Hausdorff Distance for Arbitrary Measurable Sets

Johan M. Bogoya [1,*]**, Andrés Vargas** [1]**, Oliver Cuate** [2] **and Oliver Schütze** [2]

[1]   Department of Mathematics, Pontificia Universidad Javeriana, 110231 Bogotá, Colombia;
    a.vargasd@javeriana.edu.co
[2]   Computer Science Department, Cinvestav-IPN, 07360 Mexico City, Mexico;
    ocuate@computacion.cs.cinvestav.mx (O.C.); schuetze@cs.cinvestav.mx (O.S.)
*   Correspondence: jbogoya@javeriana.edu.co; Tel.: +57-312-5588102

Received: 22 July 2018; Accepted: 16 September 2018; Published: 18 September 2018

Abstract: The Hausdorff distance is a widely used tool to measure the distance between different sets. For the approximation of certain objects via stochastic search algorithms this distance is, however, of limited use as it punishes single outliers. As a remedy in the context of evolutionary multi-objective optimization (EMO), the averaged Hausdorff distance Δ_p has been proposed that is better suited as an indicator for the performance assessment of EMO algorithms since such methods tend to generate outliers. Later on, the two-parameter indicator $\Delta_{p,q}$ has been proposed for finite sets as an extension to Δ_p which also averages distances, but which yields some desired metric properties. In this paper, we extend $\Delta_{p,q}$ to a continuous function between general bounded subsets of finite measure inside a metric measure space. In particular, this extension applies to bounded subsets of $\mathbb{R}^k$ endowed with the Euclidean metric, which is the natural context for EMO applications. We show that our extension preserves the nice metric properties of the finite case, and finally provide some useful numerical examples that arise in EMO.

Keywords: averaged Hausdorff distance; evolutionary multi-objective optimization; power means; metric measure spaces; performance indicator; Pareto front

---

1. Introduction

The Hausdorff distance d_H (see [1]) is an established and widely used tool to measure the proximity of different sets. It is, among others, used in several research fields such as image matching (e.g., [2–4]), the approximation of manifolds in dynamical systems ([5–7]), in fractal geometry ([8]), or in the context of convergence analysis in multi-objective optimization ([9–13]). One major reason for the use of d_H is that it defines a metric on the set of all nonempty bounded closed sets in a metric space. However, one characteristic of the Hausdorff distance is that it heavily punishes single outliers which is a severe drawback in many cases. For instance, it is known that stochastic search algorithms are generally quite effective in the (global) approximation of certain objects, however, it is also known that these approximations may come with a few outliers (e.g., [14]). For those cases, the approximation quality is not reflected by the value of the Hausdorff distance.

As a remedy, in the context of evolutionary multi-objective optimization, Schütze et al. [14] have made a first effort to propose the averaged Hausdorff distance Δ_p. As opposed to d_H, this indicator averages the distances involved in the proximity measure of the given sets and is hence much more suitable in the context of stochastic search as single (or few) outliers in a candidate solution set are not punished hard any more. On the other hand, compared to d_H, Δ_p has two shortcomings: (*i*) it only defines an inframetric instead of a metric; and (*ii*) it is only defined for *finite* approximations of the solution set. In the particular context of continuous multi-objective optimization, it is known that the solution set, the so-called Pareto set, and its image, the Pareto front, form manifolds of certain dimensions. Hence, it is natural that the candidate solution set (i.e., the set computed by a given solver) is not restricted to finitely many points, but may also form a continuous set. This is in fact already the case for set-based optimization techniques such as the cell-to-cell mappings ([15–17]) and the subdivision techniques ([10,18,19]). In the context of evolutionary multi-objective optimization, typically a finite set of candidate solutions (a *population*) is generated ([20–23]). However, also here it is a rather natural approach to construct a continuous set out of the final population using, e.g., interpolation techniques (see [24,25]).

In [26], a modification of the Δ_p indicator called the (p,q)-averaged Hausdorff distance $\Delta_{p,q}$ has been introduced by the first two authors. This indicator generalizes the averaged Hausdorff distance Δ_p, is strongly related to the Hausdorff distance d_H, and admits an expression in terms of the matrix $\ell_{p,q}$-norm $\| \cdot \|_{p,q}$. Moreover, when $1 \leqslant p,q < \infty$ it is a proper metric, while for the remaining cases where $|p|, |q| \geqslant 1$ it is still an inframetric. In addition, when finding optimal archives the parameters p and q play crucial geometrical roles. More precisely, in the context of EMO, p handles the closeness to the Pareto front and q handles the dispersion. The indicator, however, is restricted to finite sets.

In this work, we propose a more general version of the $\Delta_{p,q}$ indicator that can be applied to general measurable subsets and that preserves the useful advantages of the finite case. Consideration is also given to the Pareto-compliance of an intermediate indicator $GD_{p,q}$ that is employed to define $\Delta_{p,q}$. The indicator is hence the first one that can be used in the context of multi-objective optimization using continuous approximations of the Pareto set/front as described above. Numerical results on two well known evolutionary algorithms will show the benefit of such continuous archives compared to discrete ones that have been used so far in lack of a suitable performance indicator.

This paper is organized as follows: In Section 2, we briefly state the background required for the understanding of this work. In Section 3, we introduce the extended version of the $GD_{p,q}$ and $\Delta_{p,q}$ indicators, discussing their properties and providing some sufficient criteria for the Pareto compliance of the first one. In Section 4, we present some numerical results that show the applicability and the benefit of the novel indicator in particular in the context of multi-objective optimization. Finally, we draw our conclusions and present possible paths for future research in Section 5.

2. Preliminaries

In this section, we briefly present the required background on integral power means and multi-objective optimization that will be needed for our purposes. Throughout the document we employ the notation $\mathbb{R}^\times := \mathbb{R} \setminus \{0\}$ and $\overline{\mathbb{R}} := [-\infty, \infty]$ for simplicity.

2.1. Integral Power Means

The theory can be presented in the general setting of metric measure spaces, briefly outlined below, but for simplicity the reader may assume that the specific context of our interest is that of well-behaved bounded subsets of the n-dimensional Euclidean space $\mathbb{R}^n$ endowed with its standard Lebesgue measure which gives rise to the conventional notion of volume (when it is defined). For a quick review of measure spaces see [27] (Section 1.4), and for a simple explanation of the Lebesgue measure see [28] (Chapter 2). Integral means appear already in [29] (Chapter 6). A comprehensive account on the properties of means can be found in [30].

For greater generality, we recall that (Σ, d, μ) is called a metric measure space if (Σ, d) is a metric space with a measure μ defined on its Borel σ-algebra $\mathfrak{M}(\Sigma)$, i.e., the smallest σ-algebra containing all the open subsets of the metric topology of (Σ, d). A measure μ is said to be finite if $\mu(\Sigma) < \infty$, and in this case Σ is called a finite-measure space.

Now, given $p \in \mathbb{R}^\times$ and any measurable function $f \colon X \subset \Sigma \to [0, \infty)$ over a finite-measure set X, we can define the *p-average* of f over X (or the *p-power mean* of f over X), by

$$\underset{x \in X}{\mathcal{M}^p}(f(x)) := \left(\frac{1}{\mu(X)} \int_X f(x)^p \, d\mu \right)^{\frac{1}{p}}. \tag{1}$$

Henceforth, the integral at the RHS will be abbreviated as

$$\fint_X f^p \, d\mu := \frac{1}{\mu(X)} \int_X f(x)^p \, d\mu.$$

If necessary, when the measure μ is clear from the context, the element $d\mu$ will be written as dx to emphasize the variable of integration x. In addition, the notation $\mathcal{M}^p(f(X)) \equiv \underset{x \in X}{\mathcal{M}^p}(f(x))$ and $|X| \equiv \mu(X)$ will also be employed to simplify expressions whenever appropriate.

Let us note that for $p \geqslant 1$ we have $\mathcal{M}^p(f(X)) = \mu(X)^{-\frac{1}{p}} \|f\|_p$, where $\| \cdot \|_p$ denotes the *p-norm* of the Lebesgue space $L^p(X, \mu)$. Furthermore, it is not difficult to show, with the aid of L'Hôpital's rule, that the integral power mean $\mathcal{M}^p$ can be extended to the cases $p = \pm\infty$. Indeed, if $f \not\equiv 0$, denoting the essential supremum and essential infimum of f on X by $\|f\|_\infty := \text{ess sup}_{x \in X} f(x)$ and $\|1/f\|_\infty^{-1} := \text{ess inf}_{x \in X} f(x)$, respectively, it follows that

$$\underset{x \in X}{\mathcal{M}^\infty}(f(x)) := \lim_{p \to \infty} \left(\fint_X f^p \, d\mu \right)^{\frac{1}{p}} = \|f\|_\infty \lim_{p \to \infty} \left(\fint_X \left(\frac{f(x)}{\|f\|_\infty} \right)^p d\mu \right)^{\frac{1}{p}} = \|f\|_\infty,$$

because the last integrand is smaller than 1 and the limit is 1. Similarly,

$$\underset{x \in X}{\mathcal{M}^{-\infty}}(f(x)) := \lim_{p \to -\infty} \left(\fint_X f^p \, d\mu \right)^{\frac{1}{p}} = \lim_{p \to \infty} \left(\fint_X \left(\frac{1}{f} \right)^p d\mu \right)^{-\frac{1}{p}} = \left\| \frac{1}{f} \right\|_\infty^{-1}.$$

We recall that $\| \cdot \|_\infty$ is precisely the norm of the Lebesgue space $L^\infty(X, \mu)$. We can also define $\mathcal{M}^p$ when $p = 0$ as follows:

$$\underset{x \in X}{\mathcal{M}^0}(f(x)) := \exp \left(\fint_X \log f \, d\mu \right).$$

It can be considered the integral generalization of the notion of geometric mean for finitely many elements.

2.2. Multi-objective Optimization

As an application of the (p, q)-distances, we will consider in this work continuous multi-objective optimization problems (MOPs). Problems of this kind can be expressed mathematically as

$$\min \{ F(x) \colon x \in Q \subset \mathbb{R}^n \}, \tag{2}$$

where the function F is defined as a vector of objective functions

$$F \colon Q \subset \mathbb{R}^n \to \mathbb{R}^k, \qquad F(x) := (f_1(x), \ldots, f_k(x)).$$

We will assume here that all objectives $f_i \colon X \to \mathbb{R}$, for $i \in \{1, \ldots k\}$, are continuous. The optimality of MOPs is typically defined via the concept of dominance (see [31]).

Definition 1. *In the context of MOPs the following are standard notions:*

(i) *Let $v = (v_1, \ldots, v_k)$ and $w = (w_1, \ldots, w_k) \in \mathbb{R}^k$. Then the vector v is less than w (denoted $v <_p w$), if $v_i < w_i$ for all $i \in \{1, \ldots, k\}$. The relation $\leqslant_p$ is defined analogously.*

(ii)   A vector $y \in Q$ is dominated by a vector $x \in Q$ (in short: $x \prec y$) with respect to (2) if $F(x) \leqslant_P F(y)$ and $F(x) \neq F(y)$, i.e., there exists a $j \in \{1, \ldots, k\}$ such that $f_j(x) < f_j(y)$.

(iii)   A point $x \in Q$ is called Pareto optimal or a Pareto point if there is no $y \in Q$ which dominates x.

(iv)   The set of all Pareto optimal solutions is called the Pareto set, denoted by P_Q.

(v)   The image of the Pareto set, $F(P_Q)$, is called the Pareto front.

It is known that under certain mild smoothness assumptions the Pareto set and the Pareto front define $(k-1)$-dimensional objects [32]. Hence, for set oriented solvers such as cell mapping, subdivision techniques, and evolutionary algorithms, the question naturally arises as to how to measure the approximation quality of the obtained solution set with respect to the Pareto set/front. To accomplish this task, several *performance indicators* have been proposed in the specialized literature. There exist, for instance, the hypervolume indicator [21,33], the R2 indicator [34], the IGD$^+$ [35], and the DOA [36]. Moreover, in the context of multi-criteria decision-making processes, the properties of some distance measures, as the Hamming, Euclidean, and Hausdorff metrics, is investigated in [37,38]. In this work, we will focus on a new variant of the Hausdorff distance [6]. For convenience of the reader, we recall in the following the most important definitions.

Definition 2. *Let $u, v \in \mathbb{R}^n$, arbitrary $A, B \subset \mathbb{R}^n$, and $\|\cdot\|$ be a vector norm. The Hausdorff distance $d_H(\cdot, \cdot)$ is defined as follows:*

(i) $\text{dist}(u, A) := \inf\{\|u - v\| : v \in A\}$,

(ii) $\text{dist}(B, A) := \sup\{\text{dist}(u, A) : u \in B\}$,

(iii) $d_H(A, B) := \max\{\text{dist}(A, B), \text{dist}(B, A)\}$.

The Hausdorff distance d_H is widely used in many fields. It is, however, of limited practical use when measuring the distance of the outcome of a stochastic search method such as an evolutionary algorithm to the Pareto set/front. The main reason for this is that evolutionary algorithms may generate outliers that are punished too strongly by d_H. As a remedy, the *averaged* Hausdorff distance has been proposed in [14]. In this study the vector norm is the 2-norm, i.e., the Euclidean norm.

Definition 3 (Schütze et al. [14]). *For $p \in \mathbb{N}$, and finite sets $A, B \subset \mathbb{R}^n$ the value*

$$\Delta_p(A, B) := \max\{\text{GD}_p(A, B), \text{IGD}_p(A, B)\},$$

where

$$\text{GD}_p(A, B) := \left(\frac{1}{|A|}\sum_{a \in A} d(a, B)^p\right)^{\frac{1}{p}} \quad \text{and} \quad \text{IGD}_p(A, B) := \left(\frac{1}{|B|}\sum_{b \in B} d(b, A)^p\right)^{\frac{1}{p}},$$

is called the averaged *Hausdorff distance between A and B.*

The indicator Δ_p can be viewed as a composition of slight variations of the Generational Distance (GD, see [39]) and the Inverted Generational Distance (IGD, see [40]). It is $\Delta_\infty = d_H$, but for finite values of p the indicator Δ_p averages the distances considered in d_H. More precisely, the larger the value of p, the harder single outliers will be punished by Δ_p. Hence, as opposed to d_H, the distance Δ_p does not punish single (or few) outliers in a candidate set. For more discussion about Δ_p and its relation to other indicators we refer to [14,41].

Definition 4 (Vargas–Bogoya [26]). *For $p, q \in \mathbb{R}^\times$, and finite sets $A, B \subset \mathbb{R}^n$ the value*

$$\Delta_{p,q}(A, B) := \max\{\text{GD}_{p,q}(A, B \setminus A), \text{GD}_{p,q}(B, A \setminus B)\},$$

where $\mathrm{GD}_{p,q}(A, B) := \left(\frac{1}{|A|} \sum_{a \in A} \left(\frac{1}{|B|} \sum_{b \in B} d(a,b)^q \right)^{\frac{p}{q}} \right)^{\frac{1}{p}}$ *is called the* (p, q)-*averaged Hausdorff distance between A and B.*

For finite sets, the indicator $\Delta_{p,q}$, introduced in [26] was also defined for p or $q = 0$, and even for p or $q = \pm\infty$. It is a generalization of Δ_p in the sense that between disjoint subsets we have

$$\lim_{q \to -\infty} \Delta_{p,q} = \Delta_p.$$

The parameters p and q can be independently modified in order to produce customary spread archives (depending on q) located with customary closeness (depending on p) to the Pareto front.

Finally, let us recall that one characteristic of a performance indicator is *Pareto compliance*: for two subsets A and B we say that $A \preceq B$ if for every $b \in B$ there exists an element $a \in A$ such that $a \preceq b$. If this does not hold, we write $A \npreceq B$. We say that a performance indicator $\mathcal{I}$ is Pareto compliant if for any two sets A and B with $A \preceq B$ and $B \npreceq A$ it follows $\mathcal{I}(A) \leqslant \mathcal{I}(B)$. We refer to [42] for details.

3. The (p, q)-Averaged Hausdorff Distance for Measurable Sets

3.1. Properties of Integral Power Means

We start summarizing some fundamental properties of integral power means that we will need for our subsequent calculations.

Theorem 1. *Let X and Y denote finite-measure spaces, $f, g: X \to [0, \infty)$ non-negative measurable functions, and $d: X \times Y \to [0, \infty)$ a measurable function with respect to the product measure on $X \times Y$. The integral power mean $\mathcal{M}$ satisfies the following properties:*

(i)   *If $p \in \overline{\mathbb{R}}$ and $k \in [0, \infty)$, then $\underset{x \in X}{\mathcal{M}^p}(k) = k$ and $\underset{x \in X}{\mathcal{M}^p}(kf(x)) = k \underset{x \in X}{\mathcal{M}^p}(f(x))$.*

(ii)   *For any $p \in \overline{\mathbb{R}}$, we have $\underset{x \in X}{\mathcal{M}^p}\left(\underset{y \in Y}{\mathcal{M}^p}(d(x,y)) \right) = \underset{y \in Y}{\mathcal{M}^p}\left(\underset{x \in X}{\mathcal{M}^p}(d(x,y)) \right)$.*

(iii)   *If $1 \leqslant p \leqslant \infty$, then $\underset{x \in X}{\mathcal{M}^p}(f(x) + g(x)) \leqslant \underset{x \in X}{\mathcal{M}^p}(f(x)) + \underset{x \in X}{\mathcal{M}^p}(g(x))$.*

(iv)   *If $p \in \overline{\mathbb{R}}$ and $f(x) \leqslant g(x)$ for all $x \in X$, then $\underset{x \in X}{\mathcal{M}^p}(f(x)) \leqslant \underset{x \in X}{\mathcal{M}^p}(g(x))$.*

(v)   *For $p, q \in \overline{\mathbb{R}}$ with $0 < p \leqslant q$, we have that $\underset{x \in X}{\mathcal{M}^p}(f(x)) \leqslant \underset{x \in X}{\mathcal{M}^q}(f(x))$.*

Proof. The proofs of (i) and (ii) are straightforward. To prove (iii) we only need the Minkowski inequality,

$$\underset{x \in X}{\mathcal{M}^p}(f(x) + g(x)) = \mu(X)^{-\frac{1}{p}} \|f + g\|_p$$

$$\leqslant \mu(X)^{-\frac{1}{p}} \left(\|f\|_p + \|g\|_p \right) = \underset{x \in X}{\mathcal{M}^p}(f(x)) + \underset{x \in X}{\mathcal{M}^p}(g(x)).$$

The proof of (iv) is also straightforward from the definitions and a simple proof of (v) can be given as a particular case of [43] (Theorem 3) which we recall here for completeness. For a positive real v, consider the function

$$\omega_r(v) := \int_1^v t^{r-1} \, dt = \begin{cases} \dfrac{v^r - 1}{r}, & r \neq 0; \\[2mm] \log v, & r = 0. \end{cases}$$

Since the function t^u, with t a positive constant and $u \geqslant 0$ is increasing with respect to u, we easily get $\omega_r(v) \leqslant \omega_s(v)$ for $0 \leqslant r \leqslant s$ and every $v \geqslant 0$. Consider the following linear integral operator

$$\mathcal{J}[f] := \frac{1}{\mu(X)} \int_X f \, d\mu = \fint_X f \, d\mu.$$

Assume first, that $p \neq 0$, then for any $x \in X$,

$$\mathcal{J}\left[\omega_p\left(\frac{f(x)}{\mathcal{M}^p(f(X))}\right)\right] = \frac{1}{p}\left(\mathcal{J}\left[\frac{f(x)}{\mathcal{M}^p(f(X))}\right]^p - \mathcal{J}[1]\right)$$

$$= \frac{1}{p}\left(\frac{1}{(\mathcal{M}^p(f(X)))^p}\fint_X f^p \, d\mu - \fint_X d\mu\right) = 0.$$

Similarly, if $p = 0$ we have

$$\mathcal{J}\left[\omega_0\left(\frac{f(x)}{\mathcal{M}^0(f(X))}\right)\right] = \mathcal{J}[\log(f(x)) - \log(\mathcal{M}^0(f(X)))]$$

$$= \fint_X \log f \, d\mu - \log(\mathcal{M}^0(f(X)))\mathcal{J}[1] = 0.$$

Suppose that $0 \leqslant p < q$. Since $\omega_p(\cdot) \leqslant \omega_q(\cdot)$ implies $\mathcal{J}(\omega_p(\cdot)) \leqslant \mathcal{J}(\omega_q(\cdot))$, we obtain

$$0 \leqslant \mathcal{J}\left[\omega_q\left(\frac{f(x)}{\mathcal{M}^p(f(X))}\right)\right] = \frac{1}{q}\left[\left(\frac{\mathcal{M}^q(f(X))}{\mathcal{M}^p(f(X))}\right)^q - 1\right],$$

from which it follows that $\mathcal{M}^p_{x \in X}(f(x)) \leqslant \mathcal{M}^q_{x \in X}(f(x))$. $\quad\square$

3.2. Definition of $\Delta_{p,q}$ for Measurable Sets

With the aid of Theorem 1 we generalize the results of [26] (Section 3). For easy reference, we provide here slightly abbreviated but complete proofs. Given a metric measure space (Σ, d, μ), let $\mathfrak{M}(\Sigma)$ denote the σ-algebra of all measurable subsets of Σ and let $\mathfrak{M}_{<\infty}(\Sigma)$ refer to those elements of $\mathfrak{M}(\Sigma)$ having finite measure. As it should be expected from the context, any set relation obtained from calculations involving an underlying measure μ should be understood to hold in a measure-theoretic sense, i.e., almost everywhere (*a.e.*). For example, for $X, Y \in \mathfrak{M}_{<\infty}(\Sigma)$, a result saying $X = Y$, or $X \subset Y$ actually holds almost everywhere, which means that $\mu\{X \neq Y\} = 0$, or $\mu\{X \not\subset Y\} = 0$, respectively. Thus, it is convenient in this setting to identify a set $X \in \mathfrak{M}_{<\infty}(\Sigma)$ with the whole equivalence class $[X] := \{Y \mid X = Y, \, a.e.\}$, and think of these classes as the elements of $\mathfrak{M}_{<\infty}(\Sigma)$ to remove the need for the *a.e.* abbreviation. Also, to avoid an overload of parentheses in the forthcoming expressions, the distance $d(x, y)$ between $x, y \in \Sigma$ will be abbreviated by $d_{x,y}$.

Definition 5. *For $p, q \in \mathbb{R}^\times$, the generational (p, q)-distance $\mathrm{GD}_{p,q}(X, Y)$ between two sets $X, Y \in \mathfrak{M}_{<\infty}(\Sigma)$ is given by*

$$\mathrm{GD}_{p,q}(X, Y) := \mathcal{M}^p_{x \in X}\left(\mathcal{M}^q_{y \in Y}(d_{x,y})\right) = \left(\fint_X \left(\fint_Y d^q_{x,y} \, dy\right)^{\frac{p}{q}} dx\right)^{\frac{1}{p}},$$

where the sets X and Y are implicitly assumed to be disjoint when $p < 0$ or $q < 0$.

As in the finite case, the definition of $\mathrm{GD}_{p,q}$ can be easily extended for $p, q \in \overline{\mathbb{R}}$, but still has two undesirable drawbacks, first $\mathrm{GD}_{p,q}(X, X)$ can be different from zero, and second, in general the values of $\mathrm{GD}_{p,q}(X, Y)$ and $\mathrm{GD}_{p,q}(Y, X)$ can be different, thus this indicator does not define a metric. To obtain a proper metric we introduce the following modification.

Math. Comput. Appl. **2018**, 23, 51

Definition 6. *The (p,q)-averaged Hausdorff distance is the map $\Delta_{p,q}\colon \mathfrak{M}_{<\infty}(\Sigma) \times \mathfrak{M}_{<\infty}(\Sigma) \to [0,\infty)$ given by*

$$\Delta_{p,q}(X,Y) := \max\left\{ \mathrm{GD}_{p,q}(X, Y \setminus X), \mathrm{GD}_{p,q}(Y, X \setminus Y)\right\}.$$

Remark 1. *For finite subsets $X, Y \subset \mathbb{R}^n$ endowed with the standard counting measure μ, the previous notions of $\mathrm{GD}_{p,q}$ and $\Delta_{p,q}$ coincide with the ones in Definition 4.*

Figure 1 illustrates how the shape of $\Delta_{p,q}$-metric balls $B_\varepsilon := \{x \in \mathbb{R}^2 : \Delta_{p,q}(A,x) \leqslant \varepsilon\}$ around a discrete set A of ten points (that approximates a segment of negative slope in the plane) varies as p and q take several different values. Notice that for negative values of p and q the balls' shape resemble the shape of A and enclose all of its points.

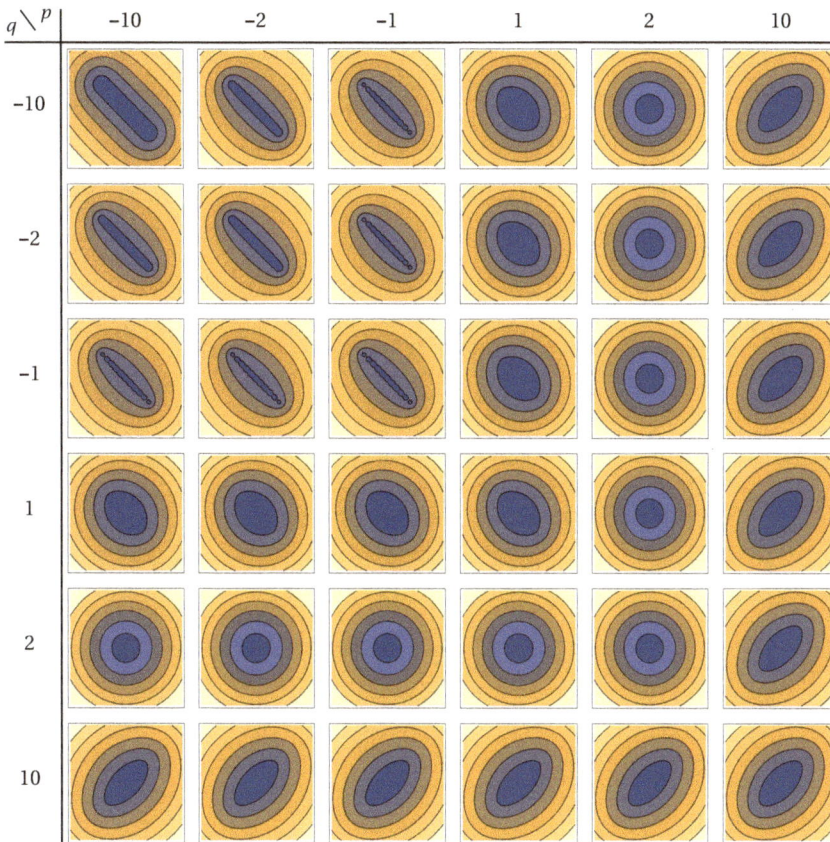

Figure 1. Table of $\Delta_{p,q}$-neighborhoods of increasing radius around a discrete set of ten equidistant points along the line $y = -x$ in $\mathbb{R}^2$, showing how their shape change for different values of p and q.

3.3. Metric Properties

The extension of $\Delta_{p,q}$ to measurable sets given in Definition 6 preserves the nice metric properties of the finite version considered in [26] (Section 3). In particular, Theorem 1 enables us to show a result analogous to [26] (Theorem 3.3).

Theorem 2. *Suppose that $1 \leqslant p, q < \infty$. Then the generational (p,q)-distance $\mathrm{GD}_{p,q}$ satisfies the triangle inequality, namely*

$$\mathrm{GD}_{p,q}(X, Z) \leqslant \mathrm{GD}_{p,q}(X, Y) + \mathrm{GD}_{p,q}(Y, Z)$$

for any sets $X, Y, Z \in \mathfrak{M}_{<\infty}(\Sigma)$.

Proof. From the triangle inequality for the metric $d(\cdot, \cdot)$ we have

$$d_{x,z} \leqslant d_{x,y} + d_{y,z} \quad (x \in X, \ y \in Y, \ z \in Z).$$

Taking the q-average over Z at both sides and using Theorem 1 *(i)–(iii)*, yields

$$\underset{z \in Z}{\mathcal{M}^q}(d_{x,z}) \leqslant \underset{z \in Z}{\mathcal{M}^q}(d_{x,y} + d_{y,z}) \leqslant \underset{z \in Z}{\mathcal{M}^q}(d_{x,y}) + \underset{z \in Z}{\mathcal{M}^q}(d_{y,z}) = d_{x,y} + \underset{z \in Z}{\mathcal{M}^q}(d_{y,z}). \tag{3}$$

Now, we consider two cases for the parameters $1 \leqslant p, q < \infty$, independently.

Case $p \leqslant q$: Taking at both sides of (3) the p-average over X and using Theorem 1 *(i)*, *(iii)*, and *(iv)*, we get

$$\underset{x \in X}{\mathcal{M}^p}\left(\underset{z \in Z}{\mathcal{M}^q}(d_{x,z}) \right) \leqslant \underset{x \in X}{\mathcal{M}^p}\left(d_{x,y} + \underset{z \in Z}{\mathcal{M}^q}(d_{y,z}) \right) = \underset{x \in X}{\mathcal{M}^p}(d_{x,y}) + \underset{z \in Z}{\mathcal{M}^q}(d_{y,z}). \tag{4}$$

In this expression, the LHS is precisely $\mathrm{GD}_{p,q}(X, Z)$ which does not depend on Y. We now take the p-average over Y at both sides of (4) and use Theorem 1 *(i)*, *(iii)*, and *(iv)*, to obtain

$$\mathrm{GD}_{p,q}(X, Z) \leqslant \underset{y \in Y}{\mathcal{M}^p}\left(\underset{x \in X}{\mathcal{M}^p}(d_{x,y}) + \underset{z \in Z}{\mathcal{M}^q}(d_{y,z}) \right) = \underset{y \in Y}{\mathcal{M}^p}\left(\underset{x \in X}{\mathcal{M}^p}(d_{x,y}) \right) + \mathrm{GD}_{p,q}(Y, Z).$$

To finish this case note that from Theorem 1 *(ii)*, *(iv)*, and *(v)*, we have that

$$\underset{y \in Y}{\mathcal{M}^p}\left(\underset{x \in X}{\mathcal{M}^p}(d_{x,y}) \right) = \underset{x \in X}{\mathcal{M}^p}\left(\underset{y \in Y}{\mathcal{M}^p}(d_{x,y}) \right) \leqslant \underset{x \in X}{\mathcal{M}^p}\left(\underset{y \in Y}{\mathcal{M}^q}(d_{x,y}) \right) = \mathrm{GD}_{p,q}(X, Y)$$

which proves the claim.

Case $q \leqslant p$: Here, we note that the LHS of (3) does not depend on Y, and take at both sides of (3) the q-average over Y. Hence, Theorem 1 *(i)*, *(iii)–(v)* yield

$$\underset{z \in Z}{\mathcal{M}^q}(d_{x,z}) \leqslant \underset{y \in Y}{\mathcal{M}^q}\left(d_{x,y} + \underset{z \in Z}{\mathcal{M}^q}(d_{y,z}) \right) \leqslant \underset{y \in Y}{\mathcal{M}^q}(d_{x,y}) + \mathrm{GD}_{p,q}(Y, Z).$$

Lastly, we take the p-average over X and use Theorem 1 *(ii)–(iv)*, to obtain

$$\mathrm{GD}_{p,q}(X, Z) \leqslant \underset{x \in X}{\mathcal{M}^p}\left(\underset{y \in Y}{\mathcal{M}^q}(d_{x,y}) + \mathrm{GD}_{p,q}(Y, Z) \right) = \mathrm{GD}_{p,q}(X, Z) + \mathrm{GD}_{p,q}(Y, Z),$$

which is the required result. $\square$

Corollary 1. *For $p, q \in \mathbb{R}^\times$ the (p,q)-averaged Hausdorff distance $\Delta_{p,q}$ is a semimetric on the collection $\mathfrak{M}_{<\infty}(\Sigma)$ of all measurable subsets of Σ with finite measure. Moreover, between disjoint sets, $\Delta_{p,q}$ is a proper metric on $\mathfrak{M}_{<\infty}(\Sigma)$ for $1 \leqslant p, q < \infty$.*

Proof. Definition 6 easily implies that $\Delta_{p,q}(\cdot, \cdot) \geqslant 0$ as well as $\Delta_{p,q}(X, Y) = \Delta_{p,q}(Y, X)$, for every pair $X, Y \in \mathfrak{M}_{<\infty}(\Sigma)$ and all $p, q \in \mathbb{R}^\times$. From Definition 5 we can see that $\mathrm{GD}_{p,q}(X, Y \setminus X) = 0$ if and only if $X = \varnothing$ or $Y \subseteq X$ (and hence $Y \setminus X = \varnothing$). We thus find, for $X, Y \neq \varnothing$, that

$$\Delta_{p,q}(X, Y) = 0 \ \text{ if and only if } \ X = Y.$$

We have shown that $\Delta_{p,q}$ is a semimetric on $\mathfrak{M}_{<\infty}(\Sigma)$, and since the maximum of two functions satisfying the triangle inequality also satisfies it, Theorem 2 shows that $\Delta_{p,q}$ satisfies the triangle inequality when $1 \leqslant p, q < \infty$. $\square$

Theorem 3. *Suppose that for any sets* $X, Y, Z \in \mathfrak{M}_{<\infty}(\Sigma)$ *there exist some constants* $0 < r < R$ *such that* $r \leqslant d_{u,v} \leqslant R$ *holds for all pairs* (u, v) *in* $X \times Y$, $X \times Z$, *or* $Y \times Z$. *Then, for all non-simultaneously positive* $p, q \in \mathbb{R}^\times$ *with* $|p|, |q| \geqslant 1$ *the generational* (p, q)-*distance* $\mathrm{GD}_{p,q}$ *satisfies the following relaxed triangle inequality*

$$\mathrm{GD}_{p,q}(X, Z) \leqslant \frac{R^2}{r^2}\Big(\mathrm{GD}_{p,q}(X, Y) + \mathrm{GD}_{p,q}(Y, Z)\Big).$$

Proof. We prove the theorem in three steps.

Step 1: Take $p \in \mathbb{R}^\times$ and assume that $q < 0$, we will show that

$$\mathrm{GD}_{p,|q|}(X, Y) \leqslant \frac{R}{r}\,\mathrm{GD}_{p,q}(X, Y). \tag{5}$$

For any $x \in X$ and all $y_1, y_2 \in Y$ we have $\frac{r}{R} \leqslant \frac{d_{x,y_1}}{d_{x,y_2}} \leqslant \frac{R}{r}$, thus

$$\frac{R}{r} \geqslant \left(\fint_Y \fint_Y \left[\frac{d_{x,y_1}}{d_{x,y_2}}\right]^{|q|} dy_1\, dy_2\right)^{\frac{1}{|q|}} = \left(\fint_Y d_{x,y_1}^{|q|}\, dy_1\right)^{\frac{1}{|q|}}\left(\fint_Y d_{x,y_2}^{-|q|}\, dy_2\right)^{\frac{1}{|q|}}.$$

Using the fact that $q = -|q|$, we get

$$\left(\fint_Y d_{x,y_1}^{|q|}\, dy_1\right)^{\frac{1}{|q|}} \leqslant \frac{R}{r}\left(\fint_Y d_{x,y_2}^{q}\, dy_2\right)^{\frac{1}{q}},$$

which by (1), proves that $\underset{y \in Y}{\mathcal{M}^{|q|}}(d_{x,y}) \leqslant \frac{R}{r}\underset{y \in Y}{\mathcal{M}^{q}}(d_{x,y})$. Calculating the p-average $\underset{x \in X}{\mathcal{M}^{p}}$ of both sides, and from Theorem 1 (*i*) and (*iv*), we finally get $\underset{x \in X}{\mathcal{M}^{p}}\Big(\underset{y \in Y}{\mathcal{M}^{|q|}}(d_{x,y})\Big) \leqslant \frac{R}{r}\underset{x \in X}{\mathcal{M}^{p}}\Big(\underset{y \in Y}{\mathcal{M}^{q}}(d_{x,y})\Big)$, which, by Definition 5, is precisely (5).

Step 2: Now, take $q \in \mathbb{R}^\times$ and assume that $p < 0$, we will show that

$$\mathrm{GD}_{|p|,q}(X, Y) \leqslant \frac{R}{r}\,\mathrm{GD}_{p,q}(X, Y). \tag{6}$$

By hypothesis, for any $y \in Y$ and all $x_1, x_2 \in X$ we have $\frac{r}{R} \leqslant \frac{d_{x_1,y}}{d_{x_2,y}} \leqslant \frac{R}{r}$. Therefore, proceeding as before and applying again Theorem 1 (*i*) and (*iv*) we conclude that $\underset{y \in Y}{\mathcal{M}^{q}}(d_{x_1,y}) \leqslant \frac{R}{r}\underset{y \in Y}{\mathcal{M}^{q}}(d_{x_2,y})$. Hence,

$$\left(\fint_X \left(\underset{y \in Y}{\mathcal{M}^{q}}(d_{x_1,y})\right)^{|p|} dx_1\right)^{\frac{1}{|p|}}\left(\fint_X \left(\underset{y \in Y}{\mathcal{M}^{q}}(d_{x_2,y})\right)^{p} dx_2\right)^{\frac{1}{|p|}} = \left(\fint_X \fint_X \left[\frac{\underset{y \in Y}{\mathcal{M}^{q}}(d_{x_1,y})}{\underset{y \in Y}{\mathcal{M}^{q}}(d_{x_2,y})}\right]^{|p|} dx_1\, dx_2\right)^{\frac{1}{|p|}} \leqslant \frac{R}{r},$$

from which we deduce

$$\left(\fint_X \left(\underset{y \in Y}{\mathcal{M}^{q}}(d_{x_1,y})\right)^{|p|} dx_1\right)^{\frac{1}{|p|}} \leqslant \frac{R}{r}\left(\fint_X \left(\underset{y \in Y}{\mathcal{M}^{q}}(d_{x_2,y})\right)^{p} dx_2\right)^{\frac{1}{p}}.$$

Using (1), the previous inequality can be written as

$$\underset{x \in X}{\mathcal{M}^{|p|}}\Big(\underset{y \in Y}{\mathcal{M}^{q}}(d_{x,y})\Big) \leqslant \frac{R}{r}\underset{x \in X}{\mathcal{M}^{p}}\Big(\underset{y \in Y}{\mathcal{M}^{q}}(d_{x,y})\Big),$$

which, by Definition 5, is precisely (6).

Step 3: From the previous two steps we easily obtain

$$\text{GD}_{|p|,|q|}(X,Y) \leqslant \frac{R}{r}\,\text{GD}_{|p|,q}(X,Y) \leqslant \frac{R^2}{r^2}\,\text{GD}_{p,q}(X,Y). \tag{7}$$

Theorem 1 *(iv)* and Definition 5 imply that $\text{GD}_{p,q}(X,Z) \leqslant \text{GD}_{|p|,|q|}(X,Z)$. Finally, the triangle inequality for $\text{GD}_{|p|,|q|}$ (Theorem 2) and (7), produces the desired relation

$$\text{GD}_{p,q}(X,Z) \leqslant \text{GD}_{|p|,|q|}(X,Y) + \text{GD}_{|p|,|q|}(Y,Z) \leqslant \frac{R^2}{r^2}\Big(\text{GD}_{p,q}(X,Y) + \text{GD}_{p,q}(Y,Z)\Big). \quad \square$$

Remark 2. *When the pair (p,q) lies in the light-gray or violet regions of Figure 2, the distance $\text{GD}_{p,q}$ satisfies a relaxed triangle inequality, with the drawback that the constant R^2/r^2 depends on the condition that $r \leqslant d_{u,v} \leqslant R$, for all pairs $(u,v) \in X \times Y, X \times Z$, or $Y \times Z$. For bounded and separated sets this condition always holds, and on those sets the associated (p,q)-averaged Hausdorff distance $\Delta_{p,q}$ becomes an inframetric as the following corollary implies.*

Corollary 2. *Under the same hypothesis of Theorem 3 we have*

$$\Delta_{p,q}(X,Z) \leqslant \frac{R^2}{r^2}\Big(\Delta_{p,q}(X,Y) + \Delta_{p,q}(Y,Z)\Big).$$

Proof. It follows immediately from Theorem 3 and Definition 6. $\square$

Theorem 4. *Let $X,Y \in \mathfrak{M}_{<\infty}(\Sigma)$ and suppose that $p,p',q,q' \in \overline{\mathbb{R}}$ satisfy $p \leqslant p'$ and $q \leqslant q'$. Then*

$$\Delta_{p,q}(X,Y) \leqslant \Delta_{p',q}(X,Y) \quad \text{and} \quad \Delta_{p,q}(X,Y) \leqslant \Delta_{p,q'}(X,Y).$$

Proof. It follows easily from Theorem 1 *(v)* and Definition 6. $\square$

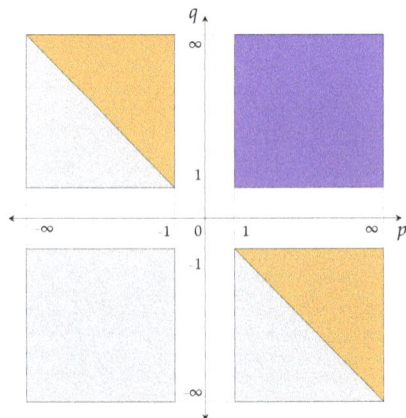

Figure 2. Representation of key regions on the (p,q)-plane. Corollary 1 shows that $\Delta_{p,q}$ is a proper metric in the violet region and Corollary 2 shows that it is an inframetric in the orange and light-gray ones. Numerical evidence suggests that $\Delta_{p,q}$ is still a proper metric in the orange regions.

3.4. Pareto-Compliance

We return now to the setting of MOPs to consider the behavior of the generalized $\mathrm{GD}_{p,q}$ and $\Delta_{p,q}$ distances as performance indicators by studying their Pareto-compliance. A discussion of the Pareto-compliance for the indicators GD_p and Δ_p appeared in [14] (Section 3). Similar observations are valid for these new (p,q)-indicators, but a detailed and complete account of the details is part of ongoing research and will appear elsewhere. Here, as a first approach to the compliance question we present a basic result that describes the behavior of the indicator $\mathrm{GD}_{p,q}$ under stronger assumptions than the compliance notion mentioned at the end of Section 2.2.

Let us assume that given a decision space $Q \subset \mathbb{R}^n$, a MOP has an associated objective function $F: Q \to \mathbb{R}^k$, with objective space $F(Q) \subset \mathbb{R}^k$ endowed with the Euclidean distance $d(\cdot, \cdot)$ and the inherited Lebesgue measure μ. Furthermore, let P_Q denote the Pareto set and $F(P_Q) \subset \mathbb{R}^k$ the corresponding Pareto front. If $X \subset Q$ denotes an approximating subset (or archive), the explicit $\mathrm{GD}_{p,q}$-performance indicators assigned to X is given by

$$\mathcal{I}_{p,q}^{\mathrm{GD}}(X) := \mathrm{GD}_{p,q}(F(X), F(P_Q)).$$

For the following statement, let us recall here that a partition of a set X is a collection of disjoint and non-empty subsets of X whose union is the whole of X. Furthermore, for any $q \in \overline{\mathbb{R}}$ we abbreviate the q-averaged distance of $F(u) \in F(Q)$ to the Pareto front $F(P_Q)$ by $\delta_q(u) := \mathcal{M}_{v \in P_Q}^q(d(F(u), F(v)))$.

Theorem 5. *Suppose that for fixed $p, q \in \overline{\mathbb{R}}$ a pair of measurable archives $X, Y \subset Q$, satisfy that:*

(i)   *X and Y admit finite partitions $X = \bigsqcup_{i=1}^m X_i$ and $Y = \bigsqcup_{i=1}^m Y_i$ such that for each $i \in \{1, \ldots, m\}$:*

    (a)   *$X_i \subset X$ and $Y_i \subset Y$ are subsets of non-null finite measure.*

    (b)   *$\forall x \in X_i, \forall y \in Y_i: x \preceq y$,*

(ii)   *$\forall x \in X, \forall y \in Y$: if $x \preceq y \implies \delta_q(x) \leqslant \delta_q(y)$.*

Then $\mathcal{I}_{p,q}^{\mathrm{GD}}(X) \leqslant \mathcal{I}_{p,q}^{\mathrm{GD}}(Y)$.

Proof. From condition (i) the archives X and Y admit partitions into the same number m of subsets and from (ii) it is clear that for any $i \in \{1, \ldots, m\}$ if $x \in X_i$ and $y \in Y_i$ then $\delta_q(x) \leqslant \delta_q(y)$. Hence, taking integral p-averages over X_i, and then over Y_i of the quantities at both sides of this inequality we obtain for each i that

$$a_i^p := \frac{1}{|X_i|} \int_{X_i} \delta_q(x)^p \, dx \leqslant \frac{1}{|Y_i|} \int_{Y_i} \delta_q(y)^p \, dy =: b_i^p. \tag{8}$$

Now, for each $i \in \{1, \ldots, m\}$ for which the inequality $\frac{|X_i|}{|X|} \leqslant \frac{|Y_i|}{|Y|}$ does not hold, we can further subdivide X_i into a sufficiently large partition of m_i non-null finite measure subsets $X_{i,1}, X_{i,2}, \ldots, X_{i,m_i}$, so that for all $j \in \{1, \ldots, m_i\}$ we can guarantee that

$$w_{i,j} := \frac{|X_{i,j}|}{|X|} \leqslant \frac{|Y_i|}{|Y|} =: \widetilde{w}_i. \tag{9}$$

Please note that this should be possible due to the assumption that X_i has non-null finite measure. Since part (b) of condition (i) still holds for these subsets, (i.e., $\forall x \in X_{i,j}, \forall y \in Y_i: x \preceq y$), an analogous relation to Inequality (8) is valid for them. Explicitly, for each $i \in \{1, \ldots, n\}$ and all $j \in \{1, \ldots, m_i\}$ we have

$$a_{i,j}^p := \frac{1}{|X_{i,j}|} \int_{X_{i,j}} \delta_q(x)^p \, dx \leqslant \frac{1}{|Y_i|} \int_{Y_i} \delta_q(y)^p \, dy =: b_i^p.$$

Due to the chosen partitions of X and Y, it is clear that $|X| = \sum_{i=1}^m |X_i|$, where $|X_i| = \sum_{j=1}^{m_i} |X_{i,j}|$, and $|Y| = \sum_{i=1}^m |Y_i|$. Therefore, with the notation of (9) it follows $\sum_{i=1}^m \sum_{j=1}^{m_i} w_{i,j} = \sum_{i=1}^m \widetilde{w}_i = 1$, which implies that the quantities $w_{i,j}$ and $\tilde{w}_i$ can be regarded as normalized weights appropriate for taking weighted averages. Using that $0 \leqslant a_{i,j} \leqslant b_i$ and $0 \leqslant w_{i,j} \leqslant \tilde{w}_i \leqslant 1$, simple properties of (discrete) weighted power means ensure that $\sum_{i=1}^{m} \sum_{j=1}^{m_i} w_{i,j} a_{i,j}^p \leqslant \sum_{i=1}^{m} \tilde{w}_i b_i^p$. Thus, we can finally write

$$
\mathcal{I}_{p,q}^{GD}(Y)^p = \frac{1}{|X|} \sum_{i=1}^{m} \sum_{j=1}^{m_i} \int_{X_{i,j}} \delta_q(x)^p \, dx = \sum_{i=1}^{m} \sum_{j=1}^{m_j} \frac{|X_{i,j}|}{|X|} a_{i,j}^p = \sum_{i=1}^{m} \sum_{j=1}^{m_i} w_{i,j} a_{i,j}^p
$$

$$
\leqslant \sum_{i=1}^{m} \tilde{w}_i b_i^p = \sum_{i=1}^{m} \frac{|Y_i|}{|Y|} b_i^p = \frac{1}{|Y|} \sum_{i=1}^{m} \int_{Y_i} \delta_q(y)^p \, dy = \mathcal{I}_{p,q}^{GD}(Y)^p,
$$

proving the statement. □

Remark 3. *Condition (i) of Theorem 5 implies the simpler (and weaker) dominance conditions:*

(a') $X \preceq Y$ *(i.e., $\forall y \in Y$, $\exists x \in X$ such that $x \preceq y$), and*
(b') $\forall x \in X$, $\exists y \in Y$ such that $x \preceq y$.*

> In many simple examples for which (a') and (b') hold, it is not difficult to find the partitions needed for Theorem 5 (i), however this is not always possible, and the question of when such partitions exist will not be considered here. Figure 3, show examples where (a') and (b') hold and the inequality $\mathcal{I}_{p,q}^{GD}(X) \leqslant \mathcal{I}_{p,q}^{GD}(Y)$ is both, true (left) and false (right). In these cases it can be shown that X and Y satisfy (left), and do not satisfy (right) the requirements of Theorem 5 (i), respectively.

Remark 4. *Another important observation is that condition (ii) of Theorem 5 allows for some freedom in the choice of an appropriate $q \in \overline{\mathbb{R}}$ such that the inequality $\delta_q(x) \leqslant \delta_q(y)$ holds for $x \preceq y$, ensuring the compliance to Pareto optimality. This freedom is not available for the indicator GD_p because in that case $\delta_q(x)$ should be replaced by the corresponding quantity when $q \to \infty$ which is the standard distance from a set to a point $d(F(x), F(P_Q))$. The possibility to choose a value of q according to the problem is clearly an advantage, and provides an argument in favor of the generalized version $GD_{p,q}$.*

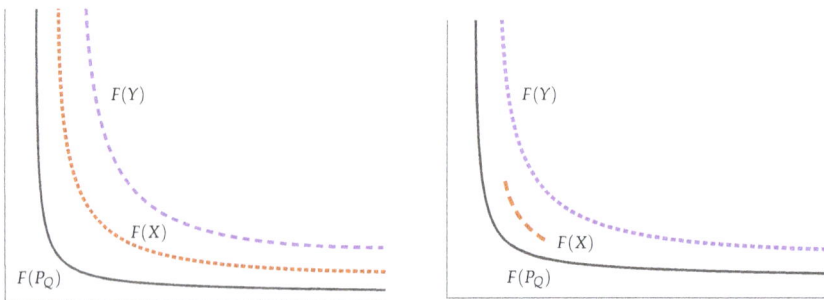

Figure 3. (Left) Example of a Pareto front $F(P_Q)$ with two archives satisfying condition (*i*) of Theorem 5 for which $\mathcal{I}_{p,q}^{GD}(X) \leqslant \mathcal{I}_{p,q}^{GD}(Y)$. (Right) Example of a Pareto front $F(P_Q)$ with two archives satisfying conditions (*a'*) and (*b'*) of Remark 3 but for which $\mathcal{I}_{p,q}^{GD}(X) \nleqslant \mathcal{I}_{p,q}^{GD}(Y)$. In this case partitions of the archives satisfying Theorem 5 (*i*) do not exist.

4. Numerical Examples

In this section, we demonstrate the applicability and usefulness of the new distance measure on two examples.

4.1. General Example

As a first example we consider the following sets within the Euclidean plane $\mathbb{R}^2$: the first set, A, is a line segment connecting two points $a = (-1, 0)$ and $b = (1, 0)$, i.e.,

$$A = \overline{ab}. \tag{10}$$

Next, for some given $\delta > 0$ and any fixed value of $\varepsilon > 0$ we consider sets B_δ defined as the union of line segments

$$B_\delta = \overline{cd_\delta} \cup \overline{e_\delta f_\delta} \cup \overline{g_\delta h}, \tag{11}$$

where $c = (-1, \varepsilon)$, $d_\delta = (-\delta, \varepsilon)$, $e_\delta = (-\delta, 1)$, $f_\delta = (\delta, 1)$, $g_\delta = (\delta, \varepsilon)$, and $h = (1, \varepsilon)$ are the segment end-points in $\mathbb{R}^2$. Hereby, a set B_δ can be seen as a certain approximation of A, where the segment $\overline{e_\delta f_\delta}$ can be considered to be the outlier in the approximation.

Figure 4 shows the sets A and B_δ for the values $\delta \in \{0.05, 0.10, 0.20, 0.40\}$ and $\varepsilon = 0.10$. Apparently, for smaller δ, the outlier region gets smaller, and hence, the approximation B_δ of A gets "better". This is reflected by the values of the (p, q)-distance in Table 1.

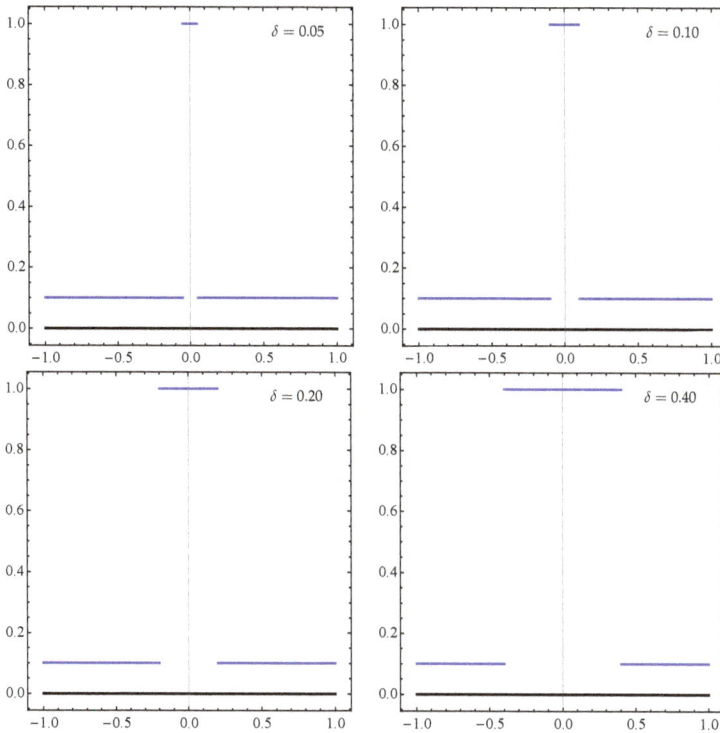

Figure 4. Example of four approximations of A (black horizontal segment) with B_δ (blue piecewise function) for four different values of δ and fixed $\varepsilon = 0.10$.

Table 1. $\Delta_{p,q}$ values for A and B_δ in (10) and (11), for different values of p, q, and δ, with fixed $\varepsilon = 0.10$.

p	q	$\Delta_{pq}(A, B_{0.05})$	$\Delta_{pq}(A, B_{0.10})$	$\Delta_{pq}(A, B_{0.20})$	$\Delta_{pq}(A, B_{0.40})$
1	1	0.7149	0.7464	0.8091	0.9324
1	−1	0.4105	0.4506	0.5311	0.6945
1	−100	0.1503	0.1961	0.2878	0.4711
1	−200	0.1479	0.1934	0.2844	0.4663
1	−10,000	0.1451	0.1901	0.2802	0.4602

On the other hand, if choosing the classical Hausdorff distance, all values of $d_H(A, B_\delta)$ are equal to 1, regardless of the choice of $\delta > 0$. Hence, the (p, q)-distance is more appropriate in this example to identify "better" approximations.

4.2. Approximation of Pareto Sets/Fronts

As a second example we consider the approximation of the Pareto set and front of a given MOP. For this, we define the following bi-objective problem that is known as the *Lamé super-sphere function* [32]:

$$\min_x F \colon \mathbb{R}^n \to \mathbb{R}^2, \tag{12}$$

where $F(x) = (f_1(x), f_2(x))$ is given by

$$f_1(x) = \left(\frac{1}{n} \sum_{i=1}^{n} x_i^2 \right)^{\frac{\gamma}{2}} \quad \text{and} \quad f_2(x) = \left(\frac{1}{n} \sum_{i=1}^{n} (x_i - 1)^2 \right)^{\frac{\gamma}{2}}$$

for $x \in \mathbb{R}^n$ and $\gamma \in \mathbb{R}$. Figures 5 and 6 show the Pareto sets and fronts for the special cases $n = 2$ with $\gamma = 2$ and $\gamma = 1/2$, respectively.

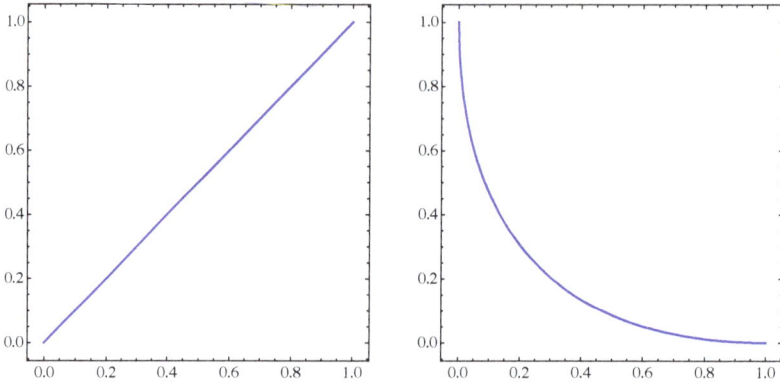

Figure 5. Pareto set (left) and front (right) of MOP (12) for $n = 2$ with $\gamma = 2$.

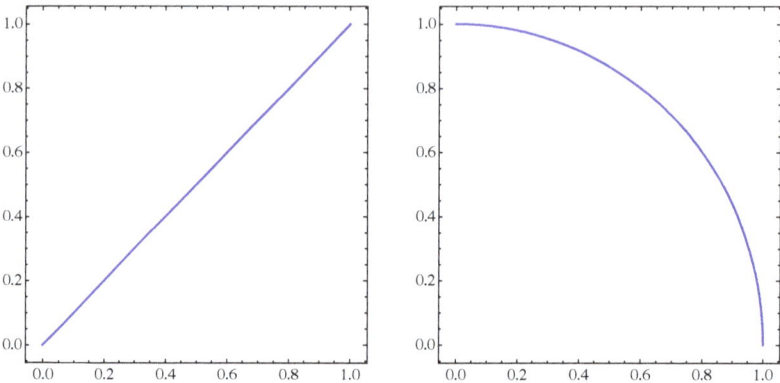

Figure 6. Pareto set (left) and front (right) of MOP (12) for $n = 2$ with $\gamma = 1/2$.

For the first step, we consider a simple hypothetical example to illustrate the concept of continuous archives in the context of evolutionary multi-objective optimization. For these, assume we are given the discrete archive $A = \{x_1, \ldots, x_5\} \subset \mathbb{R}^2$, where

$$x_1 = (-0.0129, -0.0421), \quad x_2 = (0.2525, 0.2912), \quad x_3 = (0.4903, 0.4035),$$
$$x_4 = (0.6258, 0.6912), \quad x_5 = (1.0212, 0.9930).$$

The set A is hence consisting of only five candidate solutions. Analogously, the image $F(A)$ of A can be considered as an approximation of the Pareto front that consists as well of five candidate solutions. Now, in order to improve the quality of the approximation, instead of A one can consider the polygon that is defined by the elements of A, namely

$$B := \overline{x_1 x_2} \cup \cdots \cup \overline{x_4 x_5}. \tag{13}$$

In what follows, we will call this polygon *the continuous archive*. The approximations A, B, $F(A)$, and $F(B)$ can be seen in the Figures 7 and 8. By visual inspection, the approximation qualities increase significantly when using the linear interpolation, in particular in objective space. This is reflected by the (p, q)-distances which are shown in Table 2 where we can find the following general behavior: first, the distances within decision and objective spaces, decreases from finite to continuous archives, and this phenomena is stronger in the objective space; and second, following the result of Theorem 4, the distances decreases as q decreases.

In a next step, we consider discrete and continuous archives resulting from two of the most famous EMO algorithms: NSGA-II [44] and MOEA/D [45], see Table 3 for the parameter setting of these algorithms. To this end, we first consider the result of NSGA-II with a population size of 12 after 500 generations, see Figures 9 and 10 and Table 4 for the numerical results. Finally, we consider the MOEA/D generational algorithm to get 500 finite archives of 12 elements each, see Figures 11 and 12 and Table 5 for the numerical results.

For both EMO algorithms, it can be observed that the indicator values for the continuous archives are significantly better than for the respective discrete archives. Next, note that the $\Delta_{p,q}$ values oscillate for NSGA-II which is a typical behavior for this dominance-based algorithm. These oscillations, however, are less distinct for the continuous archives.

Table 2. $\Delta_{p,q}$ values for the Pareto set/front approximations for MOP (12).

p	q	Decision Space		Objective Space	
		Finite Archive	Continuous Archive	Finite Archive	Continuous Archive
$\gamma = 2$					
1	1	0.5314	0.4841	0.4369	0.3943
1	-1	0.2732	0.1750	0.2095	0.0945
1	-100	0.1140	0.0213	0.0893	0.0018
1	-200	0.1131	0.0208	0.0886	0.0017
1	$-10,000$	0.1122	0.0202	0.0879	0.0017
$\gamma = \frac{1}{2}$					
1	1	0.5314	0.4841	0.5629	0.5024
1	-1	0.2732	0.1750	0.2807	0.1072
1	-100	0.1140	0.0213	0.1202	0.0015
1	-200	0.1131	0.0208	0.1192	0.0015
1	$-10,000$	0.1122	0.0202	0.1183	0.0014

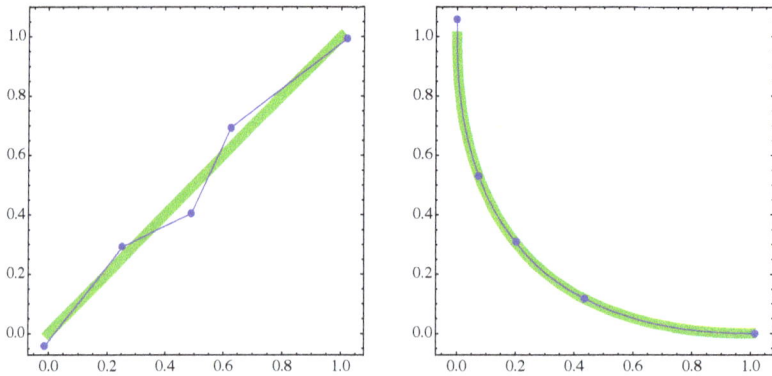

Figure 7. Left: Approximations A (blue dots) and B (blue polygon line) of the Pareto set (green thick line) of MOP (12) for $n = 2$. Right: corresponding approximations $F(A)$ and $F(B)$ of the Pareto front, for $\gamma = 2$.

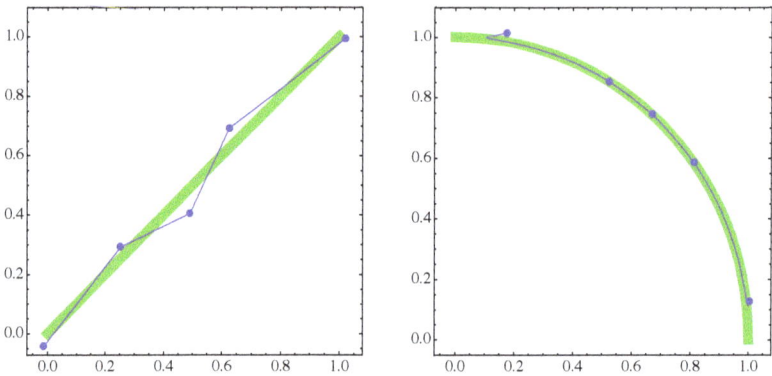

Figure 8. Left: Approximations A (blue dots) and B (blue polygon line) of the Pareto set (green thick line) of MOP (12) for $n = 2$. Right: corresponding approximations $F(A)$ and $F(B)$ of the Pareto front, for $\gamma = 1/2$.

Table 3. Parameter setting for NSGA-II and MOEA/D.

Algorithm	Parameter	Value
NSGA-II	Population size	12
	Number of generations	500
	Crossover probability	0.8
	Mutation probability	$1/n$
	Distribution index for crossover	20
	Distribution index for mutation	20
MOEA/D	Population size	12
	# weight vectors	12
	Number of generations	500
	Crossover probability	1
	Mutation probability	$1/n$
	Distribution index for crossover	30
	Distribution index for mutation	20
	Aggregation function	Tchebycheff
	Neighborhood size	3

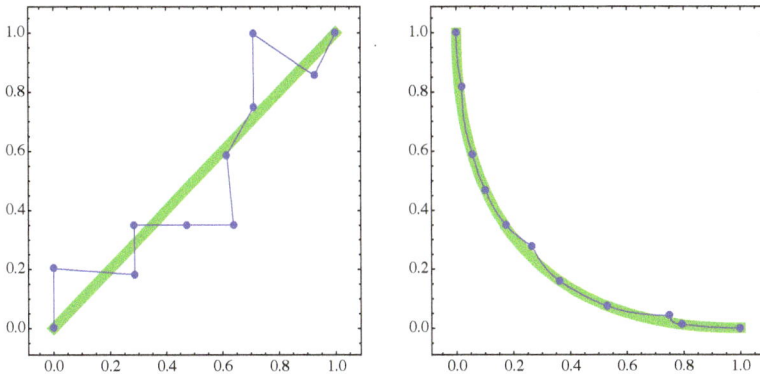

Figure 9. Left: approximations A (blue dots) corresponding to the 500th generation of the NSGA-II algorithm, and B (blue polygon line) of the Pareto set (green thick line) of MOP (12) for $n = 2$. Right: respective approximations $F(A)$ and $F(B)$ of the Pareto front for $\gamma = 2$.

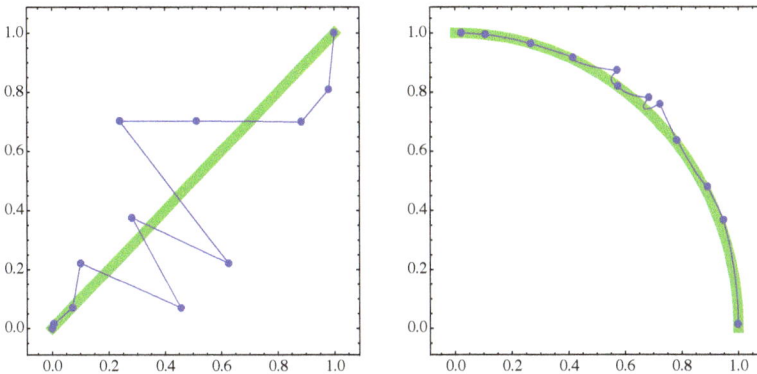

Figure 10. Left: approximations A (blue dots) corresponding to the 500th generation of the NSGA-II algorithm, and B (blue polygon line) of the Pareto set (green thick line) of MOP (12) for $n = 2$. Right: respective approximations $F(A)$ and $F(B)$ of the Pareto front for $\gamma = 1/2$.

Table 4. $\Delta_{p,q}$ values for the Pareto front approximations for MOP (12) using the NSGA-II archives and with $p = 1, q = -10$.

Generation	$\gamma = 1/2$		$\gamma = 2$	
	Finite Archive	Continuous Archive	Finite Archive	Continuous Archive
50	0.0439	0.0147	0.0696	0.0160
100	0.0498	0.0109	0.0540	0.0102
200	0.0613	0.0118	0.0716	0.0207
250	0.0651	0.0265	0.0572	0.0061
400	0.0602	0.0102	0.0723	0.0276
450	0.0630	0.0154	0.0584	0.0088
460	0.0612	0.0154	0.0658	0.0098
470	0.0523	0.0102	0.0566	0.0083
480	0.0754	0.0269	0.0684	0.0241
490	0.0510	0.0091	0.0584	0.0118
500	0.0722	0.0097	0.0560	0.0103

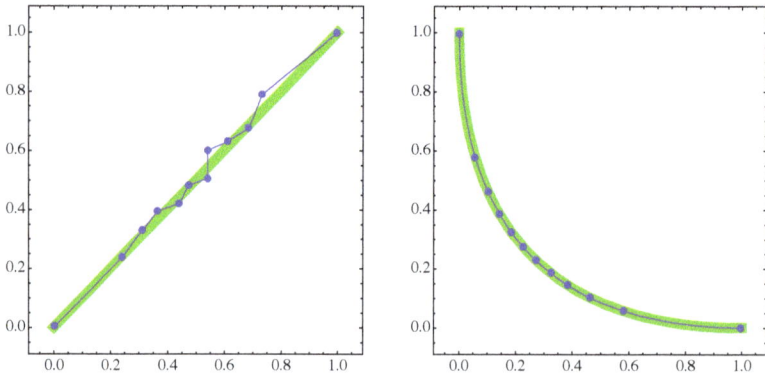

Figure 11. Left: approximations A (blue dots) corresponding to the 500th generation of the MOEA/D algorithm), and B (blue polygon line) of the Pareto set (green thick line) of MOP (12) for $n = 2$. Right: respective approximations $F(A)$ and $F(B)$ of the Pareto front for $\gamma = 2$.

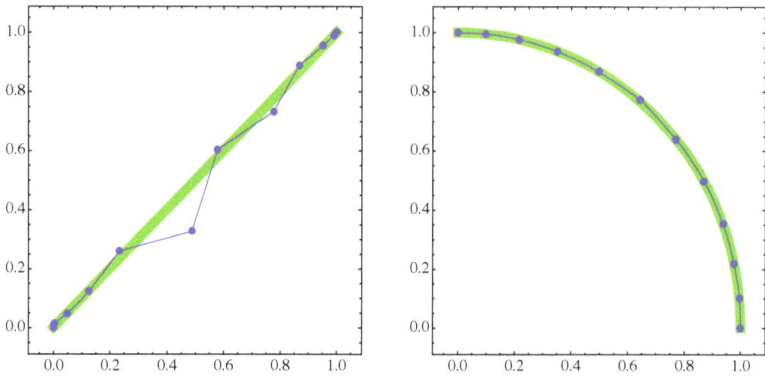

Figure 12. Left: approximations A (blue dots) corresponding to the 500th generation of the MOEA/D algorithm), and B (blue polygon line) of the Pareto set (green thick line) of MOP (12) for $n = 2$. Right: respective approximations $F(A)$ and $F(B)$ of the Pareto front for $\gamma = 1/2$.

Table 5. $\Delta_{p,q}$ values for the Pareto front approximations for MOP (12) using the MOEA/D archives and with $p = 1, q = -10$.

Generation	$\gamma = 1/2$		$\gamma = 2$	
	Finite Archive	Continuous Archive	Finite Archive	Continuous Archive
50	0.0610	0.0171	0.0648	0.0119
100	0.0519	0.0051	0.1093	0.0016
200	0.0536	0.0037	0.0781	0.0009
250	0.0522	0.0037	0.0790	0.0008
400	0.0511	0.0017	0.0784	0.0009
450	0.0511	0.0017	0.0784	0.0009
460	0.0509	0.0012	0.0784	0.0009
470	0.0509	0.0012	0.0784	0.0009
480	0.0509	0.0010	0.0783	0.0009
490	0.0509	0.0010	0.0783	0.0009
500	0.0509	0.0010	0.0783	0.0009

To further investigate the last statement, we consider finally the convex bi-objective problem $f_1, f_2 \colon \mathbb{R}^3 \to \mathbb{R}$, where $x = (x_1, x_2, x_3)$ and

$$f_1(x) = (x_1 + 1)^2 + x_2^2 + x_3^2$$
$$f_2(x) = (x_1 - 1)^2 + x_2^2 + x_3^2. \tag{14}$$

The Pareto set of MOP (14) is the line segment connecting the points $(0,0,0)$ and $(1,0,0)$, and the Pareto front is as shown in Figure 13.

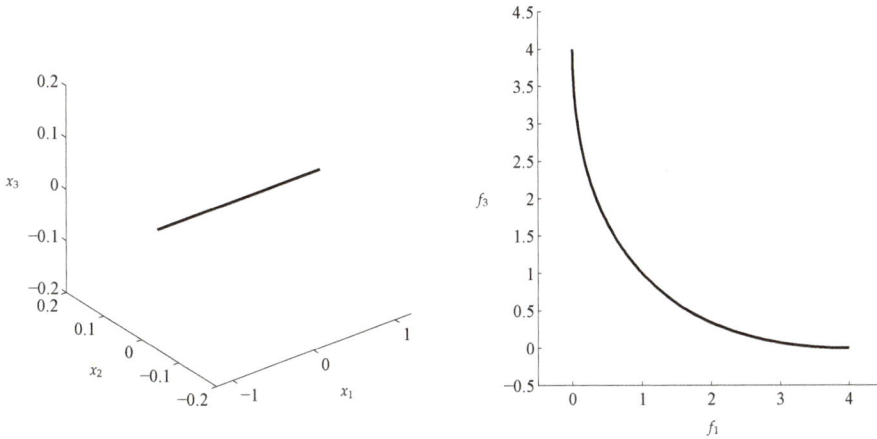

Figure 13. Pareto set (left) and front (right) of MOP (14).

Figure 14 and Table 6 show the $\Delta_{p,q}$ values for both the discrete and continuous archives obtained by NSGA-II using a population size of 20. As it can be seen, again the continuous archives achieve much better indicator values, and the amplitudes of the oscillations are significantly smaller compared to the discrete archives. This is confirmed by Figures 15–17 that show the results of the discrete and continuous archives after 300, 400, and 500 generations, respectively. As it can be seen, NSGA-II is able to compute solutions along the Pareto front, however, with varying distribution along this set (In fact, it is known that there is no "limit archive" for NSGA-II since this algorithm is not indicator-based). In turn, for each of the results of NSGA-II, all of the continuous archives represent—at least by visual inspection—perfect approximations of the Pareto front, which is reflected by the good $\Delta_{p,q}$ values.

Concluding, the results presented in this section strongly indicate the convenience of the new indicator that is able to assess the performance of continuous archives. Though in principle also other indicators can be extended to continuous sets, this has not been done so far, and this is not a straightforward task. Hence, no comparisons to other indicators can be considered here. The presented results further indicate the benefit of the use of continuous archives instead of discrete ones that are being used classically. This would, among others, allow for the usage of smaller population sizes which would in turn allow to reduce the computational burden of the evolutionary algorithms (note that the time complexity for all MOEAs in each generation is quadratic in the population size). The verification of this statement, however, is left for future work as this goes beyond the scope of this study.

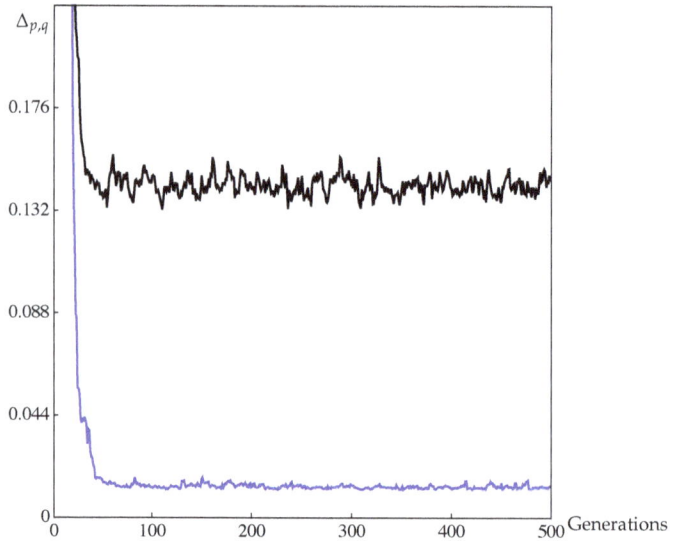

Figure 14. $\Delta_{p,q}$ values for the discrete (black curve) and the continuous archives (blue curve) of NSGA-II for MOP (14).

Table 6. $\Delta_{p,q}$ values for the discrete and continuous archives of NSGA-II for MOP (14). The results are averaged over 20 independent runs.

Generation	Continuous Archive	Finite Archive
20	0.1333	0.2401
40	0.0176	0.1451
60	0.0090	0.1561
80	0.0088	0.1355
100	0.0065	0.1472
120	0.0074	0.1412
140	0.0081	0.1395
160	0.0075	0.1549
180	0.0092	0.1468
200	0.0074	0.1429
220	0.0066	0.1408
240	0.0075	0.1397
260	0.0066	0.1460
280	0.0074	0.1439
300	0.0084	0.1421
320	0.0070	0.1352
340	0.0070	0.1373
360	0.0081	0.1454
380	0.0079	0.1413
400	0.0066	0.1388
420	0.0063	0.1400
440	0.0097	0.1384
460	0.0067	0.1418
480	0.0067	0.1421
500	0.0076	0.1426

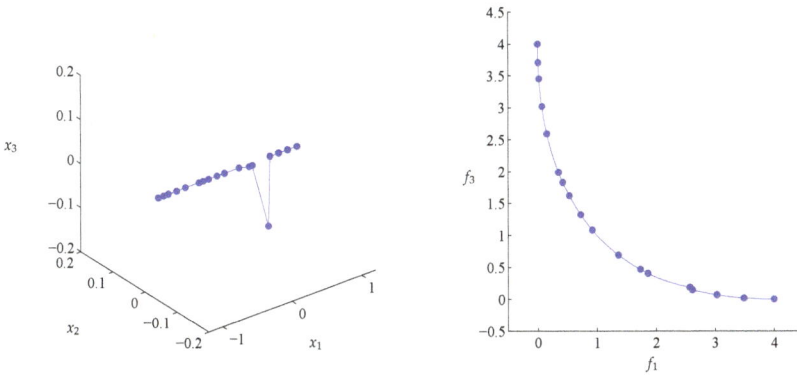

Figure 15. Left: Approximations A (blue dots) and B (blue continuous polygon line) of the Pareto set of MOP (14) in the 300th generation. Right: corresponding approximations $F(A)$ and $F(B)$ of the Pareto front.

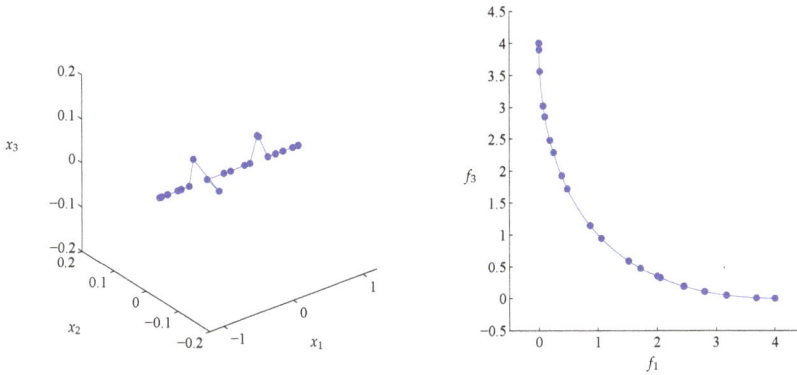

Figure 16. Left: Approximations A (blue dots) and B (blue continuous polygon line) of the Pareto set of MOP (14) in the 400th generation. Right: corresponding approximations $F(A)$ and $F(B)$ of the Pareto front.

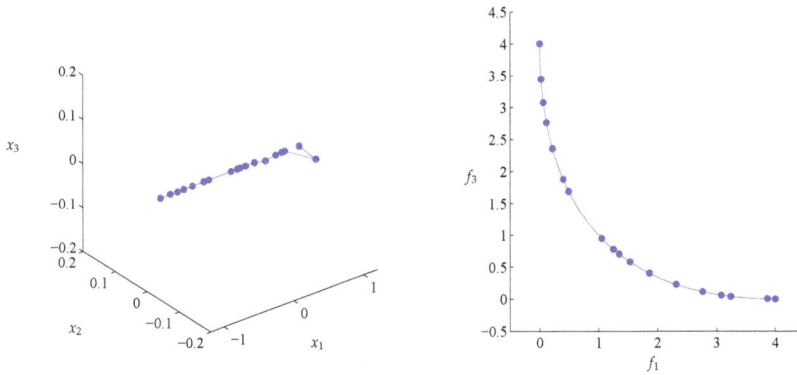

Figure 17. Left: Approximations A (blue dots) and B (blue continuous polygon line) of the Pareto set of MOP (14) in the 500th generation. Right: corresponding approximations $F(A)$ and $F(B)$ of the Pareto front.

5. Conclusions and Future Work

In this work, we have proposed extensions of the existing $GD_{p,q}$ and $\Delta_{p,q}$ performance indicators that allow to compute the distance between two general measurable sets. In particular, this is a natural setting in multi-objective optimization because the solution of such a problem typically forms a set of certain dimension (and is thus not given by finitely many points). We have shown that the extended indicators keep the nice metric properties from its finite-version predecessors (see [14,26]). Moreover, for $GD_{p,q}$, sufficient conditions have been provided ensuring that certain compliance to Pareto optimality of this indicator can be guaranteed. Further study is needed to determine the precise relation between these conditions and other ones appearing in the literature.

We have demonstrated the applicability and usefulness of the novel indicator on examples related to evolutionary multi-objective optimization.

As part of future work, we intend to further investigate the use of $\Delta_{p,q}$ within evolutionary multi-objective optimization. For instance, it might be interesting to integrate this performance indicator within an evolutionary multi-objective optimization algorithm as it was done, e.g., in [46] for its predecessor Δ_p. Although it is clear that the individual roles of p and q are related with the convexity of the metric neighborhoods of point and sets, further research is needed to elucidate more precisely useful ways to take advantage of their joint behavior in concrete situations. Additionally, to understand the behavior of $\Delta_{p,q}$ in relation to Pareto compliance and to complete the partial results that have been established in Section 3.4 for $GD_{p,q}$, consideration should be given to the inverted generational indicator $IGD_{p,q}$. Finally, one interesting aspect is to see if the indicator can be used as a proximity measure in other research fields.

Author Contributions: J.B. and A.V. obtained the theoretical results concerning the (p,q)-averaged Hausdorff distance. O.S. and O.C. conceived and designed the experiments; J.B. and O.C. performed the experiments and provided the related figures and tables; O.S. analyzed the data and revised the text. J.B. and A.V. wrote the paper.

Acknowledgments: We would like to thank the referees for the for their useful comments and suggestions that helped to improve our manuscript. The first two authors were partially supported by the research project ID-PRY: 7919 of the Faculty of Sciences, Pontificia Universidad Javeriana, Bogotá, Colombia. The last two authors acknowledge support from the Conacyt Basic Science Group project no. 285599.

Conflicts of Interest: The authors declare no conflict of interest.

References

1. Heinonen, J. *Lectures on Analysis on Metric Spaces*; Springer: New York, NY, USA, 2001.
2. Huttenlocher, D.P.; Klanderman, G.A.; Rucklidge, W.A. Comparing Images Using the Hausdorff Distance. *IEEE Trans. Pattern Anal. Mach. Intell.* **1993**, *15*, 850–863. [CrossRef]
3. Yi, X.; Camps, O.I. Line-Based Recognition Using A Multidimensional Hausdorff Distance. *IEEE Trans. Pattern Anal. Mach. Intell.* **1999**, *21*, 901–916.
4. De Carvalho, F.; de Souza, R.; Chavent, M.; Lechevallier, Y. Adaptive Hausdorff distances and dynamic clustering of symbolic interval data. *Pattern Recognit. Lett.* **2006**, *27*, 167–179. [CrossRef]
5. Dellnitz, M.; Hohmann, A. A subdivision algorithm for the computation of unstable manifolds and global attractors. *Numerische Mathematik* **1997**, *75*, 293–317. [CrossRef]
6. Aulbach, B.; Rasmussen, M.; Siegmund, S. Approximation of attractors of nonautonomous dynamical systems. *Discret. Contin. Dyn. Syst. Ser. B* **2005**, *5*, 215–238.
7. Emmerich, M.; Deutz, A.H. Test Problems Based on Lamé Superspheres. In Proceedings of the 4th International Conference on Evolutionary Multi-criterion Optimization, Matsushima, Japan, 5–8 March 2007; Springer: Berlin/Heidelberg, Germany, 2007; pp. 922–936.
8. Falconer, K. *Fractal Geometry: Mathematical Foundations and Applications*, 2nd ed.; John Wiley & Sons, Inc.: Hoboken, NJ, USA, 2003.
9. Schütze, O. Set Oriented Methods for Global Optimization. Ph.D. Thesis, University of Paderborn, Paderborn, Germany, 2004.

10. Dellnitz, M.; Schütze, O.; Hestermeyer, T. Covering Pareto Sets by Multilevel Subdivision Techniques. *J. Optim. Theory Appl.* **2005**, *124*, 113–155. [CrossRef]

11. Padberg, K. Numerical Analysis of Transport in Dynamical Systems. Ph.D. Thesis, University of Paderborn, Paderborn, Germany, 2005.

12. Schütze, O.; Coello Coello, C.A.; Mostaghim, S.; Talbi, E.G.; Dellnitz, M. Hybridizing Evolutionary Strategies with Continuation Methods for Solving Multi-Objective Problems. *Eng. Optim.* **2008**, *40*, 383–402. [CrossRef]

13. Schütze, O.; Laumanns, M.; Coello Coello, C.A.; Dellnitz, M.; Talbi, E.G. Convergence of Stochastic Search Algorithms to Finite Size Pareto Set Approximations. *J. Glob. Optim.* **2008**, *41*, 559–577. [CrossRef]

14. Schütze, O.; Esquivel, X.; Lara, A.; Coello Coello, C.A. Using the averaged Hausdorff distance as a performance measure in evolutionary multiobjective optimization. *IEEE Trans. Evol. Comput.* **2012**, *16*, 504–522. [CrossRef]

15. Hernández, C.; Naranjani, Y.; Sardahi, Y.; Liang, W.; Schütze, O.; Sun, J.Q. Simple Cell Mapping Method for Multi-objective Optimal Feedback Control Design. *Int. J. Dyn. Control* **2013**, *1*, 231–238. [CrossRef]

16. Siwel, J.; Yew-Soon, O.; Jie, Z.; Liang, F. Consistencies and contradictions of performance metrics in multiobjective optimization. *IEEE Trans. Evol. Comput.* **2014**, *44*, 2329–2404.

17. Sun, J.Q.; Xiong, F.R.; Schütze, O.; Hernández, C. *Cell Mapping Methods—Algorithmic Approaches and Applications*; Springer: Singapore, 2019.

18. Jahn, J. Multiobjective search algorithm with subdivision technique. *Comput. Optim. Appl.* **2006**, *35*, 161–175. [CrossRef]

19. Schütze, O.; Vasile, M.; Junge, O.; Dellnitz, M.; Izzo, D. Designing optimal low thrust gravity assist trajectories using space pruning and a multi-objective approach. *Eng. Optim.* **2009**, *41*, 155–181. [CrossRef]

20. Deb, K. *Multi-Objective Optimization Using Evolutionary Algorithms*; John Wiley & Sons: Chichester, UK, 2001.

21. Beume, N.; Naujoks, B.; Emmerich, M. SMS-EMOA: Multiobjective selection based on dominated hypervolume. *Eur. J. Oper. Res.* **2007**, *181*, 1653–1669. [CrossRef]

22. Garg, H. Solving structural engineering design optimization problems using an artificial bee colony algorithm. *J. Ind. Manag. Optim.* **2014**, *10*, 777–794. [CrossRef]

23. Garg, H. A hybrid PSO-GA algorithm for constrained optimization problems. *Appl. Math. Comput.* **2016**, *274*, 292–305. [CrossRef]

24. Zapotecas-Martínez, S.; López-Jaimes, A.; García-Nájera, A. LIBEA: A Lebesgue Indicator-Based Evolutionary Algorithm for Multi-objective Optimization. *Swarm Evolut. Comput.* **2018**. [CrossRef]

25. Hartikainen, M.; Miettinen, K.; Wiecek, M. PAINT: Pareto front interpolation for nonlinear multiobjective optimization. *Comput. Optim. Appl.* **2012**, *52*, 845–867. [CrossRef]

26. Vargas, A.; Bogoya, J.M. A generalization of the averaged Hausdorff distance. *Computación y Sistemas* **2018**, *22*, 331–345. [CrossRef]

27. Tao, T. *An Introduction to Measure Theory (Graduate Studies in Mathematics)*; American Mathematical Society: Providence, RI, USA, 2011.

28. Jones, F. *Lebesgue Integration on Euclidean Space*; Jones and Bartlett Publishers: Boston, MA, USA, 2001.

29. Hardy, G.H.; Littlewood, J.E.; Pólya, G. *Inequalities*, 2nd ed.; Cambridge University Press: Cambridge, UK, 1952.

30. Bullen, P.S. *Handbook of Means and Their Inequalities*; Kluwer Academic Publishers Group: Dordrecht, The Netherlands, 2003.

31. Pareto, V. *Manual of Political Economy*; The Macmillan Press: London, UK, 1971.

32. Hillermeier, C. *Nonlinear Multiobjective Optimization: A Generalized Homotopy Approach*; Springer Science & Business Media: Berlin, Germany, 2001.

33. Zitzler, E.; Thiele, L. Multiobjective evolutionary algorithms: A comparative case study and the strength Pareto approach. *IEEE Trans. Evol. Comput.* **1999**, *3*, 257–271. [CrossRef]

34. Brockhoff, D.; Wagner, T.; Trautmann, H. On the Properties of the R2 Indicator. In Proceedings of the 14th Annual Conference on Genetic and Evolutionary Computation, Philadelphia, PA, USA, 7–11 July 2012; ACM: New York, NY, USA, 2012; pp. 465–472.

35. Ishibuchi, H.; Masuda, H.; Nojima, Y. A Study on Performance Evaluation Ability of a Modified Inverted Generational Distance Indicator. In Proceedings of the 2015 Annual Conference on Genetic and Evolutionary Computation, Madrid, Spain, 11–15 July 2015; ACM: New York, NY, USA, 2015; pp. 695–702.

Math. Comput. Appl. **2018**, *23*, 51

36. Dilettoso, E.; Rizzo, S.A.; Salerno, N. A Weakly Pareto Compliant Quality Indicator. *Math. Comput. Appl.* **2017**, *22*, 25.

37. Garg, H.; Kumar, K. Distance measures for connection number sets based on set pair analysis and its applications to decision-making process. *Appl. Intell.* **2018**, *48*, 1–14. [CrossRef]

38. Singh, S.; Garg, H. Distance measures between type-2 intuitionistic fuzzy sets and their application to multicriteria decision-making process. *Appl. Intell.* **2017**, *46*, 788–799. [CrossRef]

39. Van Veldhuizen, D.A.; Lamont, G.B. Multiobjective evolutionary algorithm test suites. In Proceedings of the 1999 ACM Symposium on Applied Computing, San Antonio, TX, USA, 28 February–2 March 1999; ACM: New York, NY, USA, 1999; pp. 351–357.

40. Coello Coello, C.A.; Cruz Cortés, N. Solving Multiobjective Optimization Problems using an Artificial Immune System. *Genet. Program. Evolvable Mach.* **2005**, *6*, 163–190. [CrossRef]

41. Rudolph, G.; Schütze, O.; Grimme, C.; Domínguez-Medina, C.; Trautmann, H. Optimal averaged Hausdorff archives for bi-objective problems: Theoretical and numerical results. *Comput. Optim. Appl.* **2016**, *64*, 589–618. [CrossRef]

42. Zitzler, E.; Thiele, L.; Laumanns, M.; Fonseca, C.M.; da Fonseca, V.G. Performance assessment of multiobjective optimizers: An analysis and review. *IEEE Trans. Evol. Comput.* **2003**, *7*, 117–132. [CrossRef]

43. Witkowski, A. A new proof of the monotonicity property of power means. *JIPAM. J. Inequal. Pure Appl. Math.* **2004**, *5*, 73.

44. Deb, K.; Pratap, A.; Agarwal, S.; Meyarivan, T. A fast and elitist multiobjective genetic algorithm: NSGA-II. *IEEE Trans. Evol. Comput.* **2002**, *6*, 182–197. [CrossRef]

45. Zhang, Q.; Li, H. MOEA/D: A Multiobjective Evolutionary Algorithm Based on Decomposition. *IEEE Trans. Evol. Comput.* **2007**, *11*, 712–731. [CrossRef]

46. Schütze, O.; Domínguez-Medina, C.; Cruz-Cortés, N.; de la Fraga, L.G.; Sun, J.Q.; Toscano, G.; Landa, R. A scalar optimization approach for averaged Hausdorff approximations of the Pareto front. *Eng. Optim.* **2016**, *48*, 1593–1617. [CrossRef]

© 2018 by the authors. Licensee MDPI, Basel, Switzerland. This article is an open access article distributed under the terms and conditions of the Creative Commons Attribution (CC BY) license (http://creativecommons.org/licenses/by/4.0/).

Mathematical
and Computational
Applications

MDPI

Article

Variation Rate to Maintain Diversity in Decision Space within Multi-Objective Evolutionary Algorithms [†]

Oliver Cuate [1],* and Oliver Schütze [2],*

[1]   Computer Science Department, Cinvestav-IPN, Mexico City 07360, Mexico
[2]   Dr. Rodolfo Quintero Ramirez Chair, UAM Cuajimalpa, Mexico City 05370, Mexico
*   Correspondence: ocuate@computacion.cs.cinvestav.mx (O.C.); schuetze@cs.cinvestav.mx (O.S.)
†   This paper is an extended version of our paper published in the 10th International Conference on
    Evolutionary Multi-Criterion Optimization (EMO 2019), East Lansing, MI, USA, in 10–13 March 2019.

Received: 30 July 2019; Accepted: 11 September 2019; Published: 13 September 2019

Abstract: The performance of a multi-objective evolutionary algorithm (MOEA) is in most cases measured in terms of the populations' approximation quality in objective space. As a consequence, most MOEAs focus on such approximations while neglecting the distribution of the individuals of their populations in decision space. This, however, represents a potential shortcoming in certain applications as in many cases one can obtain the same or very similar qualities (measured in objective space) in several ways (measured in decision space). Hence, a high diversity in decision space may represent valuable information for the decision maker for the realization of a given project. In this paper, we propose the Variation Rate, a heuristic selection strategy that aims to maintain diversity both in decision and objective space. The core of this strategy is the proper combination of the averaged distance applied in variable space together with the diversity mechanism in objective space that is used within a chosen MOEA. To show the applicability of the method, we propose the resulting selection strategies for some of the most representative state-of-the-art MOEAs and show numerical results on several benchmark problems. The results demonstrate that the consideration of the Variation Rate can greatly enhance the diversity in decision space for all considered algorithms and problems without a significant loss in the approximation qualities in objective space.

Keywords: evolutionary computation; multi-objective optimization; decision space diversity

---

1. Introduction

In many areas such as Economy, Finance, or Industry, the problem arises naturally that several conflicting objectives have to be optimized concurrently [1,2]. Such problems are called multi-objective optimization problems (MOPs) in literature [3–5]. The solution set of an MOP (in decision space) is called the Pareto set and its image (defined in objective space) the Pareto front. One important characteristic of continuous MOPs is that their Pareto sets and fronts typically form $(k-1)$-dimensional objects, where k is the number of objectives considered in the problem. In many applications, the Pareto front is of primary interest as it contains information about the desired qualities of each selected solution. As a consequence, almost all existing MOEAs focus on approximations in objective space while entirely neglecting the distribution of the individuals in decision space. There exist, however, also applications where the values of the solutions in decision variable space are of great importance. As an example, consider that the amount of a certain resource (i.e., the value of a variable x_i) used to obtain the desired quality (measured in objective space) is important. For two solutions that are equal or similar in objective space, one may prefer the one that has a lower value of x_i. Another example is that the variable x_i could represent the launch date of a project, as, e.g., in [6] in the context of space mission design. For such problems, different values of x_i directly relate to different timescales in the realization of the project. In that case, the decision maker may select one "optimal" realization of the project, and keep solutions with similar objective values but later launch dates as back-up solutions. For such problems, the sole consideration of the approximation quality in objective space represents a potential shortcoming. On the one hand, it is of course possible to formulate all such problems via additional constraints and objectives. On the other hand, such re-modellations of the problem lead in general to a higher complexity compared to the original problem. For instance, by each added objective, the dimension of the Pareto set/front increases by one. Hence, in this context, there exists an additional challenge in solving a given MOP, since we have to find a suitable approximation to the optimal set both in objective and decision space, in order to provide a satisfying overview of the possible solutions to the decision maker. Another application can be found in [7], where data analysis techniques are used to discover patterns and rules. Here, authors conclude that such process of innovation by optimization ("innovization") has an enormous potential to revolutionize the engineering of the design process in the industry.

In this paper, we propose a framework for equipping MOEAs with a mechanism that performs an exploration of both decision and objective spaces. The underlying idea of this proposal, called Variation Rate, is to measure the spread via using the averaged distance in decision space for elements for which function values are close in objective space. After discussion of the general framework, we present implementations of variants of NSGA-II [8] (elitist algorithm), SMS-EMOA [9] (indicator based algorithm), MOEA/D [10] (decomposition based algorithm), and NSGA-III [11] (designed for the treatment of MOPs with many objectives). Further on, we demonstrate the effectiveness of our approach on several benchmark problems, where we show that, compared to the base variants, we greatly improve the approximation qualities in decision space without any significant loss in the qualities in objective space. A preliminary study of this work can be found in [12], where the discussion on the proposed method is reduced, and where it has only been applied to NSGA-III.

The remainder of the paper is organized as follows: in Section 2, we briefly present the background required for the understanding of this paper and discuss the related work. In Section 3, we first present the general framework of the Variation Rate and further on provide particular implementations for four different MOEAs that are representative of the state-of-the-art. In Section 4, we present some numerical results on selected benchmark problems using both the four base MOEAs as well as their variants that use the Variation Rate. Finally, in Section 5, we discuss the advantages of the proposed approach and discuss possible paths for future research.

2. Background and Related Work

Optimization refers to finding the best possible solution to a problem given a set of constraints [4]. Multi-objective optimization refers to the simultaneous optimization of multiple and usually conflicting objectives. More precisely, a multi-objective optimization problem (MOP) with k objectives is mathematically defined as follows:

$$\min_{x \in D} F(x), \tag{1}$$

where $D \subset \mathbb{R}^n$ is the domain and $F : D \subset \mathbb{R}^n \to \mathbb{R}^k$ is called the objective function.

The optimality of an MOP is defined by the concept of *strict dominance*. Let $v, w \in \mathbb{R}^k$, the vector v is *less than* w ($v <_p w$), if $v_i < w_i$ for all $i \in \{1, \ldots, k\}$; the relation $\leq_p$ is defined analogously. A vector $y \in D$ is *dominated* by a vector $x \in D$ ($x \prec y$) with respect to (1) if $F(x) \leq_p F(y)$ and $F(x) \neq F(y)$, else y is called non-dominated by x. A point $x^* \in \mathbb{R}^n$ is Pareto optimal to (1) if there is no $y \in D$ that dominates x. The set of all the Pareto optimal points P_D is called the Pareto set and its image $F(P_D)$ is called the Pareto front. Typically, i.e., under certain mild smoothness assumption on the model, both Pareto set and front form at least locally $(k-1)$-dimensional objects.

Math. Comput. Appl. **2019**, 24, 82

Unlike evolutionary algorithms for single objective optimization problems (SOPs), maintaining diversity in decision space is not a priority for most MOEAs; most of the performance indicators are developed in order to measure the accuracy based only on the objective function (e.g., the hypervolume [13] and the Degree of Approximation [14]). As exceptions, we have some of the measures used for multimodal optimization (see [15]) and two particular examples. The first one is the Δ_p indicator [16,17], which can be viewed as an averaged Hausdorff distance and which actually measures the distance between two general sets and we can use it as an indicator both in objective space as well as in decision space. The other work that can deal with this aspect is the diversity integrating multiobjective optimizer (DIOP) [18], which is user-defined and concurrently optimizes two set-based diversity measures, one in decision space and the other in objective space. In this work, the relationship between the two set-based diversity measures is conceived as a bi-objective optimization problem and it is solved via a weighted sum of the two diversity indicators. In particular, the authors consider the Solow–Polasky measure as it satisfies three requirements: monotonicity in varieties, twinning, and monotonicity in the distance.

Although works that explicitly consider at the same time variables and objectives are scarce, one can find some related work on this topic in [19] and some specific algorithms. For instance, the NSGA [20] (the algorithm that precedes the well-known NSGA-II) uses fitness sharing in decision space. In [21], some possible techniques are proposed to spread out solutions both in objective and decision decision space: *pointwise expansion, threshold sharing, sequential sharing, simultaneous sharing multiplicative*, and *simultaneous sharing additive*. It is important to point out that the above approaches are only part of the discussion of the paper and they were not implemented; the implemented algorithm was the Niched Pareto GA, a method with phenotypic sharing. In addition, all of the described techniques depend on the normal fitness sharing method, that is, two additional parameters must be provided or approximated (the niche radius σ_{share} in both decision and objective spaces).

The omni-optimizer algorithm [22] is proposed as a procedure that aims at solving a wide variety of optimization problems (single or multi-objective and uni- or multi-modal problems). The authors argue that, to solve different kinds of problems, it is necessary to know different specialized algorithms. Thus, it is desirable to have an algorithm that adapts itself for handling any number of conflicting objectives, constraints, and variables. The omni-optimizer is important in the context of this work as it uses a two-tier fitness assignment scheme based on the crowding distance of the NSGA-II. The primary fitness is computed using the phenotypes (objectives and constraint values) and the secondary fitness is computed using both phenotypes and genotypes (decision variables). The modified crowding distance computes the average crowding distance of the population in objective and decision spaces. If the crowding value for some individual is above average (at any space), it is assigned the larger of the two distances; otherwise, the smaller of the two distances is assigned. However, we must not lose sight of the fact that omni-optimizer was developed not only to maintain diversity in decision space but with a more general purpose (it adapts itself to solve different kinds of problems).

An algorithm that explicitly promotes the diversity of the decision space is the *MOEA/D with Enhanced Variable-Space Diversity (MOEA/D-EVSD)*, proposed in [23]. This method is an extension of the MOEA/D [10] but with an enhanced variable-space diversity control. In the first generations, the MOEA/D-EVSD tries to induce a larger diversity via promoting the mating of dissimilar individuals. Similarly to MOEA/D, a new individual is created for each subproblem. Then, instead of randomly selecting two individuals of the neighborhood, a pool of α candidate parents is randomly filled from the neighborhood with probability δ, and with probability $1 - \delta$ from the whole population. The two selected parents are the ones in the pool that have the largest distance among them. As the δ parameter is dynamically set, a gradual change between exploration and exploitation can be induced. Additionally, a final phase to further promote intensification is included, which is essentially a traditional MOEA/D coupled with Differential Evolution (DE) operators. For the last generations of MOEA/D-EVSD, the traditional mating selection of MOEA/D is conserved together with the

Rand/1/bin scheme for the DE operators. The authors of this paper show that, by inducing a gradual loss of diversity in the decision space, the state-of-the-art of MOEAs can be improved.

Finally, in [24], the diversity integrating hypervolume-based search algorithm (DIVA) is proposed. Here, the authors proposed a modified hypervolume indicator, which is integrated into an evolutionary algorithm. They employ a so-called diversity function, which fulfills certain requirements such that the modified hypervolume indicator remains compliant with the underlying preference relation.

3. Proposed Framework

State-of-the-art MOEAs that measure the approximation quality of their outcome entirely in objective space work typically well if there is a 1:1 relationship between Pareto set and Pareto front (that is, if, for every $y \in F(P_D)$, there exists exactly one $x \in P_D$ such that $F(x) = y$). That is, if a good finite size approximation of the Pareto front is found by the MOEA (the goodness can be measured, e.g., by any existing performance indicator), the respective finite size approximation of the Pareto set is in many cases also satisfying. This, however, does not hold any more if there is an $m : 1$ relationship between Pareto set and front (i.e., if there are multiple $x_i \in P_D$ such that $F(x_i) = y$ for a $y \in F(P_D)$). If, for instance, there are several connected components of the Pareto set that map to the same part of the Pareto front, a good Pareto front approximation does not imply a good (or at least satisfying) approximation of the Pareto set. To see this, consider the hypothetical bi-objective problem that is shown in Figure 1. The Pareto set of this problem consists of two disjunct connected components that map both to the same Pareto front (that is, every $y \in F(P_D)$ has exactly two pre-images). Figure 2 shows four possible approximations in decision and objective space. As it can be seen, the approximation quality is very high for all sets in objective space, while this is not the case for the Pareto set approximations. Out of them, only the last one is "complete" according to the given discretization. MOEAs that merely measure their outcomes in objective space cannot distinguish between those solutions, and, consequently, the Pareto set approximation is left to chance. MOPs of this kind are termed Type III problems in [25].

To overcome this problem, we propose to perform a density estimator that aims to obtain a good distribution both in objective and decision space. Usually, a classical density estimator groups the population considering only the objective values. Such classification is commonly used to define selection criteria for its elements, giving them certain **reference value** based on its distribution in objective space. According to the design of each algorithm, the individual with either lower or higher reference value is chosen. The idea is to define a relationship between this reference value in objective space and a certain measurement in decision space. In this way, the first grouping phase identifies promising solutions in objective space; meanwhile, the second phase favors solutions with the most different values in the decision space. Our goal is to properly represent the trade-off between these two aspects.

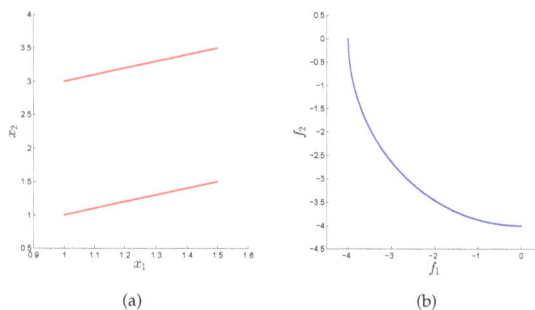

(a)   (b)

Figure 1. (**a**) Pareto set and (**b**) Pareto front of a hypothetical bi-objective problem where the Pareto set consists of two connected components that both map to the entire Pareto front.

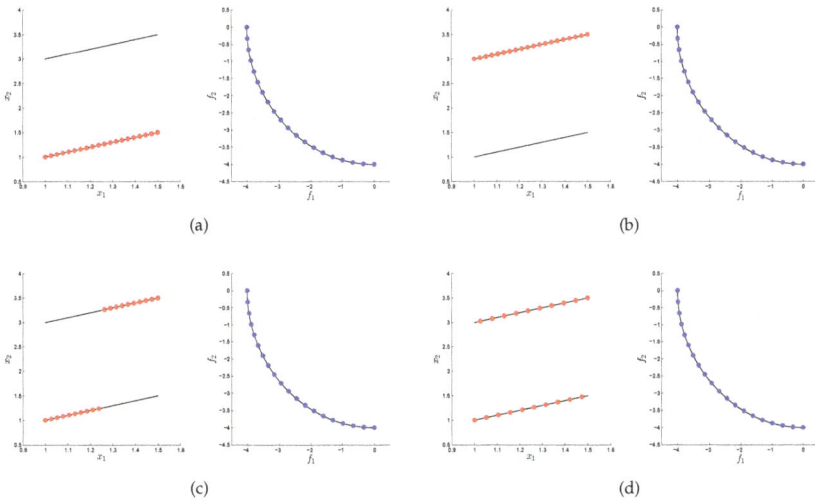

Figure 2. Four different Pareto set/front approximations, where all Pareto front approximations are good (e.g, in the Hausdorff sense), but where only in case (**d**) the Pareto set approximation is complete.

3.1. Using the Averaged Distance in Variable Space

In the following, we discuss why we think that the usage of the averaged distance is an adequate measure in decision space that serves our purpose.

Let $I = \{x_1, \ldots, x_s\} \subset \mathbb{R}$ be a finite set, then the averaged distance $\bar{d}$ between each element $x_i \in I$ and the rest of the elements in I is given by

$$\bar{d}(x_i, I) = \frac{1}{s-1} \sum_{\substack{j=1 \\ j \neq i}}^{s} d(x_i, x_j), \tag{2}$$

where $d(x_i, x_j)$ is the desired metric for the distance between the two elements x_i and x_j in the decision space that can vary according to the codification or the used norm. In this work, we consider the Euclidean distance, i.e., $d(x_i, x_j) = \|x_i - x_j\|_2$.

Though the averaged distance is defined for every finite set $I \in \mathbb{R}^n$, we will apply it on sets where the values of their images, $F(x_i)$, $i = 1, \ldots, s$, are close to each other.

As an illustrative example, consider again the Type III bi-objective problem whose Pareto set and front are shown in Figure 3. The set I is given by the three points ▼, ◆, and ★. All three images are relatively close to the given reference point Z point Z in objective space; assume, for simplicity, that the distances of all three images to Z are given by one (i.e., $\|Z - F(▼)\| = \|Z - F(◆)\| = \|Z - F(★)\| = 1$). Furthermore, we assume that, for the distances in variable space, we have $d(◆, ★) = 0.4$, $d(◆, ▼) = 2$, and $d(▼, ★) = 2.2$. Then, we obtain

$$\bar{d}(★, I) = \frac{2.6}{2} = 1.3, \qquad \bar{d}(◆, I) = \frac{2.4}{2} = 1.2, \qquad \text{and} \qquad \bar{d}(▼, I) = \frac{4.2}{2} = 2.1.$$

Notice that the point with the *biggest* average distance in variable space is also the most different individual in this space. In other words, elements with the maximum average distance in decision space have the desired behavior for Type III problems.

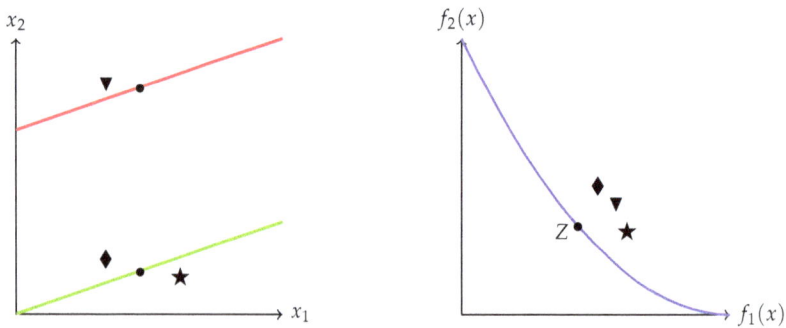

Figure 3. Illustrative example.

However, it is not sufficient to only take into account the average distance of the variables as a selection criterion. Our problem now is how to select an individual that has both a good quality in objective space as well as a good distribution in variable space. We discuss this issue in the following.

3.2. Variation Rate

As explained before, the generic selection criterion of most MOEAs prefers individuals with "the best" reference value in objective space, and this could be a maximum or a minimum value according to the selection procedure. For instance, the selection criterion of the NSGA-II prefers individuals with the biggest crowding distance, while the niching procedure of the NSGA-III favors individuals with the least distance to an induced line. On our part, we have to consider both the reference value provided by the classical selection criterion, as well as the average distance in variables to solve Type III problems.

We first consider selection mechanisms that prefer small reference values. For this, let I be a set of points in decision space whose images are close to each other, let v_i be the reference value in objective space for each $x_i \in I$ and let $\bar{d}(x_i, I)$ as defined as in (2) (in decision space). Then, the variation rate r_i for each element x_i is stated as follows:

$$r_i = \frac{v_i}{\bar{d}(x_i, I)}. \tag{3}$$

This makes sense because for Type III problems the elements of the neighborhood I will have a similar reference value in objectives while the average distances will be larger for the most different solutions in decision space. Hence, its quotient (the variation rate) will tend to be smaller than for the rest of the quotients. Thus, through the variation rate, we have a way to relate the objective and the decision spaces in order to choose the best individual in each group.

Next, we address selection mechanisms that prefer large reference values. For this, we have two options. The first one is to edit the selection criterion to prefer small values in order to use the variation rate. The second alternative is to use a product instead of a quotient. We decide to conserve the essence of each MOEA; for this reason, we implement the second option in this work, which we call *the inverse variation rate*. More precisely, for a set I as above with reference values v_i for each $x_i \in I$ and $\bar{d}(x_i, I)$ as defined in (2), the inverse variation rate $\tilde{r}_i$ is defined as follows:

$$\tilde{r}_i = v_i \cdot \bar{d}(x_i, I). \tag{4}$$

Using this definition, elements with the largest values are hence preferred.

In order to illustrate these two approaches, we go back to the example shown in Figure 3 where we already have the values of $\bar{d}$ and v_i for all the three elements of the set I. Suppose that we have to select two out of the three individuals ★, ▼, and ◆.

If we work with an MOEA that has a selection criterion that prefers individuals with the least distance to Z in objective space, then the only option we have is randomly choosing them (since $v_\blacktriangledown = \|Z - F(\blacktriangledown)\| = 1$, $v_\blacklozenge = \|Z - F(\blacklozenge)\| = 1$, and $v_\bigstar = \|Z - F(\bigstar)\| = 1$). However, if we use the variation rate, then the values change to:

$$r_\bigstar = \frac{1}{1.3} \approx 0.7692 \qquad r_\blacklozenge = \frac{1}{1.2} \approx 0.8333, \qquad \text{and} \qquad r_\blacktriangledown = \frac{1}{2.1} \approx 0.4762.$$

In this way, we select ▼ as the preferred solution and the second one is ★, which preserves individuals in both of the disconnected regions in decision space.

The desired two-element population is hence given by

$$P = \{\blacktriangledown, \bigstar\}.$$

Otherwise, if we work with an MOEA that has a selection criterion that prefers individuals with the largest distance to Z in objective space (maybe in order to preserve diversity), then the option we have again is randomly choosing two of them. However, if we use the inverse variation rate, that is:

$$\tilde{r}_\bigstar = 1 \cdot 1.3 = 1.3, \qquad \tilde{r}_\blacklozenge = 1 \cdot 1.2 = 1.2, \qquad \text{and} \qquad \tilde{r}_\blacktriangledown = 1 \cdot 2.1 = 2.1,$$

then it also leads ▼ and ★ as the selected individuals.

The desired two-element population is again given by

$$P = \{\blacktriangledown, \bigstar\}.$$

Observe that, in both cases, we conserve one individual in each disconnected component of the Pareto set. This is something that we can not guarantee with the use of a standard approach.

Notice that, for all MOPs with a 1:1 relationship of the Pareto set and the Pareto front, it is expected that solutions in the same neighborhood have similar reference values in objective space and also a similar average distance in decision space. Thus, it is also likely that making the quotient or the product of these values does not significantly affect the original selection criterion. This will be shown in Section 4 on several classical benchmark problems.

We can now state a general framework. A pseudocode of the Variation Rate is shown in Algorithm 1.

Algorithm 1 Framework to include the Average Distance in Variables within any MOEA

Require: Parameters of the selected MOEA
Ensure: Final population P_t
1: $t \leftarrow 0$
2: $P_0 \leftarrow \texttt{InitializePopulation}()$
3: **while** the stop criterion is not satisfied **do**
4: \qquad $M_t \leftarrow \texttt{VariationOperator}(P_t)$
5: \qquad $V_t \leftarrow \texttt{SelectProcedure}(P_t \cup M_t)$
6: \qquad $P_t \leftarrow \texttt{SelectByVariationRate}(V_t)$
7: \qquad $t \leftarrow t + 1$
8: **end while**

In Algorithm 1, the procedure $\texttt{SelectByVariationRate}$ takes the reference values V_t in objective space provided by $\texttt{SelectProcedure}$ (we assume it is based on a classical selection criterion), and then it updates such values according to the variation rate or the inverse variation rate in order to improve the selection mechanism to deal with Type III problems.

As we can see, this framework can be used in principle within any MOEA; however, the particular use of the variation rate or the inverse variation rate will depend on the given MOEA. In the following, we explain how to adapt the Variation Rate for four of the most representative MOEAs. The reader can find the pseudocode of these four variants in Appendix A.

3.3. Integration into NSGA-II

The first algorithm that we consider is the classical NSGA-II, which has been used successfully for the treatment of a large number of applications. This is a domination-based multi-objective evolutionary algorithm; that is, this method directly applies the Pareto dominance relation and an elitism strategy to preserve the best individuals along the optimization process. The elitism operator is incorporated via a special parent selection based on two mechanisms: **fast-non-dominated-sorting** and **crowding distance**. The former conserves best individuals based on the Pareto dominance relation, whereas the latter is used to promote the preservation of the diversity.

We consider the classification in fronts performed by the fast-non-dominated-sorting as our neighborhood structure because the crowding distance is applied only in the last front that can contribute elements to the next population. This means that we have to integrate the diversity into decision space into the crowding distance. The crowding distance procedure sorts the elements in the last front according to the values of objectives; then, the crowding distance of an individual p_i is the average distance in objective space from the previous and the next individuals (according to the induced order), that is, the individuals p_{i-1} and the p_{i+1}. In order to preserve the extreme individuals, the crowding distance of the first and the last element is set as a big value. This means that the crowding distance prefers elements with big values, and, hence, we use the inverse variation rate.

The pseudocode of the modification of the NSGA-II with the variation rate (VR-NSGA-II) is shown in Algorithm A1 of Appendix A.1.

3.4. Integration into NSGA-III

We consider this algorithm here because it is able to properly deal with MOPs with many objectives. This algorithm is similar to its predecessor, the NSGA-II in the variation operators and in the classification of the fronts via the fast-non-dominated-sorting; however, the crowding distance is replaced by a more sophisticated procedure.

Here, the idea is to take advantage of the **association** method of the NSGA-III, which defines a "neighborhood" structure in a very convenient way for our purpose. The association method assigns each element of F_j (the last front classified after the fast-non-dominated-sorting) to the nearest induced line by some weight $w_i \in Z$, where Z is a set of reference points. Each weight can have more than one associated element, forming a neighborhood.

In the original NSGA-III, the **niching** is realized by sorting the obtained groups in the association stage according to its cardinality in ascending order. The element with the least distance to the induced line in each group is selected, and the algorithm continues with the next group until the population is filled. Thus, we modify the **niching** method. To include the diversity in decision space, our new niching procedure does not prefer the element with the least distance value. Instead, it prefers the one with the smallest *variation rate*.

The pseudocode of an iteration of the VR-NSGA-III algorithm with variation rate is shown in Algorithm A2 of Appendix A.2.

3.5. Integration into MOEA/D

MOEA/D is part of the Decomposition-Based Evolutionary Algorithms, which transform the original multi-objective optimization problem into a set of single-objective optimization problems that are simultaneously solved. In particular, this method takes a set of weights to define neighborhoods. The set of nearest weights defines one neighborhood and the best individuals are selected based on the value of a certain aggregative function. MOEA/D considers the weighted aggregation of objectives as an elitism mechanism. Furthermore, the neighborhood structure promotes the mating of close solutions. Different aggregative functions can be used in the MOEA/D framework, However, individuals with the least values are selected. In this work, we employ the Tchebycheff function, which is the most popular approach.

In order to include variation rate to the MOEA/D, we modify the selection criterion. Instead of preferring individuals with the least aggregative function value, we use the least value of the variation rate. For this, we employ the neighborhood structure of the original MOEA/D.

The pseudocode of VR-MOEA/D is shown in Algorithm A3 of Appendix A.3.

3.6. Integration into SMS-EMOA

SMS-EMOA is an indicator based algorithm; this means that it uses as a selection criterion the value of a certain performance indicator. In the case of SMS-EMOA, it is the hypervolume indicator.

This algorithm is similar to the NSGA-II, but it replaces the crowding distance by the contribution to the hypervolume of each individual in the last front. That is, the individuals of the last front with the biggest contribution to the hypervolume are preferred.

In this case, to adapt the SMS-EMOA, we consider the inverse variation rate, as the original mechanism criterion of this algorithm prefers high values. Again, we use the last front as our neighborhood structure.

The pseudocode of this method is shown in Algorithm A4 of Appendix A.4.

4. Numerical Results

In this section, we show some numerical results and comparisons to the state-of-the-art to demonstrate the benefit and strength of the variation rate. To this end, we first compare the original version of each algorithm against its corresponding version that uses the variation rate (respectively, the inverse variation rate) on some widely used (non-Type III) benchmark problems. This is done in order to show that the performance of each algorithm is not significantly affected for standard problems. In the next step, we test again the original and variation rate versions of the selected MOEAs on some Type III problems, where the advantage of the variation rate becomes apparent.

The benchmark problems that we use for the first part of these experiments are the well known test problems DTLZ 1-4 [26], IDTLZ 1 and IDTLZ2 [27], as well as the test problems WFG 1-5 [28]. For the second part, we use following six Type III problems.

The first Type III problem is taken from [22], which is defined as follows:

$$f_1(x) = \sum_{i=1}^{n} \sin(\pi x_i), \quad f_2(x) = \sum_{i=1}^{n} \cos(\pi x_i), \tag{5}$$

where $0 \leq x_i \leq 6$. This problem, denoted as OMNI1 in this work, has a total of 243 different disconnected components that form the Pareto set, and all of these components map to the same Pareto front.

The second problem, also taken from [22], is defined as follows:

$$f_1(x) = \sin\left(\pi \sum_{i=1}^{n} x_i\right), \quad f_2(x) = \cos\left(\pi \sum_{i=1}^{n} x_i\right), \tag{6}$$

where $0 \leq x_i \leq 1$, and $i = 1, 2, \ldots, 6$. This problem is denoted as OMNI2 in this work. Let $y = \sum_{i=1}^{6} x_i$ be the sum of the variables, then the Pareto set consists of the points x where $1 \leq y \leq 1.5$ or $3 \leq y \leq 3.5$. In addition, here, both connected components map to the same Pareto front. That is, every point on the Pareto front can be obtained in different infinite ways via the combinations of the variables mentioned.

The third Type III problem is the application stated in [29,30], where subdivision techniques have been used to tackle the problem. It is stated as follows: for f_1, f_2 $\mathbb{R}^5 \rightarrow \mathbb{R}$, it is:

$$
\begin{aligned}
f_1(x) &= \sum_{i=1}^{n} x_i, \\
f_2(x) &= 1 - \prod_{i=1}^{n}(1 - w_i(x_i)),
\end{aligned}
\tag{7}
$$

where

$$
w_i(z) = \begin{cases}
0.01 \cdot \exp\left(-\left(\dfrac{z}{20}\right)^{2.5}\right), & \text{for } i = 1, 2, \\
0.01 \cdot \exp\left(-\dfrac{z}{15}\right), & \text{for } 3 \le i \le 5.
\end{cases}
\tag{8}
$$

Finally, we consider the methodology from [25] to construct three more problems, denoted in this paper as RPH1, RPH2, and RPH3. These are bi-objective problems with two variables. In order to properly define them, we use the following functions.

First, we define the objective functions for the RPH1-3 problems

$$
\begin{aligned}
f_1(x) &= (x_1 + a)^2 + x_2^2, \\
f_2(x) &= (x_1 - a)^2 + x_2^2,
\end{aligned}
\tag{9}
$$

where $x \in \mathbb{R}^2$ and $a \in \mathbb{R}^+$. The variants of the RPH problems are obtained with the following functions. Let $t_1(x)$ and $t_2(x)$, with $x \in \mathbb{R}^2$, be the tile identifiers that are determined via:

$$
\begin{aligned}
\hat{t}_1(x) &= \text{sgn}(x_1) \cdot \left\lceil \frac{|x_1| - \left(a + \frac{c}{2}\right)}{2a + c} \right\rceil, \\
\hat{t}_2(x) &= \text{sgn}(x_2) \cdot \left\lceil \frac{|x_2| - \frac{b}{2}}{b} \right\rceil,
\end{aligned}
\tag{10}
$$

which restrict the problem to nine tiles using the relation $t_i = \text{sgn}(\hat{t}_i(x)) \cdot \min\{|\hat{t}_i|, 1\}$, with $i = 1, 2$.

Then, RPH1 is defined as $f_i^{(1)}(x) = f(\hat{x}(x))$, where $\hat{x} : \mathbb{R}^2 \rightarrow \mathbb{R}^2$ is defined by the following transformation:

$$
\begin{aligned}
\hat{x}_1(x_1) &= x_1 - t_1 \cdot (c + 2a), \\
\hat{x}_2(x_2) &= x_2 - t_2 \cdot b.
\end{aligned}
\tag{11}
$$

For the RPH1-3 problems, we fix the constant values $a = 4$, $b = 10$, and $c = 4$.

The RPH2 problem is defined as the RPH1, but it rotates the variables. That is, for an angle θ, we have

$$
r(x) = \begin{pmatrix} \cos\theta & -\sin\theta \\ \sin\theta & \cos\theta \end{pmatrix} x,
\tag{12}
$$

and then $f_i^{(2)}(x) = f_i^{(1)}(r(x))$. In this paper, we use $\theta = \frac{\pi}{4}$.

Finally, via the following transformation $d : \mathbb{R}^2 \rightarrow \mathbb{R}$

$$
d(x) = x_1 \cdot \left(\frac{x_2 - L + \epsilon}{U - L} \right),
\tag{13}
$$

for some small $\epsilon > 0$ and where U and L denote the upper and lower bound of the search space, respectively; we can define the RPH3 as: $f_i^{(3)}(x) = f_i^{(2)}(d(x), x_2)$, which is a rotated and transformed problem. In this paper, we use $\epsilon = 0.1$ for the RPH3 problem, while $L = -20$ and $U = 20$ are the the the upper and lower bounds of each variable for the RPH1-3 problems.

We use the PlatEMO platform [31] to make our test. The parameter settings of all the used algorithms are shown in Table 1. For all experiments, we have executed 30 independent runs. The numerical results with the mean and standard deviation of the hypervolume and Δ_p indicators are shown in Tables 2 and 3. In these tables, we have put in bold the best value between each pair of algorithms (the original version and the version with variation rate). We also performed the Wilcoxon test [32] as statistical significance proof to validate the results. For this, we consider the value $\alpha = 0.05$. We put in gray the cell where such difference has statistical significance according to this test.

Table 1. Parameter configuration for each algorithm. Mutation probability m_p, crossover probability c_p, neighborhood size T, and number of reference points #Z.

Parameter	NSGA-II	NSGA-III	MOEA/D	SMS-EMOA
m_p	$1/n$	$1/n$	0.1	$1/n$
c_p	0.8	0.8	1.0	0.8
T	-	-	20	-
#Z	-	200	-	-

From the tables, we obtain that, for the classical benchmark problems, the original version of the selected MOEAs has a better Δ_p value than the variation rate version in 27 out of the 44 combinations, where only 19 out of these values have statistical significance, which is an expected result. However, it is important to notice that the variation rate versions do not always lose, according to the Δ_p indicator values, the variation rate versions are better in 17 out of 44 cases, but only in two with statistical significance.

For the hypervolume indicator, something similar happens. Here, the original version of the MOEAs is better than the variation rate version in 33 out of 44 runs, with statistical significance in 21 cases, while the variation rate version wins in 11 out of 44 cases for this indicator, where three of them have statistical significance.

In total, from the 88 possible combinations (algorithms, indicators and problems), we have statistical significance in 45 cases; this means that, almost 50% of the time, it is not possible to say that the original version is different than the variation rate version. Moreover, in the cases when we have statistical significance, we can see in Table 2 that the averaged values are very similar.

We observe the advantages of the variation rate versions with the Type III problems (see Table 3). For these problems, we use the Δ_p both in objective and decision space (we denote this in the table as Obj. Δ_p and Var. Δ_p, respectively). We observe a similar behavior in objective space, here 16 out of 24 possible combinations are better for the original MOEAs (but only 6 out of these 34 have statistical significance). However, in decision space, the variation rate versions are better than the original versions; we have that 18 out of 24 combinations have better Δ_p values, almost all of them with statistical significance (only, in one case, we can not reject the null hypothesis).

Graphical results are shown in Figures 4–9; we plot the original MOEA and its corresponding variation rate version with the best value for each problem (according to the median of all runs using the Var. Δ_p).

Table 2. Numerical results for the original and variation rate version of some MOEAs in standard benchmark test problems. We show the mean and standard deviation (up and down in the cell, respectively). We put in bold the best value and in gray the cells with statistical significance according to the Wilcoxon test.

Problem	Ind.	NSGA-II Original	NSGA-II VR	NSGA-III Original	NSGA-III VR	MOEAD Original	MOEAD VR	SMSEMOA Original	SMSEMOA VR
DTLZ1	Δ_p	**0.0155** (0.0004)	0.0181 (0.0006)	0.0105 (0.0000)	0.0105 (0.0000)	0.0105 (0.0000)	0.0105 (0.0000)	0.0358 (0.0639)	0.0719 (0.1213)
	HV	**0.8473** (0.0013)	0.8454 (0.0012)	0.8576 (0.0003)	0.8575 (0.0006)	0.8571 (0.0005)	0.8572 (0.0004)	0.8359 (0.0081)	0.8351 (0.0079)
DTLZ2	Δ_p	0.0410 (0.0008)	0.0410 (0.0014)	0.0277 (0.0000)	0.0277 (0.0000)	**0.0277** (0.0000)	0.0280 (0.0000)	0.0539 (0.0041)	**0.0536** (0.0041)
	HV	0.5630 (0.0018)	**0.5641** (0.0015)	0.5813 (0.0000)	0.5813 (0.0001)	**0.5814** (0.0000)	0.5793 (0.0001)	0.5593 (0.0013)	0.5588 (0.0020)
DTLZ3	Δ_p	0.0666 (0.0780)	**0.0572** (0.0855)	0.0301 (0.0021)	0.0295 (0.0015)	0.0316 (0.0027)	0.0308 (0.0019)	0.0687 (0.0522)	0.1386 (0.2410)
	HV	0.5577 (0.0052)	**0.5598** (0.0074)	0.5714 (0.0060)	0.5730 (0.0044)	0.5660 (0.0068)	0.5677 (0.0053)	0.5526 (0.0066)	**0.5456** (0.0213)
DTLZ4	Δ_p	**0.0400** (0.0007)	0.0450 (0.0015)	0.0277 (0.0000)	0.0277 (0.0000)	0.1359 (0.2069)	0.0293 (0.0000)	0.0719 (0.1096)	0.0725 (0.1095)
	HV	0.5650 (0.0013)	**0.5676** (0.0013)	0.5812 (0.0001)	0.5813 (0.0001)	0.5287 (0.0971)	**0.5721** (0.0002)	0.5521 (0.0483)	**0.5522** (0.0481)
IDTLZ1	Δ_p	**0.0156** (0.0005)	0.0187 (0.0011)	0.1542 (0.2459)	0.0774 (0.1318)	**0.0177** (0.0000)	0.0181 (0.0001)	0.0371 (0.0708)	0.0422 (0.0984)
	HV	**0.2301** (0.0017)	0.2268 (0.0018)	**0.2331** (0.0007)	0.2326 (0.0006)	**0.2287** (0.0001)	0.2275 (0.0002)	0.2218 (0.0035)	0.2217 (0.0038)
IDTLZ2	Δ_p	**0.0394** (0.0008)	0.0406 (0.0012)	0.0412 (0.0008)	0.0412 (0.0010)	0.0475 (0.0008)	**0.0456** (0.0007)	0.0513 (0.0035)	0.0510 (0.0048)
	HV	**0.5495** (0.0011)	0.5425 (0.0019)	**0.5550** (0.0020)	0.5533 (0.0017)	**0.5565** (0.0001)	0.5548 (0.0002)	**0.5410** (0.0023)	0.5408 (0.0018)
WFG1	Δ_p	**0.1626** (0.0224)	0.2488 (0.0405)	**0.2143** (0.0266)	0.2392 (0.0342)	**0.2495** (0.0176)	0.7113 (0.0929)	0.3413 (0.0353)	0.3408 (0.0283)
	HV	**0.9257** (0.0095)	0.8582 (0.0249)	**0.8740** (0.0157)	0.8629 (0.0198)	**0.8892** (0.0237)	0.5933 (0.0428)	0.8697 (0.0170)	**0.8725** (0.0152)
WFG2	Δ_p	**0.1270** (0.0072)	0.1329 (0.0083)	0.0830 (0.0012)	0.0830 (0.0013)	**0.2166** (0.0335)	0.2222 (0.0320)	0.1405 (0.0069)	**0.1377** (0.0067)
	HV	**0.9356** (0.0009)	0.9334 (0.0010)	0.9393 (0.0006)	0.9392 (0.0006)	**0.9203** (0.0061)	0.9200 (0.0049)	**0.9309** (0.0015)	0.9307 (0.0011)
WFG3	Δ_p	**0.5261** (0.0241)	0.7354 (0.0234)	0.7229 (0.0180)	0.7200 (0.0224)	**0.8110** (0.0261)	0.8788 (0.0462)	0.4541 (0.0488)	0.4633 (0.0653)
	HV	**0.4131** (0.0013)	0.4080 (0.0023)	0.4013 (0.0016)	0.4006 (0.0018)	**0.3826** (0.0163)	0.3613 (0.0142)	**0.3672** (0.0052)	0.3649 (0.0033)
WFG4	Δ_p	**0.1641** (0.0044)	0.1968 (0.0053)	**0.1142** (0.0007)	0.1147 (0.0007)	**0.1361** (0.0028)	0.2240 (0.0053)	0.2349 (0.0099)	0.2301 (0.0136)
	HV	**0.5502** (0.0026)	0.5328 (0.0029)	**0.5751** (0.0008)	0.5741 (0.0008)	**0.5550** (0.0028)	0.5180 (0.0029)	0.5336 (0.0026)	0.5339 (0.0031)
WFG5	Δ_p	**0.1857** (0.0038)	0.1927 (0.0041)	**0.1385** (0.0004)	0.1388 (0.0003)	**0.1515** (0.0013)	0.1549 (0.0014)	0.2191 (0.0149)	0.2181 (0.0131)
	HV	**0.5178** (0.0032)	0.5148 (0.0029)	**0.5396** (0.0003)	0.5393 (0.0002)	**0.5225** (0.0023)	0.5200 (0.0029)	0.5168 (0.0021)	0.5166 (0.0027)

Table 3. Numerical results for the original and variation rate version of some MOEAs in Type III test problems. We show the mean and standard deviation (up and down in the cell, respectively). We put in bold the best value and in gar the cells with statistical significance according to the Wilcoxon test.

Problem	Ind.	NSGA-II		NSGA-III		MOEAD		SMSEMOA	
		Original	VR	Original	VR	Original	VR	Original	VR
OMNI1	Obj Δ_p	**0.0218** 0.0006	0.0220 0.0005	0.0293 0.0030	0.0306 0.0046	**0.1605** 0.0145	0.2363 0.0547	0.0210 0.0017	**0.0206** 0.0012
	Var Δ_p	2.2071 0.2928	**1.8804** 0.1603	1.8876 0.1852	1.9313 0.2480	4.3767 0.6295	**3.2976** 0.7931	2.2525 0.2978	**2.0409** 0.2544
OMNI2	Obj Δ_p	**0.0038** 0.0001	0.0039 0.0001	0.0048 0.0000	0.0048 0.0000	**0.0237** 0.0000	0.0278 0.0076	0.0039 0.0001	0.0039 0.0000
	Var Δ_p	1.1649 0.0360	**0.7206** 0.0075	0.8714 0.0956	**0.7576** 0.0312	1.2561 0.1039	**0.9800** 0.1302	0.9073 0.0853	**0.7366** 0.0071
SCM1	Obj Δ_p	0.3149 0.0135	**0.3138** 0.0123	0.3599 0.0198	0.3594 0.0303	**81.3852** 0.0043	81.3853 0.0042	0.2852 0.0297	**0.2681** 0.0252
	Var Δ_p	**3.7266** 0.1461	3.7751 0.1440	**3.5737** 0.1586	3.6209 0.1468	**38.5910** 0.0020	38.5912 0.0022	2.5058 0.0309	**2.4572** 0.0261
RPH1	Obj Δ_p	**0.1636** 0.0042	0.1653 0.0040	0.4323 0.0028	0.4334 0.0042	**2.6365** 0.0027	3.3847 0.7119	0.1551 0.0023	**0.1535** 0.0026
	Var Δ_p	6.8061 2.5349	**4.2771** 0.7180	1.6001 1.3496	**1.0450** 0.8738	8.0157 2.8617	**5.2802** 0.7714	**2.7075** 1.9102	5.3446 0.4255
RPH2	Obj Δ_p	**0.1725** 0.0027	0.1745 0.0051	0.4302 0.0109	0.4365 0.0141	**2.6459** 0.0175	3.1017 0.6411	0.1580 0.0038	**0.1574** 0.0035
	Var Δ_p	**2.6822** 0.9276	3.0836 0.7577	1.4661 0.5815	**1.2051** 0.6150	11.3906 3.0919	**6.6796** 1.5272	1.2333 0.5085	**1.1097** 0.5499
RPH3	Obj Δ_p	**0.1701** 0.0040	0.1732 0.0047	0.4314 0.0182	0.4340 0.0140	**2.6572** 0.0360	3.1012 0.6169	**0.1555** 0.0038	0.1564 0.0032
	Var Δ_p	6.3361 1.7299	**2.8003** 0.8529	3.0233 1.9181	**2.1521** 0.9408	8.6873 2.0764	**4.7500** 1.5926	2.8545 1.1084	**1.6417** 0.5711

In Figure 4, we can see that the obtained distribution is better for the variation rate version; it looks similar to that obtained by the original algorithm. However, we have to recall that this problem has 243 different disconnected components in the Pareto set, and some of them are overlapping in the plot. In Figure 5, we notice that the variation rate version can obtain points at the two regions in this representation (we plot y on both axes, where $y = \sum_{i=1}^{6} x_i$), while the original version is only able to compute points in one of them. On the other hand, in Figure 6, we can see that both the original and the variation rate versions can approximate the disconnected components of this problem well (except by the MOEA/D algorithm). However, the variation rate version is better in this problem, according to the values of Table 3.

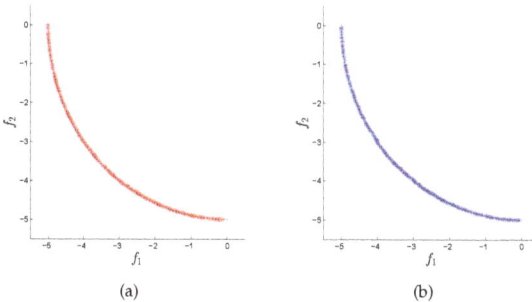

(a)

(b)

Figure 4. *Cont.*

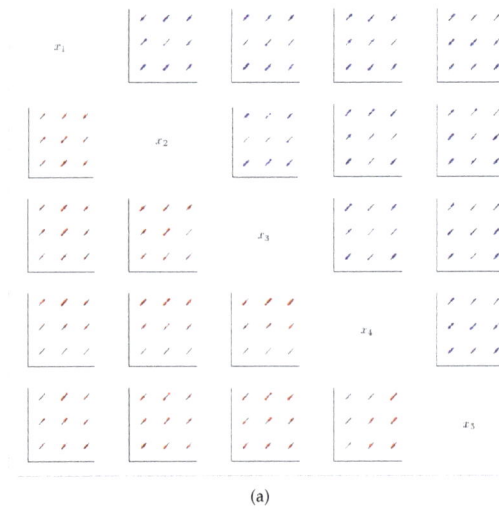

(a)

Figure 4. Graphical results of the run with the median values for the OMNI1 function for NSGA-II.
(a) Objective Space Original; (b) Objective Space VR; (c) Decision Space, pairwise plot of each variable.
The left-down and red marks correspond to the original algorithm, while the right-up and blue ones
are the VR version.

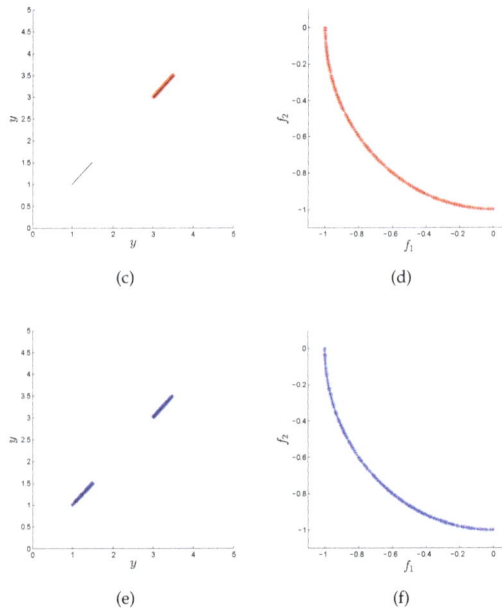

Figure 5. Graphical results of the run with the median values for the OMNI2 function for the
NSGA-II algorithm. (a) Decision Space Original; (b) Objective Space Original; (c) Decision Space
VR; (d) Objective Space VR.

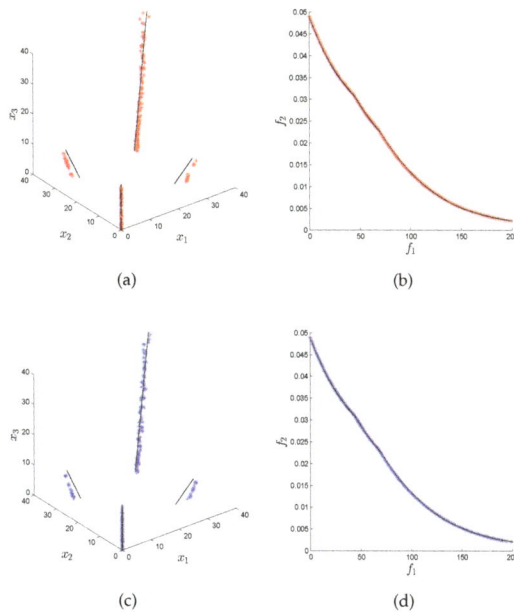

(a)   (b)

(c)   (d)

Figure 6. Graphical results of the run with the median values for the SCM1 function for the SMS-EMOA algorithm. (**a**) Decision Space Original; (**b**) Objective Space Original; (**c**) Decision Space VR; (**d**) Objective Space VR.

For the RPH problems, we can see in Figure 7 that the variation rate version of the NSGA-II can obtain four out of the nine disconnected components of the Pareto set, while the original version only gets three out of them; moreover, the distribution in decision space is also improved. In Figure 8, we can see again a similar behavior between the variation rate and the original version of the NSGA-III algorithm, which is also confirmed by the values of Table 3, where the variation rate version wins in three out of the four baseline algorithms for the RPH2 problem, but only in one does it have statistical significance. Finally, for the RPH3 problem, we can see in Figure 9 how the variation rate can significantly improve the performance of the MOEA/D algorithm in its distribution in decision space, as the original version only obtains points in one out of the nine disconnected components of the Pareto set, while the variation rate version obtains five out of them.

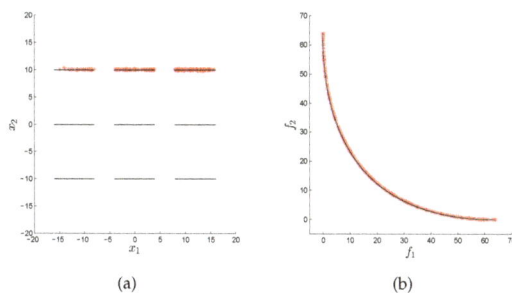

(a)   (b)

Figure 7. *Cont.*

(a)

(b)

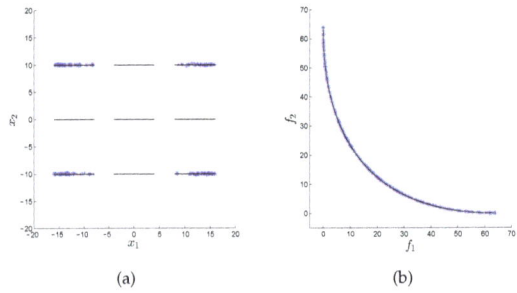

Figure 7. Graphical results of the run with the median values for the RPH1 function for the NSGA-II algorithm. (**a**) Decision Space Original; (**b**) Objective Space Original; (**c**) Decision Space VR; (**d**) Objective Space VR.

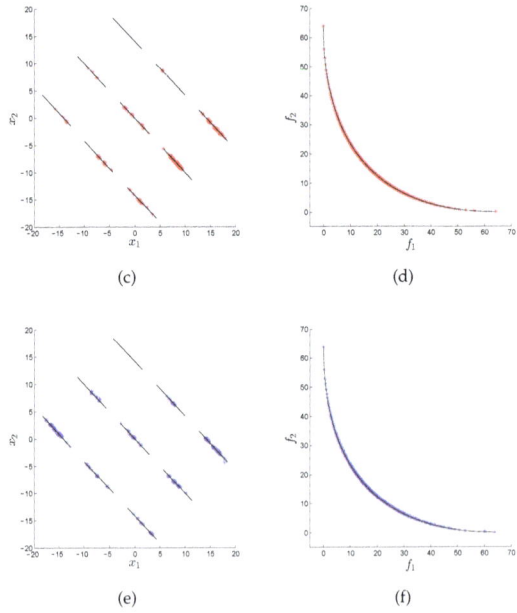

(c)

(d)

(e)

(f)

Figure 8. Graphical results of the run with the median values for the RPH2 function for the NSGA-III algorithm. (**a**) Decision Space Original; (**b**) Objective Space Original; (**c**) Decision Space VR; (**d**) Objective Space VR.

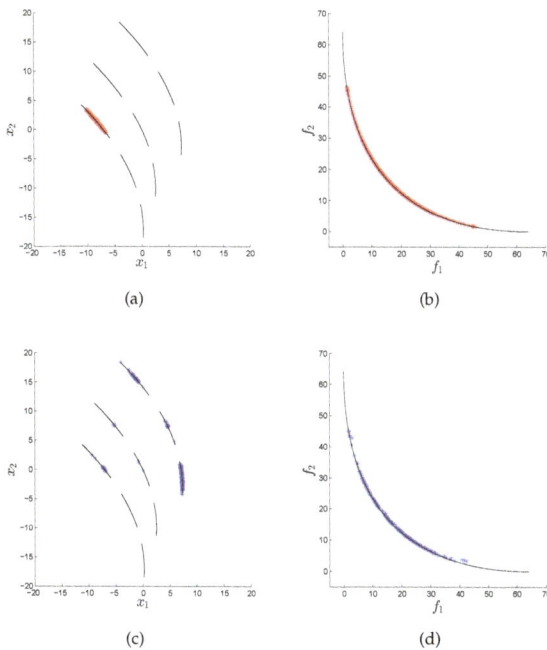

Figure 9. Graphical results of the run with the median values for the RPH3 function for the MOEA/D algorithm. (**a**) Decision Space Original; (**b**) Objective Space Original; (**c**) Decision Space VR; (**d**) Objective Space VR.

5. Conclusions and Future Work

In this work, we have addressed the problem of computing diverse solutions both in decision and objective space for a given multi-objective optimization problem via specialized evolutionary strategies. While so far quite a few good diversity mechanisms exist to obtain a spread in objective space, the consideration of the Pareto set approximations has been mainly neglected so far. This represents a possible shortcoming in particular for Type III problems where points in the Pareto front may have multiple pre-images. To achieve this goal, we have first presented the general framework of the variation rate that combines the usage of the averaged distance in variable space with the selection operator that is given by the multi-objective evolutionary algorithm (MOEAs). We have demonstrated further on possible integration of the variation rate into four MOEAs that represent the state-of-the-art.

Numerical results have shown that the use of the variation rate improves the performance of the standalone algorithms for Type III problems, while the variation rate algorithms are not significantly worse for the standard benchmark problems (i.e., problems with a 1:1 relationship between Pareto set and Pareto front), even in some cases variation rate improves the performance of the original algorithm. Of course, the behavior of variation rate by itself is not enough for the treatment of every kind of MOP as this depends on the operators of every algorithm. That is, the variation rate can enhance the overall performance of a certain algorithm, but if, for instance, such algorithm is not conceived to deal with MOPs with many objectives, the addition of the variation rate would be not enough to solve them. We can, for instance, consider some of the algorithms and test functions of [33], in order to show if the variation rate provides some advantages when dealing with MOPs with many objectives.

As future work, it will be mandatory to adapt the genetic operators of the evolutionary algorithms to exploit the diversity in decision space. As the above results have shown, the diversity in decision space has already increased significantly; however, the variation rate is only a selection mechanism. In order to obtain

Math. Comput. Appl. **2019**, *24*, 82

optimal solutions in particular in decision space, the exploration will have to be increased. Furthermore, it will be necessary to develop a particular indicator for problems of Type III. In general, performance indicators evaluate an approximation based on the value of the objectives, but, for problems such as OMNI1, this does not provide enough information. Once we have such an indicator, we can validate better performance of our methods in decision space. However, it is ad hoc not clear which property has to be satisfied with this approximation. We also need to test this approach in problems with different properties in decision space, in particular problems with disconnected Pareto sets.

Author Contributions: O.C. and O.S. conceived and designed the experiments; O.C. performed the experiments; O.C. and O.S. analyzed the data; O.C. and O.S. wrote the paper.

Funding: The authors acknowledge funding from the Conacyt Basic Science project No. 285599 and SEP Cinvestav project No. 231.

Acknowledgments: The first author acknowledges support from the Conacyt to pursue his Ph.D. studies.

Conflicts of Interest: The authors declare no conflict of interest.

Appendix A. Pseudo-Codes

The appendix contains the pseudocode of all the variation rate algorithms that were used in this work.

Appendix A.1. VR-NSGA-II

The differences between the original NSGA-II and the VR-NSGA-II are lines 12 and 20. In line 12, we compute the inverse variation rate values of the last front, while, on line 20, we perform the selection according to these values.

Algorithm A1 Pseudocode of VR-NSGA-II

Require: Population size (P_s), crossover probability (P_c), mutation probability (P_m)
Ensure: Final Population
1: Population ← InitializePopulation(P_s)
2: FastNondominatedSorting(Population)
3: Selected ← SelectParentsByRank(Population, P_s)
4: Children ← CrossoverAndMutation(Selected, P_c, P_m)
5: **while** StopCondition() **do**
6:     Union ← Merge(Population, Children)
7:     Fronts ← FastNondominatedSorting(Union)
8:     Parents ← ∅
9:     $Front_L$ ← ∅
10:     **for** $Front_i$ ∈ Fronts **do**
11:         V_t ← CrowdingDistanceAssignment($Front_i$)
12:         R_t ← InverseVariationRateAssigment(V_t)
13:         **if** Size(Parents)+Size($Front_i$) > P_s **then**
14:             $Front_L$ ← i
15:             **break**
16:         **else**
17:             Parents ← Merge(Parents, $Front_i$)
18:         **end if**
19:         **if** Size(Parents) < P_s **then**
20:             $Front_L$ ← SortByRankAndInverseVariationRate($Front_L$)
21:             **for** P_1 to $P_{P_s-\text{Size}(Front_L)}$ **do**
22:                 Parents ← P_i
23:             **end for**
24:         **end if**
25:         Selected ← SelectParentsByRankAndDistance(Parents, P_s)
26:         Population ← Children
27:         Children ← CrossoverAndMutation(Selected, P_c, P_m)
28:     **end for**
29: **end while**
30: **return** Children

Appendix A.2. VR-NSGA-III

We only show the iteration as the complete code is basically identical to NSGA-II with a different selection mechanism. Here, the main difference of the variation rate version, compared to the original NSGA-III, is on line 14; while the original NSGA-III uses the ninching as selection criterion, the VR-NSGA-III employs the variation rate, as it is stated on line 14.

Algorithm A2 Iteration of the VR-NSGA-III

Require: Reference points Z, current population P_t
Ensure: Next, population P_{t+1}
  1: $S_t = \varnothing$, $i = 1$
  2: Q_t = apply variation operators to P_t
  3: $M_t = P_t \cup Q_t$
  4: $(F_1, F_2 \dots,)$ = fast-non-dominated-sorting(M_t)
  5: **while** $|S_t| \leq N$ **do**
  6:     $t = S_t \cup F_i$
  7:     $i = i + 1$
  8: **end while**
  9: Add first fronts to P_{t+1}
10: $F_i :=$ last added front
11: Normalize F_i
12: Associate elements of F_i with each Z
13: $V_t :=$ Niching of F_i
14: $R_t :=$ `SelectByVariationRate`(V_t)
15: $P_{t+1} : S_t \cup R_t$

Appendix A.3. VR-MOEA/D

The change in this algorithm, concerning the original MOEAD, is very subtle. On lines 9 and 10, we compute the variation rate of the elements of the neighborhood and the offspring ($B(i) \cup y$) instead of only computing the values of the aggregative function.

Algorithm A3 Pseudocode of VR-MOEA/D

Require: N number of solutions and weight vectors; T neighborhood size.
Ensure: Final Population
  1: Initialize N weight vectors $\lambda^1 \lambda^2, \cdots, \lambda^N$
  2: Set N subproblems defined by the N weight vectors
  3: Set N neighborhoods $B(i) = \{w_{i,1}, \cdots, w_{i,T}\}$, where $w_{i,j} = \lambda^j$ are the closest weight vectors to λ^i
  4: $\{x^1, \cdots, x^N\} \leftarrow$ `InitializePopulation`(N)
  5: **while** StopCondition() **do**
  6:     **for** $i \in N$ **do**
  7:         Randomly select two solutions from $B(i)$ to generate an offspring y
  8:         Apply using variation operators to y
  9:         Compute the values V_r of $B(i) \cup y$ via the aggregative function g.
10:         Compute Variation Rate V_t of the elements of $B(i) \cup y$
11:         **for** x in $B(i)$ **do**
12:             **if** $r_y < r_x$ **then**
13:                 Replace x with y
14:             **end if**
15:         **end for**
16:     **end for**
17: **end while**
18: **return** Children

Appendix A.4. VR-MOEA/D

Again, in this case, we only show an iteration of the method as the rest of the algorithm is basically the NSGA-II. We observe that, on line 12, we substitute the values of the hypervolume contributions with the variation rate and we use them for the selection mechanism.

Algorithm A4 Iteration of the VR-SMS-EMOA

Require: Current population P_t
Ensure: Next, population P_{t+1}
1: $S_t = \emptyset$, $i = 1$
2: Q_t = apply variation operators to P_t
3: $M_t = P_t \cup Q_t$
4: $(F_1, F_2 \dots,)$ = fast-non-dominated-sorting(M_t)
5: **while** $|S_t| \leq N$ **do**
6: $t = S_t \cup F_i$
7: $i = i + 1$
8: **end while**
9: Add first fronts to P_{t+1}
10: $F_i :=$ last added front
11: $V_t \leftarrow$ ComputeHypervolumeContributions(F_i)
12: $R_t :=$ SelectByVariationRate(V_t)
13: $P_{t+1} : S_t \cup R_t$

References

1. Coello, C.A.C.; Lamont, G.B. *Applications of Multi-Objective Evolutionary Algorithms*; World Scientific: Singapore, 2004; Volume 1.
2. Lewis, A.; Mostaghim, S.; Randall, M. *Biologically-Inspired Optimisation Methods: Parallel Algorithms, Systems and Applications*; Springer: Berlin/Heidelberg, Germany, 2009; Volume 210.
3. Deb, K. *Multi-Objective Optimization Using Evolutionary Algorithms*; John Wiley & Sons: Chichester, UK, 2001.
4. Coello, C.A.C.; Lamont, G.B.; Van Veldhuizen, D.A. *Evolutionary Algorithms for Solving Multi-Objective Problems*; Springer: New York, NY, USA, 2007; Volume 5.
5. Peitz, S.; Dellnitz, M. A Survey of Recent Trends in Multiobjective Optimal Control—Surrogate Models, Feedback Control and Objective Reduction. *Math. Comput. Appl.* **2018**, *23*, 30. [CrossRef]
6. Schütze, O.; Vasile, M.; Coello Coello, C.A. Computing the Set of Epsilon-Efficient Solutions in Multiobjective Space Mission Design. *J. Aerosp. Comput. Inf. Commun.* **2011**, *8*, 53–70. [CrossRef]
7. Basto-Fernandes, V.; Yevseyeva, I.; Deutz, A.; Emmerich, M. A Survey of Diversity Oriented Optimization: Problems, Indicators, and Algorithms. In *EVOLVE—A Bridge Between Probability, Set Oriented Numerics and Evolutionary Computation VII*; Emmerich, M., Deutz, A., Schütze, O., Legrand, P., Tantar, E., Tantar, A.A., Eds.; Springer International Publishing: Cham, Switzerland, 2017; pp. 3–23. [CrossRef]
8. Deb, K.; Pratap, A.; Agarwal, S.; Meyarivan, T. A fast and elitist multiobjective genetic algorithm: NSGA-II. *IEEE Trans. Evol. Comput.* **2002**, *6*, 182–197. [CrossRef]
9. Beume, N.; Naujoks, B.; Emmerich, M. SMS-EMOA: Multiobjective selection based on dominated hypervolume. *Eur. J. Oper. Res.* **2007**, *181*, 1653–1669. [CrossRef]
10. Zhang, Q.; Li, H. MOEA/D: A Multiobjective Evolutionary Algorithm Based on Decomposition. *IEEE Trans. Evol. Comput.* **2007**, *11*, 712–731. [CrossRef]
11. Deb, K.; Jain, H. An evolutionary many-objective optimization algorithm using reference-point-based nondominated sorting approach, part I: Solving problems with box constraints. *IEEE Trans. Evol. Comput.* **2014**, *18*, 577–601. [CrossRef]
12. Cuate, O.; Schütze, O. Variation Rate: An Alternative to Maintain Diversity in Decision Space for Multi-Objective Evolutionary Algorithms. In *Evolutionary Multi-Criterion Optimization*; Deb, K., Goodman, E., Coello Coello, C.A., Klamroth, K., Miettinen, K., Mostaghim, S., Reed, P., Eds.; Springer International Publishing: Cham, Switzerland, 2019; pp. 203–215.
13. Zitzler, E.; Thiele, L. Multiobjective evolutionary algorithms: A comparative case study and the strength Pareto approach. *IEEE Trans. Evol. Comput.* **1999**, *3*, 257–271. [CrossRef]

14. Dilettoso, E.; Rizzo, S.A.; Salerno, N. A Weakly Pareto Compliant Quality Indicator. *Math. Comput. Appl.* **2017**, *22*, 25. [CrossRef]
15. Preuss, M.; Wessing, S. Measuring Multimodal Optimization Solution Sets with a View to Multiobjective Techniques. In *EVOLVE—A Bridge Between Probability, Set Oriented Numerics, and Evolutionary Computation IV*; Emmerich, M., Deutz, A., Schuetze, O., Bäck, T., Tantar, E., Tantar, A.A., Moral, P.D., Legrand, P., Bouvry, P., Coello, C.A., Eds.; Springer International Publishing: Heidelberg, Germany, 2013; pp. 123–137.
16. Schütze, O.; Esquivel, X.; Lara, A.; Coello, C.A.C. Using the averaged Hausdorff distance as a performance measure in evolutionary multiobjective optimization. *IEEE Trans. Evol. Comput.* **2012**, *16*, 504–522. [CrossRef]
17. Bogoya, J.M.; Vargas, A.; Cuate, O.; Schütze, O. A (p,q)-Averaged Hausdorff Distance for Arbitrary Measurable Sets. *Math. Comput. Appl.* **2018**, *23*, 51. [CrossRef]
18. Ulrich, T.; Bader, J.; Thiele, L. Defining and Optimizing Indicator-Based Diversity Measures in Multiobjective Search. In *Parallel Problem Solving from Nature, PPSN XI*; Schaefer, R., Cotta, C., Kołodziej, J., Rudolph, G., Eds.; Springer: Berlin, Germany, 2010; pp. 707–717.
19. Shir, O.M.; Preuss, M.; Naujoks, B.; Emmerich, M. Enhancing Decision Space Diversity in Evolutionary Multiobjective Algorithms. In *Evolutionary Multi-Criterion Optimization*; Ehrgott, M., Fonseca, C.M., Gandibleux, X., Hao, J.K., Sevaux, M., Eds.; Springer: Berlin, Germany, 2009; pp. 95–109.
20. Srinivas, N.; Deb, K. Muiltiobjective Optimization Using Nondominated Sorting in Genetic Algorithms. *Evol. Comput.* **1994**, *2*, 221–248. [CrossRef]
21. Jeffrey, H.; Nafpliotis, N.; Goldberg, D.E. Multiobjective optimization using the Niched Pareto genetic algorithm. *IlliGAL Rep.* **1993**, 93005. Available online: https://www.researchgate.net/publication/2763393_Multiobjective_Optimization_Using_The_Niche_Pareto_Genetic_Algorithm (accessed on 13 September 2019).
22. Deb, K.; Tiwari, S. Omni-optimizer: A generic evolutionary algorithm for single and multi-objective optimization. *Eur. J. Oper. Res.* **2008**, *185*, 1062–1087. [CrossRef]
23. Castillo, J.C.; Segura, C.; Aguirre, A.H.; Miranda, G.; León, C. A Multi-Objective Decomposition-Based Evolutionary Algorithm with Enhanced Variable Space Diversity Control. In Proceedings of the GECCO 2017, Berlin, Germany, 15–19 July 2017; ACM: New York, NY, USA, 2017; pp. 1565–1571.
24. Ulrich, T.; Bader, J.; Zitzler, E. Integrating Decision Space Diversity into Hypervolume-Based Multiobjective Search. In Proceedings of the 12th Annual Conference on Genetic and Evolutionary Computation, Portland, OR, USA, 7–10 July 2010; ACM: New York, NY, USA, 2010; pp. 455–462. [CrossRef]
25. Rudolph, G.; Naujoks, B.; Preuss, M. Capabilities of EMOA to Detect and Preserve Equivalent Pareto Subsets. In Proceedings of the EMO 2017, Münster, Germany, 19–22 March 2017; Springer: Berlin, Germany, 2007; pp. 36–50.
26. Deb, K.; Thiele, L.; Laumanns, M.; Zitzler, E. Scalable Test Problems for Evolutionary Multiobjective Optimization. In *Evolutionary Multiobjective Optimization: Theoretical Advances and Applications*; Abraham, A., Jain, L., Goldberg, R., Eds.; Springer: London, UK, 2005; pp. 105–145.
27. Jain, H.; Deb, K. An evolutionary many-objective optimization algorithm using reference-point based nondominated sorting approach, part II: Handling constraints and extending to an adaptive approach. *IEEE Trans. Evol. Comput.* **2014**, *18*, 602–622. [CrossRef]
28. Huband, S.; Hingston, P.; Barone, L.; While, L. A review of multiobjective test problems and a scalable test problem toolkit. *IEEE Trans. Evol. Comput.* **2006**, *10*, 477–506. [CrossRef]
29. Schütze, O.; Witting, K.; Ober-Blöbaum, S.; Dellnitz, M. Set Oriented Methods for the Numerical Treatment of Multiobjective Optimization Problems. In *EVOLVE—A Bridge between Probability, Set Oriented Numerics and Evolutionary Computation*; Tantar, E., Tantar, A.A., Bouvry, P., Del Moral, P., Legrand, P., Coello Coello, C.A., Schütze, O., Eds.; Springer: Berlin, Germany, 2013; pp. 187–219.
30. Sun, J.Q.; Xiong, F.R.; Schütze, O.; Hernández, C. *Cell Mapping Methods*; Springer: Singapore, 2018.
31. Tian, Y.; Cheng, R.; Zhang, X.; Jin, Y. PlatEMO: A MATLAB Platform for Evolutionary Multi-Objective Optimization [Educational Forum]. *IEEE Comput. Intell. Mag.* **2017**, *12*, 73–87. [CrossRef]
32. Gibbons, J.D.; Chakraborti, S. *Nonparametric Statistical Inference*; Springer: Berlin, Germany, 2011.
33. Li, K.; Wang, R.; Zhang, T.; Ishibuchi, H. Evolutionary Many-Objective Optimization: A Comparative Study of the State-of-the-Art. *IEEE Access* **2018**, *6*, 26194–26214. [CrossRef]

© 2019 by the authors. Licensee MDPI, Basel, Switzerland. This article is an open access article distributed under the terms and conditions of the Creative Commons Attribution (CC BY) license (http://creativecommons.org/licenses/by/4.0/).

MDPI

St. Alban-Anlage 66

4052 Basel

Switzerland

Tel. +41 61 683 77 34

Fax +41 61 302 89 18

www.mdpi.com

Mathematical and Computational Applications Editorial Office

E-mail: mca@mdpi.com

www.mdpi.com/journal/mca

www.ingramcontent.com/pod-product-compliance
Lightning Source LLC
Chambersburg PA
CBHW051839210326
41597CB00033B/5709